Forecasting Forest Futures

Forecasting Forest Futures

A Hybrid Modelling Approach to the Assessment of
Sustainability of Forest Ecosystems and their Values

Hamish Kimmins, Juan A. Blanco, Brad Seely,
Clive Welham and Kim Scoullar

from Routledge

First published by Earthscan in the UK and USA in 2010

For a full list of publications please contact:

Earthscan
2 Park Square, Milton Park, Abingdon, Oxfordshire OX14 4RN
711 Third Avenue, New York, NY 10017

First issued in paperback 2015

Earthscan is an imprint of the Taylor & Francis Group, an informa business

ISBN 13: 978-1-138-86694-2 (pbk)
ISBN 13: 978-1-8440-7922-3 (hbk)

Typeset by JS Typesetting Ltd, Porthcawl, Mid Glamorgan
Cover design by Susanne Harris

A catalogue record for this book is available from the British Library

Library of Congress Cataloging-in-Publication Data

Forecasting forest futures : a hybrid modelling approach to the assessment of sustainability of forest ecosystems and their values / Hamish Kimmins ... [et al.].
        p. cm.
    Includes bibliographical references and index.
    ISBN 978-1-84407-922-3 (hardback)
    1. Forest ecology–Simulation methods. 2. Forest management. 3. Sustainable forestry. I. Kimmins, J. P.
    QK938.F6F59 2010
    634.9'2--dc22
                                        2010008425

Contents

Preface

A problem is an issue that does not get solved. An issue that gets solved is not a problem. Problem issues often persist because they are complex but only simple solutions are offered. (Kimmins 2008a, p1627)

Life is complex, and dealing with complexity is difficult. Society continually struggles to simplify the complexity of issues that it faces. We simplify the learning process in schools, colleges and universities by dividing the continuum of knowledge into individual subjects, and dividing these into individual lessons/lectures/book chapters. We simplify the creation of new knowledge and understanding through science by the application of Occam's razor and the principle of parsimony. Governments are divided into ministries, ministries into departments, and departments into programmes. We take individual snapshots (i.e. frames) from the 'movie' of life and divide them into a jigsaw puzzle of pieces so that we can describe, understand and communicate about the many sub-components of the overall picture.

This reductionism is a necessary part of the organization of the diversity of human endeavours, but it is not sufficient. If the human species is to survive and prosper, a new relationship must develop in which the jigsaw puzzles of more or less unconnected facts, understanding and experiences that populate our libraries, computers and the minds of many people are synthesized into mental images (conceptual models) of the current structure and functioning of social and environmental systems. Comparable pictures of the possible states of these systems at various times in the future must then be combined into 'movies' of the anticipated consequences of continuing on our present path of population, economic and social development, or switching to an alternative. Such snapshots of today and visions of alternative futures are prerequisites for the successful navigation of our species into an unknown and changing future in a manner that sustains the ecological systems that support us and our fellow earthly species.

This vision of how to address the seemingly overwhelmingly complex issues that face us is based on two necessities: the ability to integrate knowledge about the key components of the system(s) we are dealing with, up to the level of complexity that determines system behaviour; and the ability to make credible forecasts of future states of the system of interest. This book describes one approach to satisfying these necessities. As has been noted repeatedly, forestry is not rocket science; it is much more complex than that. This underscores the fact that most issues in forestry are complex, requiring solutions of appropriate complexity if they are not to become problems.

We assert in this book that solutions to complex issues in forestry require decision support, scenario analysis and value trade-off analysis tools that address forest processes at the ecosystem level and link this to the social values associated with forests, their conservation and utilization.

Experience has always been a valuable guide to decision making in forestry, but experience is always of the present and the past, and never of the future for which we must make decisions in our day-to-day management. The rapidly expanding understanding of ecosystem and biological processes provides a basis for predicting possible futures, but there are many difficulties faced by forecasting based solely on process-level understanding. Although these barriers are steadily being broken down, we assert that for the foreseeable future a combination of experience-based and understanding-based forecasts will provide the most practical and effective approach to choosing between alternative ways of navigating forestry into the future.

This book is about 34 years of exploration of this hybrid approach to the development of decision-support and policy-analysis tools for forestry. There is no point in this endeavour at which we can say: 'it is done'. Development of such tools is an evolutionary activity. As our knowledge of social and environmental systems increases, as the power of computers grows; and as the technology of communication changes, so will the methods – the computer-based crystal balls – used by those charged with the responsibility of managing the world's forests for multiple values and environmental services will also change. The approach that we advocate does not diminish the contributions of other, simpler models to shorter-term, smaller-scale and individual resource decisions. In developing hybrid models, we build on the contributions of these other models. However, we do assert that if these more traditional and simpler models are to be used to address the more complex issues facing forestry today and predicted for the future, they should be linked to process-based models that give them the flexibility needed to address multiple values and changing soil, climate, disturbance and management regimes: i.e. hybrid models.

We hope that this book will introduce the reader to our answers to the 'what?' and 'why?' questions about hybrid simulation modelling, and to the models that our approach has produced. We caution that while we believe that hybrid simulation modelling makes an important contribution to the design of sustainable forestry, decisions about the management of forests will always be made as much or more by ordinary citizens as by experts or the output of models. Managing forests is increasingly a bottom–up process requiring the engagement of civil society. In such public engagement processes, many of the stakeholders influencing decisions may know little about ecosystems and their management. This emphasizes that while the hybrid simulation models described here are a necessary component of scenario and value trade-off analyses, on their own they are not sufficient. They should be combined with a variety of communication tools that will help to inform decision-making processes about our best, science-based guesses about the future consequences of our decisions. Only in this way can we make the most ethical choices with respect to how much of our evolutionary inheritance of biological and ecosystem legacies be passed on to future generations.

J. P. (Hamish) Kimmins
Emeritus Professor of Forest Ecology
University of British Columbia, Vancouver

Acknowledgements

There are many individuals and organizations without whose contributions the modelling work described in this book would not have happened.

Dr Jock Carlisle, late of the Canadian Forestry Service, was instrumental in launching the work on FORCYTE as a component of the Canadian contribution to the International Energy Agency's work on renewable bioenergy. His support and enthusiasm were then taken up by Drs Lake Chatarpaul, Michael Apps and William J. Meades. This programme provided a decade of funding. Recently, additional funding has been provided by Natural Resources Canada.

The Canadian National Science and Engineering Research Council (NSERC) supported my group over the duration of my career at UBC, including the work on modelling. This involved Operating, Discovery, Collaborative Research and Development, and Strategic grants. NSERC's Canada Chairs programme funded my chair in forest ecosystem modelling and equipped our computer laboratory. The National Film Board of Canada supported the development of FORTOON.

The province of British Columbia provided major support for modelling through the Canadian–British Columbia Partnership Agreement on Forest Resource Development (FRDA), Forest Renewal BC, the BC Forest Science Program and the Science Council of BC.

Several Canadian forestry companies provided funding for various aspects of the research: Mistik Management of Meadow Lake, Saskatchewan; Western Forest Products, Canfor and Interfor of British Columbia. In addition, Syncrude Canada (in particular, Clara Qualizza) supported the application of our models of issues to oil sands reclamation, and the Cumulative Environmental Management Association (through its Terrestrial SubGroup) provided financial support for development of the FORECAST Climate version of the FORECAST model.

China contributed to the programme through the support of visiting scientists, while Spain made a major contribution through the financial support of Dr Juan Blanco. Funding from the UK Forestry Commission supported testing of FORECAST in Scotland.

The development of our models would not have happened had it not been for the dedication and loyalty of Life Science Programming, the creators of the computer code and partners in the development of the models. Kim Scoullar has spent 34 years working on the software and continued during times when there was no funding. The support of his family during those difficult times is acknowledged with gratitude.

Many other people contributed to the field and laboratory research that are an essential part of such a modelling exercise. Mr Min Tsze and Dr Michael Feller worked to develop the first calibration dataset. Drs Cindy Prescott and Robert Bradley contributed to aspects of the decomposition and soil representations. Dr Les Lavkulich assisted with the phosphorus sub-routine. Patsy Quay and Maxine Horner provided invaluable secretarial services. Numerous graduate students contributed to the development of aspects of the conceptual foundation for our models, tested assumptions and generated calibration data. They include John Yarie, Dan Binkley, Jennifer de Catanzaro, George Krumlik, Lynn Husted, Fred Nuszdorfer, Allen Banner, June Parkinson, Wayne Martin, Maria Ferreir, Phil Comeau, Dave Coopersmith, Richard Bigley, Linda Christanty, Werner Kurz, Christian Messier, Rod Keenan, Jian Rang Wang, Daniel Mailly, Anliang Zhong, Adrian Weber, Prasit Wangpakapattanawong, Clive Goodinson, Marco Albani, Sharon Hope, John Karakatsoulis, Eliot McIntyre, Robin Duchesneau, Brock Simons, Fuliang (Sam) Cao, Tanya Seebacher, Marius Boldor, Angelica Boldor, Yueh-Hsin Lo and Michael Gerzon. Many of these were supported by scholarships from UBC and a variety of grants.

Many others have been involved in the application and testing of our models. A list of publications on our website (www.forestry.ubc.ca/ecomodels) acknowledges their contributions.

We acknowledge the contributions from individuals in the Collaborative for Landscape Planning and others in the Departments of Forest Management and Forest Sciences at UBC: Drs John Nelson, Stephen Sheppard, Michael Meitner and Duncan Cavens in Forest Management, and Ralph Wells and others in Forest Sciences.

Drs Brad Seely, Clive Welham and Juan Blanco have become the major drivers of continuing model development, and I thank them for their energy, enthusiasm and commitment to the continued improvement, testing and application of the suite of models we have developed. Life Science Programming continues to provide programming and modelling services.

Finally, we would like to acknowledge with gratitude the patience and encouragement over several years by Tim Hardwick, Commissioning Editor at Earthscan, and the production work co-ordinated by Hamish Ironside. Working at a distance with a group of authors is never easy, and the Earthscan team made this a painless and enjoyable process.

List of Abbreviations

AAC	Annual allowable timber harvest
AET	Actual evapotranspiration
AM	Adaptive management
APSIM	Agricultural production systems simulator
AU	Analysis unit
BDT	Bone dry tonnes
CAD	Computer-aided design
CALP	Collaborative for Advanced Landscape Planning (UBC)
CWD	Coarse woody debris
CWH	Coastal Western Hemlock
DBDS	Design-based decision support
DBH	Diameter at breast height
DES	District energy system
DFU	Discrete forest unit
EBM	Ecosystem-based management
ENFD	Emulation of natural forest disturbance
ENFOR	Energy from the Forest programme (Canada)
ENGO	Environmental non-governmental organization
ERIN	Evaporation and radiation interception by neighbouring species
ESSF	Engelmann Spruce, Sub-Alpine Fir
FALLOW	Forest, agroforest, low-value landscape or wasteland
FIA	Forest Investment Agreement (BC)
FORCEE	Forest ecosystem complexity evaluator
FORCYTE	Forest Nutrient Cycling and Yield Trend Evaluator
FORECAST	Forestry and Environmental Change Assessment Tool
FORTOON	Forestry Cartoon
ForWaDy	Forest Water Dynamics
FRDA	Agreement on Forest Resource Development (Canada)
FSC	Forestry Stewardship Council
FVS	Forest vegetation simulator
GFI	Goodness of fit index
GHG	Greenhouse gas
GIS	Geographic information system
HB	Historical bioassay
HS	Hybrid simulation

IBP	International Biological Program
IDF	Interior Douglas-Fir
IEA	International Energy Agency
IFPA	Innovative Forest Practice Agreement (Arrow Lakes, BC)
IPCC	Intergovernmental Panel on Climate Change
LLEMS	Local Landscape Ecosystem Management Simulator
MAI	Mean annual increment
ME	Modelling efficiency
MGM	Mixed-wood growth model
MPB	Mountain pine beetle
MRM	Multiple Run Manager
NPP	Net primary production
NRCan	Natural Resources Canada
NSERC	National Science and Engineering Research Council (Canada)
OAF	Operational adjustment factor
OAPEC	Organization of Arab Petroleum-Exporting Countries
OPEC	Organization of Petroleum-Exporting Countries
OG	Old growth
OGI	Old growth index
PET	Potential evapotranspiration
PFF	Possible Forest Futures
PGCF	Prince George Community Forest (BC)
PS	Process simulation
RONV	Range of natural variation
RMSE	Root mean square error
RS	Remote sensing
SBS	Sub-boreal spruce
SCUAF	Soil changes under agroforestry
SFI	Sustainable Forestry Initiative
SFM	Sustainable forest management
SGOG	Smart Growth on the Ground (Prince George, BC)
SRIC	Short rotation, intensive culture
SSE	Sum of squares error
SVS	Stand Visualization System
TASS	Tree and stand simulator
TEM	Terrestrial ecosystem mapping
TFL	Tree Farm Licence (BC, Canada)
TPCM	Tree productivity and climate model
UBC	University of British Columbia (Canada)
VR	Variable retention
VRI	Vegetation resource inventory (BC)
WaNuLCAS	Water, nutrient and light capture in agroforestry systems
WCS	World Construction Set

Chapter 1

Introduction: Why do we Need Ecosystem-Level Models as Decision-Support Tools in Forestry?

Introduction

Why do we need ecosystem-level models as decision-support tools in forestry? The answer to this question is superficially simple: because the alternatives that have served forestry reasonably well in the past are no longer adequate today and for the future. In reality the answer is complicated and involves many issues, including: the inability of traditional reductionist science to address the social and biophysical complexities involved in managing the world's forests; the threats that are posed by continuing increases in human numbers and the even greater increase in the per capita ecological footprint as standards of living improve worldwide; the institutional barriers to managing forests as ecosystems (social, political and structural impediments to the implementation of ecosystem management); and the problems related to engaging the public in decisions about the management of public forests. These and other issues led the UBC Forest Ecosystem Management Simulation Group in 1977 to commence the development of what was then a new line of hybrid ecosystem management models, an approach that has recently attracted increasing interest amongst forest modellers, and that we predict will become the mainstream of forest modelling.

This book presents a rationale for the development path we chose, a description of the decision support and communication tools that we have developed, and examples of their application to a range of issues and processes. This chapter explores some of the issues mentioned above that require the development of a new generation of computer tools with the aim of achieving the sustainability and stewardship of our inheritance from the past and our legacy for the future.

Human population growth: the ultimate environmental threat to the world's forests

> *We have met the enemy and he is us.* (From the Foreword to The Pogo Papers, Copyright 1952–53; used in the first Earth Day poster by Walt Kelly, 1970)

Paul Ehrlich (1968) predicted imminent disaster for the world in his neo-Malthusian book *The Population Bomb*. He asserted that human population growth would soon outstrip the supply of food, energy and materials, and furthermore that human impacts on the planet would continue to increase even if population growth were to slow down. These impacts involve wealth and the power of technology, both of which are increasing faster than population size.

Following shortly after *The Population Bomb*, the Club of Rome published *The Limits to Growth* (Meadows et al, 1972). This noted that there are ultimate limits to the growth of human numbers and/or activities because of one or more factors: world population, industrialization, pollution, food production and resource depletion. Problems associated with any one of these were thought to be solvable if the factor was independent of the other factors. However, it was concluded that because these issues exist as a linked system, solving any one problem could exacerbate one or more of the other factors, which would then become limiting.

Erlich's warning was largely forgotten during the last quarter of the 20th century, partly because it coincided with a slowing of the rate of growth in human numbers (however, the annual increase in absolute numbers of people continued for many years), and because other environmental issues captured public attention. The Club of Rome's warning was rejected by many as excessively pessimistic, and because it was based on a relatively simplistic model that underestimated human creativity, ingenuity and problem solving. However, the recently published 30-year update of *The Limits to Growth*, with the support of 30 years of development in environmental sciences, is even more pessimistic than the first edition, and it points out that the message is even more urgent after squandering the opportunity to correct the course of human activity over the last 30 years (Meadows et al, 2004). The population has continued to grow from just 3 billion in 1960 – the year The Beatles formed – to 6.8 billion in 2009, and is predicted to peak at about 9.2 billion (United Nations, 2004); the 2008–2009 global economic recession has illustrated all too vividly how closely linked the different sectors of the global human system are; and climate change augmented by accelerated release of greenhouse gases through human activity echoes the Club of Rome forecasts. As predicted, increases in wealth, the use of fossil fuels and the power of technology have increased our per capita environmental impact, and the recent concern over the human ecological footprint (Wackernagel and Rees, 1996) is one of several signs that the threats associated with increasing human numbers are about to take centre stage once again. The rise in concern over the global climate is supporting this reawakening of concern over population and per capita consumption.

In addition to these neo-Malthusian predictions, societies in wealthy, developed countries have come to recognize the wide variety of environmental values and services provided by terrestrial and aquatic ecosystems. The provision of many of these values and services through fossil energy-dependent human systems is economically and/or environmentally unachievable or unsustainable. The rise of the environmental movement has focused attention on human impacts on the functioning of local and global ecosystems, and issues such as acid rain, the ozone hole, biodiversity, over-fishing, tropical deforestation, climate change and carbon budgets have become a central focus of many governments.

Not least amongst the growing concerns about the environment is the state of the world's forests and their management (FAO, 2009). Despite the origin of this concern being the rapid shrinking of species-rich tropical forests due to population growth and land-use change, much of the action by forest-related environmental non-governmental organizations (ENGOs) has until quite recently been concerned with forests at temperate and northern latitudes. In North America, issues like clearcutting and concern over iconic species, like the northern spotted owl in the north-west of the USA, have grabbed the attention of the public and put the forestry sector on notice that many of the approaches to the management of forests used in the past are no longer acceptable to the public in wealthy developed countries. Various northern forests have been labelled 'the Brazil of the north': an ecologically, socially and politically inappropriate comparison, but one that resonates with the public.

In the face of escalating public concerns, the forestry sector has two major but sometimes conflicting responsibilities:

1 to change the way in which forests are managed as the balance of values and environmental services desired from those forests changes; and
2 to reject current practices and resist proposed new practices that are inconsistent with the ecology and sociology of the new desired balance of values and services over ecologically appropriate temporal and spatial scales.

The challenge is to evaluate when and how current management practices and policies should be changed in order to deliver the desired new balance of ecologically sustainable values, and when to resist changes demanded by the public, ENGOs, government institutions or industry. There are many examples of inappropriate pressures from any one of these groups that lead, through the political process, to unsustainable policies and forestry practices (Kimmins, 1993a, 1999a). The challenge is made complex by the different spatial and temporal scales at which the sustainability of different values and services must be addressed (Kimmins, 2007b). Different answers result when issues are considered at different scales. Optimum management strategies and practices will vary between different forest ecosystem types (stand level) and different ecological regions (large landscape scale).

The choice of when and how to change forest management in response to existing or anticipated threats to desired values has traditionally been based on experience. The complexity of forest ecosystems has in the past so far exceeded our understanding of those ecosystems that prediction was of necessity based on observations of how actual ecosystems have responded to different management or natural disturbance regimes (essentially, adaptive management: Walters, 1986). Allowing the forest to inform us of the forest's response to different ways of managing it was really the only alternative: 'if in doubt, ask a tree'. However, experience is always of the past and never of the future, which is what we are concerned about. Choosing our path into this future based only on the past is a bit like rear-view mirror driving. While it may be reliable where the road ahead is straight, navigation based solely on the information provided by the rear-view mirror is fraught with risk. There is a significant probability of crashing or going off the road if the future turns out to be different from the past, or the situation is changing so rapidly that our experience is incomplete or inappropriate.

Because of the long timescales in forestry, decisions about actions today should always involve a consideration of possible outcomes far into the future ('far' usually spanning many decades or even centuries). These are timescales over which significant changes in climate, soils, biota and society's values are expected to occur. They generally exceed human lifespans, sometimes several times over. Where only short-term forecasts are needed, experience often continues to be the best foundation for prediction. But over the timescales involved in issues such as forest sustainability and stewardship, experience on its own is not enough. This lesson was learned early in the evolution of modern forestry. In the late 1700s, German foresters developed experience-based predictive growth and yield tools (yield tables) with which to calculate future supplies of timber and sustainable rates of forest harvest. By the second quarter of the 1800s, evidence was accumulating that the predictions based on these tools were inaccurate in pine forests in northern Germany, where soil fertility was being degraded by the annual collection of leaf litter by landless peasants. By 1876, the ranking German chemist of the day, Ebermayer, concluded that in the future, forecasting tree growth should incorporate our understanding of the basic ecosystem determinants of forest growth and yield, and not just experience (Ebermayer, 1876). This work continues to be an inspiration (Rennie, 1955; Assmann, 1970; Puettmann et al, 2009).

Forest management planning is all about prediction. What will the future supply of timber be? What effect will timber management and harvesting have on wildlife, habitat and measures of biodiversity? How will recreational and aesthetic values be affected by management for timber or other values? What are the possible/probable consequences for watershed and fisheries values, employment, economics, fossil fuel use and carbon budgets? Will soil fertility be sustained? These and many other questions must be addressed by the contemporary forest manager, and experience will frequently not provide the answers because our experience is of past methods of management that the public has demanded be changed. Past management frequently ignored non-timber values and ecosystem processes, so we have little experience of the effects of past management on them. Without credible, multi-value prediction tools, a forester has little basis for claiming that she or he is managing the forest sustainably, just as without such tools the critics of forestry lack a credible basis for their claims of non-sustainability. No matter which side in the forestry debate you are on, you will need the best, science-based forecasting tools available if you are to make a defensible case for your position. Such tools need to be process-based (Korzukhin et al, 1996) and at the ecosystem level of complexity (Kimmins et al, 2008a).

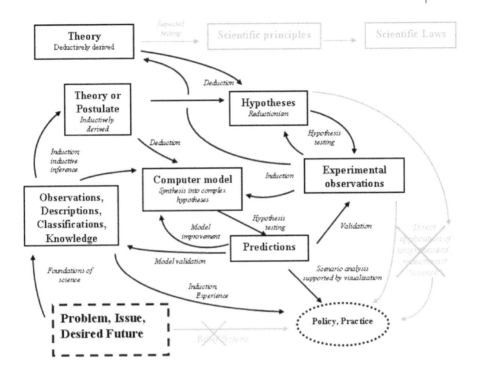

Figure 1.1 *The major components of a complete science in forestry*

Note: neither poorly informed belief systems nor the direct application of the unsynthesized results of hypothetico-deductive science (the understanding component) provide an adequate basis for forest management policy and practice.

Source: adapted from Kimmins et al (2005)

The failure of science to provide the necessary analytical and forecasting tools for resource management: the question of complexity and prediction[1]

Science is a human activity that relates to our need to *know, understand* and *predict* the objects, conditions, processes and/or events that are the focus of the science. It is generally driven by the desire of society to address persistent problems:[2] to make things better (Medawar, 1984). These problems are usually complex and difficult to understand (sometimes referred to as 'wicked problems')[3], and it is generally beyond the ability of belief systems or experience to resolve them. Herein lays the rationale for science in general, and science in forestry in particular: a way of knowing, understanding and predicting, and thereby dealing with, complexity.

Science can be reduced to three complementary components or objectives that are related to the above three needs. Each of these three parts is necessary for the solution of complex problems; none is sufficient on its own. Only when all three parts are combined are we likely to achieve the objectives of science in forestry and provide an adequate basis for guiding the relationship between people and forests (Figure 1.1). Communication is considered by some to be a fourth component of science. Science that is not communicated does not contribute to the advancement of humankind and our quest to live sustainably with our environment. Communication is required in all three of these components.

Knowing

The first stage in resolving a problem issue is to know what the problem is. Knowing about a problem, object or condition requires that it be described and classified relative to other problems, objects or conditions. The creation of this level of knowledge (which we call *experience* when knowledge is accumulated over time) by the process of induction (going from descriptions of many examples of the issue, object or condition to statements about specific examples thereof) will generally be more reliable as the basis for policy and practice than a simple belief system, if the latter lacks a solid, inductively derived foundation (i.e. an appropriate level of knowledge and experience) (Figure 1.1). However, knowledge and experience always refer to the present and the past, respectively, whereas in forestry we want to know about the future. What should we do today to ensure a desired future for both us and our descendants, and for other species and forest ecosystem functions? What policies and actions applied today will honour our intergenerational equity obligations and ensure that we practise multi-value, multi-generation, ethical forestry?

Experience on its own is an adequate (and sometimes the best) basis for predicting the future if the future is expected to be exactly like, or very similar to, the past. This is generally not the case. In fact, problems in forestry sometimes arise because the public wants to change from the policies and practices of the past to new policies and practices of which we have little or no experience. As described above, this is like driving slowly while looking only at the rear-view mirror. If there is no other traffic on the road, one might be able to navigate safely because the direction of the road is changing slowly relative to the past direction and one's speed. The knowledge gained from the rear-view mirror will show you in sufficient time that you have started to go off the road, allowing for a course correction. Thus, experience is reliable only when the future is the same as, or is changing slowly relative to, the past. Such a navigational procedure is not appropriate for driving on a winding mountain road, where the course changes rapidly and unpredictably relative to the past. It is equally inappropriate when the road ahead is crowded with obstacles, and/or one's speed is increasing. In such cases, excessive reliance on experience will probably result in a crash, or your vehicle going off the road. This has implications for the adaptive management approach in forestry (discussed in Chapter 9). If experience on its own does not give you the necessary feedback in sufficient time to avoid making significant mistakes, it cannot be relied on to prevent undesirable outcomes.

The power of the inductive component of science (knowledge, experience) is that it is based on many examples or observations of reality which incorporate implicitly the complexity of the real world. However, the complexity of the determinants that collectively result in what we have experienced is often so great that frequently we cannot be confident that we have fully described and correctly interpreted the reality that is represented in our experience. Was the gathering of the data on which induction and experience are founded biased in some way? Was it sufficiently complete to have accurately described the full complexity and variability of the focus of our experience? Are we able to perceive, observe and measure the object of our experience? Are our cognitive abilities sufficient for us to have created a sufficiently complete knowledge base? Two additional types of scientific activity are required to assue us that our induction-based interpretations can be reliable foundations for effective policy and practice. The first is the critical testing of the descriptions (i.e. experience) of, and hypotheses about, the components and processes that are the focus of our study: a process that provides *understanding*.

Understanding

Understanding involves the study of the individual parts and processes of the problem, object or condition in question. This component of science involves reductionism: the breaking down of initial theories or explanations based on induction and experience into their individual parts. The process of *deduction* is then used to derive hypotheses about these components that can be tested. A synthesis of the results of this testing provides a basis for reliable prediction, or, at the very least, a set of limited and contingent generalities. In basic forest science (that which is not related to policy and resource management practice), experimental results produced from investigations of individual components and processes in forests ecosystems are typically added to the growing body of knowledge. This body then becomes the foundation for revised or new theories or postulates (an inductive process) about these complex processes. After sufficient cycles around this inductive–deductive loop, these theories (now deductively as well as inductively derived) may be raised to the level of scientific principles. If these withstand a prolonged period of testing, they may become scientific laws. However, such principles and laws are generally sufficiently theoretical and/or reductionist that it is difficult to apply them directly to solve 'wicked' problems in forest policy and practice. They generally refer to sub-components of ecosystems – to lower levels of biological organization – and are therefore of limited value for predicting possible future conditions in real ecosystems.

For those who believe that 'understanding' is what constitutes science (i.e., that it is the only 'hard science'), the direct application of results from reductionist, hypothetico-deductive experimental science is not only justified; they assert that it should be the scientific foundation for policy and practice (Figure 1.1). However, this view is not consistent with our understanding that science has the three essential components discussed here.

Understanding the individual components and processes of a problem, object or condition is an essential component of science, but is not sufficient. The results of reductionist hypothesis testing in the context of complex, 'wicked' problems is analogous to taking a jigsaw puzzle out of the box and examining the colour, pattern, shape and texture of the individual pieces. This is an essential activity in the assembly of the puzzle, but knowledge of any single piece on its own does not give much information about the overall picture. Assembling a few pieces may help increase one's comprehension of a portion of the picture, but only when all the significant pieces have been assembled is it possible to interpret the entire picture. Using the direct results from hypothetico-deductive science as the basis for addressing complex problems in forest management is akin to predicting the completed jigsaw's picture from one or a few unassembled pieces; it results in 'jigsaw puzzle policy' that is ineffective as a basis for forest ecosystem management policy.

Building a house is another useful analogy (Forscher, 1963). To achieve the objective of a functional house, one needs bricks, timber, glass, wiring, and many other objects and materials. However, piles of bricks, pieces of wood, bits of glass and coils of wiring do not constitute a house. Only when guided by an architectural design and assembled by a builder into an appropriate spatial configuration that establishes functional linkages between the parts will the complex system of the house fulfil its purpose. Inductive science (knowing) is the scientific equivalent of the building's foundation. Hypothetico-deductive experimental science is the scientific equivalent of creating bricks and planks (creating the building blocks needed for understanding). However, not until all the components have been assembled into a system of sufficient complexity to achieve the objective (prediction in the case of science) do you have a functional house. Warehouses full of scientific 'bricks' (we call these 'libraries') do not provide society with the scientific equivalent of functional shelter and living spaces. Each component of a house is interesting in its own right, but not until it is assembled into a functional architectural system is its full value realized. The same applies to the unintegrated understanding of the individual components and processes of forest ecosystems.

Predicting

The 'architectural plan' and the 'building' of theories and hypotheses of sufficient complexity and scale to address complex and wicked problems is the focus of the third component of the science triad: *prediction*. Of course, prediction is also a component of the hypothetico-deductive activity; it is part of hypothesis testing. However, prediction at that level is not what we are discussing here. Life is like a movie; things are continually changing. Knowing is analogous to a description of the picture in one frame of the movie. Understanding is analogous to examining all the components of this frame and how they fit together to make the assembled picture. Prediction is like the making of a movie from a series of pictures of the system at various times in the future to define change over that time.

A major reason why science has not proved as useful in forestry as is often expected is because it has focused on understanding rather than prediction: it has often not been adequately grounded in knowing, has failed to deal adequately with complexity, and has focused on spatial, temporal and complexity scales far removed from the issues faced in forestry. It has provided academic bricks rather then assembled buildings. Many biologists and ecologists working on forest ecosystems have failed to comprehend these limitations, and have used population- or community-level ecology on its own to try to understand and predict forest ecosystems.

In essence, prediction is merely the process of extrapolating knowledge and understanding of the present and the past into the future. It is typically based upon defined relationships between predictor and response variables. Describing, understanding and quantifying such relationships usually occurs through the inductive–deductive hypothesis-testing loop (Figure 1.1). Many would argue that prediction is the fundamental objective of science; others would suggest that knowledge and understanding for their own sake are sufficient rationalization for this human activity.

Regardless, the application of science to address an issue or problem invariably involves some form of prediction based on a synthesis of knowledge and understanding of the components of the issue or problem.

Navigating into the future in a sustainable and ethical manner depends on our capacity to forecast the possible consequences of alternative policies and practices in complex systems. While the reductionist or analytical approach to science may work well for predicting response in relatively simple systems, it is insufficient for complex systems, where the old adage 'the whole is greater than the sum of its parts' usually applies. To adequately address the complex problems associated with the management of forest resources requires a *systems approach*, in which an understanding of the linkages and interactions of an ecosystem's key components and processes provides the basis for biophysical predictions. This biophysical (ecosystem-level) activity must then be linked to social values in order to predict (through scenario and value trade-off analysis) the interactions between the social and biophysical components of forests and forestry.

Complexity and prediction in forestry: levels of biological organization and integration

The need for adequate complexity in theories and explanations was recognized long ago. William of Occam,[4] a 14th-century English logician and Franciscan friar, asserted that explanations, theories and solutions to problems should not be more complex than necessary. This principle (called the 'principle of parsimony', and adopted by science in the form of 'Occam's razor') has become a basic tenet of reductionist science (the hypothetico-deductive *understanding* component); it encourages the use of the simplest hypothesis available. However, Occam's razor has two edges: *as simple as possible, but as complex as necessary*. This idea was echoed by Albert Einstein in his admonition to keep theories and explanations 'as simple as possible but not simpler'.

Progress in *understanding* requires the simplification of complex issues, theories etc., but the *prediction* of complex systems requires the synthesis of scientific bricks into explanations of sufficient complexity to reflect the characteristics of the object(s) of scientific activity and the determinants of complex problems. As noted earlier, *The Limits to Growth* (Meadows et al, 1972) concluded that science and society are probably capable of solving any single problem that they face if it can be isolated from other problems. However, all major problems are linked in complex ways, so solving one problem exacerbates other problems and/or reduces the possibility of solving them. Our ability to understand, predict and derive solutions to complex problems is generally limited more by our unwillingness or inability to synthesize reductionist, disciplinary knowledge up to the level of complexity of an issue than by our lack of detailed knowledge of the components and processes of the problem, the continuing incompleteness of this knowledge notwithstanding.

In the field of ecology, the issue of complexity and prediction was addressed 75 years ago by Tansley (1935), when he coined the term and described the concept of 'ecosystem'. This built on the complexity of animal population regulation recognized by Elton (1924, 1927). However, the inability of the basic tools of reductionist science to deal with the complexity implicit in these two conceptual advances led to a period of nearly half a century of ecological research that focused on *levels of biological organization*

Levels of biological organization	Key role of knowledge at each level	Levels of biological integration
Ecosystem Community Population	Understanding + PREDICTION Understanding Understanding	Ecosystem
Individual Organ systems Organs, tissues	Understanding + PREDICTION Understanding Understanding	Individual
Cell Sub-cellular	Understanding + PREDICTION Understanding	Cell

(Left vertical axis label: C O M P L E X I T Y, with upward arrow)

Figure 1.2 *Levels of biological organization and integration*

Note: Prediction is associated with true levels of integration, based on a synthesis of the understanding provided by lower levels of biological organization

Source: adapted from Kimmins (2007a)

(Rowe, 1961) below the level of ecosystem: ecophysiology/autecology, population and community. Rowe pointed out that the description and understanding of each level of biological knowledge is indispensable to the description and understanding of events and conditions at higher levels. However, the prediction of future events and conditions at any level can only be successful in the context of the next highest *level of biological integration*, up to the level of the ecosystem (Figure 1.2). An individual level of biological organization defines only a subset of the processes that affect future conditions and events at that level as it exists in real ecosystems. Only the next highest true level of integration in the hierarchy of system complexity defines the key determinants of the future for the level of interest. Thus, the fate of an individual organism in a forest cannot be defined solely on the basis of knowledge about the biology of that individual, or of the population or even of the biotic community in which it finds itself. The population level fails to identify all the biotic factors influencing that individual, and neither the population nor the community levels address the climatic and edaphic (soils) factors and the physical natural disturbance events that play such a key role in defining the future for an individual, population or community. In ecology, predictive powers for real ecosystems are vested in the ecosystem level (Figure 1.2).

Failure to understand the relationship between lower levels of biological organization and prediction has contributed to the popularity of the concept of 'emergent properties' (Anderson, 1972; Salt, 1979). These are processes and outcomes at a higher level of organization that cannot be predicted from knowledge of the lower level(s); system structures and properties that cannot be expected or predicted from knowledge of the component parts in isolation. For many decades ecology was dominated by reductionism, as ecologists and biologists at the individual organism/species level (autecology, ecophysiology) investigated the effects of physical factors on animals, microbes and within-plant processes, and species' habitat and range relationships.

Similarly, the indispensable work of population ecologists struggling to understand the effects of competition, predation, parasitism and disease on the numbers and dynamics of individual species populations was the major focus for much of the second half of 20th-century ecology. Earlier attempts at complex, whole ecosystem explanations for population-level events – such as those of Elton (1924, 1927) working with the ten-year cycle of the snowshoe hare in Canada – were widely dismissed as unscientific because they were not capable of being tested in definitive experiments involving research designs and statistical analysis techniques that were developed for agroecosystems (systems of contrived simplicity to limit spatial and temporal variability). The failure to address the complexities of ecosystems while trying to understand populations and communities has led to several controversies, including the great debate over the density-dependent vs density-independent regulation of populations, and the quest for the 'cause' of population regulation (Huffaker and Messenger, 1964). Only relatively recently has animal population ecology returned to Elton's ecosystem view of population dynamics as involving multi-trophic levels and interactions with abiotic factors (e.g. Sinclair et al, 2000; Tscharntke and Hawkins, 2002).

The role of process-based, ecosystem-level hybrid simulation models as a component of the solution to problems posed by complexity in forestry

How can the results of a synthesis of the products of biophysical sciences – complex hypotheses – be tested? One answer lies in the development and testing of mechanistic computer models (decision-support systems) that represent complex hypotheses about the structure and function of these forest ecosystems at various spatial scales. Testing complex hypotheses involves comparisons between model forecasts and the historical records of the behaviour of the systems in question (retrospective and chronosequence research), or with the results of long-term field studies (validation). This is analogous to the reductionist loop of *understanding*, but it deals with complexity rather than simplification (Figure 1.1).

As noted above, the traditional basis for prediction in forestry has been experience. Growth and yield models, and the timber supply models based thereon, are rear-view mirror forecasting systems. They are reliable, and are often the best overall basis for navigating into a future that is very similar to the past, or when the future is expected to change slowly so that adaptive management (Walters, 1986) will be an adequate navigational course corrector. However, such experience-based, inductive models (also called *historical bioassay* (HB) models; Kimmins, 1988), based on knowledge but not on understanding, are unsuitable for rapidly changing futures. They are not appropriate where society is requiring foresters to change to management policies and practices for which there is inadequate experience with which to build new HB models. Such HB decision-support systems are likely to be more useful for short-term than for long-term predictions, because the short term is generally much more similar to the past than the long-term future. They will probably continue to be valuable for short-term tactical decision making, but are generally unsuitable for long-term strategic decision support.

Purely mechanistic (*process simulation*) models that represent many of the key processes and components of forest ecosystems and can address the 'wicked' problems that confront forest policy makers and field foresters tend to be very complex. As a consequence, they have been calibrated for relatively few forests, or they are not widely and freely available, or they are so expensive to calibrate and complex to use that they have not been widely adopted by forest companies as decision-support and scenario-analysis tools. Exceptions to this generalization exist, and with the growing power of computers, the growing need for flexible decision-support tools (Korzukhin et al, 1996), and the increase in knowledge and understanding, the use of purely process models may be expected to increase in the future.

An alternative approach is to combine experience and understanding into *hybrid simulation models* that are able to predict for both unchanging futures (on the basis of inductive science), and changing futures for which we lack experience (based on our understanding of the system's components and processes). This approach to forecasting is gaining acceptance as one of the best ways to conduct scenario analysis of complex problems in forestry. It combines the strengths of the two component approaches, while reducing the problems associated with the complexity of pure process simulation prediction systems and the inability of experience-based systems to deal with future changes. Another alternative is to return to the use of 'expert' systems to construct complex conceptual models from which to make predictions. Expert input to hybrid simulation models is an expected component of developing and calibrating such models. However, expert systems on their own run the risk of reverting to rear-view mirror models.

This book describes work that has taken place over the past 33 years (1977–2010) to develop ecosystem-level hybrid simulation (HS) models for use as decision-support tools in forest policy and management, and as teaching, research and communication tools. The objective of the book is to introduce the reader to the HS approach to forecasting a range of possible forest futures, and to describe examples of this approach from the FORECAST family of HS models. More detail on these models can be found at www.forestry.ubc.ca/ecomodels. The book is organized as follows:

- Chapter 1 has introduced the need for, and the role of, prediction in science, policy and resource management, and described the limitations of HB and process-based models for prediction in forest management. It has emphasized the limitations of models at levels of integration below that of the ecosystem for providing predictions of the possible futures of forest ecosystems, and has briefly introduced the idea of a hybrid approach to modelling forests.
- Chapter 2 examines the major focus of our modelling: the analysis of the sustainability of multiple values in stand- and landscape-level forest ecosystems, and the effects of 'natural' and management-related disturbance on key ecosystem processes and structures. This chapter also examines some key concepts in the ecological and environmental literature that need to be addressed by forestry decision-support tools, and warns of the misuse of certain concepts.
- Chapter 3 provides a brief history of the development of the three major categories of forest model, emphasizing the advantages of the HS approach for applications in forest management. The major categories of HS model we have developed are introduced, more detailed descriptions being provided in subsequent chapters on FORCYTE, FORECAST, FORTOON, LLEMS, FORCEE and PFF models.

- Chapter 4 introduces the topic of models based in the simulation of individual trees. It provides the basic principles of hybrid single-tree models, and a review of the main models currently available, with examples of hybrid single trees successfully used in forest management. This chapter describes the model FORCEE (the single-tree extension of the stand-level FORECAST ecosystem management model) and its potential uses in forest management and agroforestry.
- Chapter 5 provides a description of the use of stand-level models in forest management. In this chapter the reader can find the basic principles of hybrid stand-level models and a review of this type of model as it is used in forestry, with examples of their applications. This chapter also introduces a description of FORECAST, a multi-value, ecosystem management model at the stand level.
- Chapter 6 introduces the use of landscape-level models in forest management, providing the basic ideas for this application and a review, with examples, of models used in the forest planning of large areas. This chapter also includes a description of LLEMS, a multi-value, local landscape extension of FORECAST.
- Chapter 7 discusses the important use of models as teaching tools, a role that is sometimes underestimated. This chapter describes why models are important in education, how to make an effective use of these educational tools, and includes a description of FORTOON and Possible Forest Futures, the educational game extensions of FORECAST.
- Chapter 8 provides a step-by-step guide to the development of models in forest management. It provides examples of all the steps involved in model development using the STELLA modelling environment. The chapter ends with a discussion of the importance of model evaluation, and of how much complexity should be included in forest management models, and provides examples from real forest management situations simulated with FORECAST.
- Chapter 9 discusses the increasingly important role of forest management models in adaptive management, certification and land reclamation. The rationale of using ecological forest models in these situations is described, and is supported with examples of the use of the model FORECAST in these situations.
- Chapter 10 provides a vision of the actual trends and the possible future of hybrid models in forest management. It describes two important research lines that are becoming increasingly important in forest management: meta-modelling and visualization. Some basic ideas about meta-modelling (linking different models working at different temporal or spatial scales) are provided, with the example of FORECAST and the harvest schedule model FPS-ATLAS. We discuss how visualization can facilitate the use of ecological models in forest management and public participation processes by providing user-friendly interfaces and easy-to-understand output. A description of CALP-Forester – a 3D interactive digital elevation representation of a local forested landscape driven by LLEMS – is also presented.

The reader can find additional digital content for most of the chapters on the website of the authors' research team (www.forestry.ubc.ca/ecomodels). This includes the full version of the model FORTOON, slideshows of FORECAS, LLEMS, FORCEE and PFF, colour pictures and photos and posters.

Take-home message

We close each chapter with a take-home message that is a brief summary of its content. There is some overlap between chapters, and consequently in the chapter-specific messages. We apologise if this becomes irritating for the reader, but we feel that our basic message is worth repeating.

The message from this chapter is overwhelmingly that the human species has to find a socially equitable and acceptable way of limiting human population growth while protecting the global economy, which provides people with the necessities for a safe and adequately nourished life, and with adequate health, housing and education. These basics are inadequately provided in most developing countries and amongst some indigenous and disadvantaged peoples in developed countries. Accompanying such goals, the per capita consumption of non-renewable energy and the excessive consumption of food, material things and government services in developed countries need to be moderated if the growing human impact on the planet is to be brought under control.

Science has served society poorly in providing the tools needed to address complex and often intractable social and environmental problems. We assert that this reflects fragmentation in our education systems, governments and other institutions, leading to 'jigsaw puzzle' approaches to the dynamic, complex and continually changing 'movie' of life. In the context of forestry, this argues for a new generation of science-based analytical and decision-support tools at the ecosystem level, capable of addressing complexity and conducting plausible scenario and value trade-off analyses.

This increasing need has been the driver of the forest ecosystem management simulation group at UBC, and the objective in the design of our models.

Additional material

Readers can access the complementary website (www.forestry.ubc.ca/ecomodels) to access the full versions of the following additional material:

- Slideshow on forestry, the environment and the role of science, by J. P. Kimmins, presented at the meeting with UBC Alumni in Kelowna, BC, 2005.
- Slideshow on science in forestry: why does it sometimes disappoint or even fail us?, by J. P. Kimmins, presented at the PIWAS meeting in Vancouver, BC, 2006.
- Slideshow on the management of forest complexity, by J. P. Kimmins, presented at the Faculty Research Seminar, UBC, Vancouver, BC, 2007.

Notes

1 Ideas in this section are presented in Kimmins et al, 2005.
2 A problem is an issue that does not get solved; an issue that gets solved quickly is not a problem. Problem issues often persist because they are complex, and only simple solutions are offered.
3 The term 'wicked problem' was originally proposed by Rittel and Webber (1973) in the context of social planning. Their discussion identified several characteristics of such problems, including the following:
- They are complex and not easily defined. Lacking a clear description, finding a solution is difficult.

- They have no clear stopping rules; it is often difficult to tell when the problem has been solved.
- Solutions are not right or wrong; they are better or worse. They are more dependent on value systems than on science.
- There is no immediate and objective test of a solution to a 'wicked problem'. The solution simply has to be tried, and its success or failure monitored over a considerable period of time.
- Every 'wicked problem' is more or less unique, limiting the possibility of learning from experience and applying general rules and guidelines.
- There is an almost unlimited number of potential solutions, making it difficult to evaluate and choose between different alternatives.
- Every 'wicked problem' can be considered a symptom of another problem. Linkages between problems increase the complexity of finding a workable solution (this was a major finding of Meadows et al, 1972).

For background on this topic, see Rittel and Webber, 1973; Allen and Gould, 1986; Rauscher, 1999; Shindler and Cramer, 1999; Salwasser, 2002; Conklin, 2005.

4   The writings about knowledge, logic and scientific enquiry by William of Occam (1284–1347, an English philosopher and theologian from the village of Ockham) played a major role in the transition from medieval to modern thought. Occam stressed the Aristotelian principle that entities must not be multiplied beyond what is necessary. This principle is known as Occam's razor, or 'the principle of parsimony'. It asserts that problems should be stated in their most basic and simplest terms. In science, Occam's razor asserts that the simplest theory that fits the facts of a problem is the one that should be selected. This rule is interpreted to mean that the simplest of two or more competing theories or hypotheses is preferable.

Chapter 2

Ecological and Environmental Concepts that should be Addressed in Forestry Decision-Support tools

Introduction

Language – the systematic use of sounds and written symbolic representations thereof to communicate – has played a vital role in human evolution and social development. The apocryphal story of the Tower of Babel stands as a warning of the undesirable consequences of the failure to employ a common language, especially when dealing with complex issues. The popularity and utility of dictionaries is evidence of the importance we place on agreeing about the meaning of words, and while language is a living, evolving thing, unless we can agree on the meaning of spoken and written words, we will fail to communicate. This leads to failure to understand, in turn leading to conflict (Kimmins, 1997b).

Debates about environmental issues, including forestry issues, are frequently obfuscated by the misuse, the undefined use or the uninformed use of a variety of terms related to the characteristics of ecosystems. As we struggle to break down the walls of fragmentation and reductionism, the development of interdisciplinarity often results in the transfer of concepts, ideas and terminology between disciplines without adequate analyses of their validity in the new context. This is contributing to an ecological Tower of Babel related to the uncritical transfer of concepts and terms between different levels of biological organization and different disciplines. This chapter examines some of the key concepts about forest ecosystems that should be addressed in ecosystem-based decision-support tools, and some of the disagreements about the meanings of these terms and concepts. It is important that we agree on the definition of these terms and concepts so that they can contribute to resolving issues rather than creating them.

The problem of uncritical use of, or failure to define, terms commonly used in environmental debates about forestry: the dangers posed by the anthropomorphizing of ecosystems

> *Ecosystem resilience describes the capacity of an ecosystem to cope with disturbances, such as storms, fire and pollution, without shifting into a qualitatively different state. A resilient ecosystem has the capacity to withstand shocks and surprises and, if damaged, to rebuild itself. In a resilient ecosystem, the process of rebuilding after disturbance promotes renewal and innovation. Without resilience, ecosystems become vulnerable to the effects of disturbance that previously could be absorbed.*

This quotation from a Swedish Environmental Advisory Council report (2002) poses a variety of questions about terminology:

- *What does 'disturbance' mean?* What is 'a disturbance' in the context of ecosystems that are constantly being disturbed by natural processes that are part of their ecology?
- *What is the scale, frequency and severity of disturbance implied by definitions of resilience?* Is 'disturbance' not disturbance if it is within the range of historical natural variation?
- *What does 'cope with disturbance' mean?* Do ecosystems 'cope'? Ecosystems are continually 'shifting into a qualitatively different state' as a result of stand dynamics and successional processes, sometimes involving 'disturbance' and sometimes not, so how can resilience be defined in terms of no change?
- *What are 'shocks' and 'surprises' to a forest ecosystem?* People experience shocks and surprises, but do ecosystems? Are 'shocks' and 'surprises' merely a product of human perception linked to an inadequate understanding of disturbance ecology?
- *What is 'ecosystem damage'?* Some disturbances merely accelerate stand dynamics or the successional development of stands. Is this 'damage'? Other disturbances create new successional sequences that are vital for the maintenance of biodiversity. Is successional retrogression that can occur under natural disturbance 'damage'? Does *any* change in an ecosystem constitute damage?
- *What does ecosystem resilience mean?* The term is used in two ways in this quote: to resist change, and to recover from change. Should the term have two meanings? Most usage refers to the resistance of ecosystems to change in the face of disturbance, yet a more important application of resilience in naturally disturbed and managed forests is the speed of post-disturbance ecosystem development. Policy responses will be related to which meaning is intended.
- *What is meant by 'absorbing disturbance'?* A lack of change in one or more of species composition, structure, function, or patterns of future change in the face of disturbance? But ecosystems are constantly changing in response to natural disturbance. Does this mean that they are not resilient? If post-disturbance succession follows a different pathway that does not re-establish pre-disturbance ecosystem conditions, is this 'renewal'?
- *What is ecosystem 'innovation'?* This word, which according to dictionary definitions generally implies a sentient organism, has many definitions: a creation resulting from study and experimentation; the creation of something in the mind; the

introduction of something new or different; a new idea, method or device; ideas applied successfully; and incremental, radical and revolutionary change in theory, products, processes or organization. Most of these definitions focus on human mental activity, something that is inapplicable to ecosystems. The application of 'innovation' to ecosystems implies a sentient organismal status for the ecosystem level of biological and ecological integration, which is not appropriate. It suggests that ecosystems have purpose. This is inconsistent with our current understanding of ecosystem-level forest ecology.

Clearly, this relatively simple Swedish statement, which on the face of it may seem to some to be sensible and supportable, raises so many questions that it is difficult to develop forest policy and practices that will honour its intent. There is no way that such a poorly defined and ambiguous statement can be interpreted according to our current understanding of forest ecosystems. Consequently, such statements are of little help in seeking ways to manage forest ecosystems sustainably and conserve their values.

Ecology and the environmental debate are increasingly cluttered with words taken from the social science disciplines and associated with humans. Similarly, the public debate over ecosystems is frequently obfuscated by concepts taken from the individual and population levels of biological organization and applied uncritically to the ecosystem level. When James Lovelock advanced the Gaia hypothesis (Lovelock, 1979, p10), which states that the earth is 'a complex entity involving the biosphere, atmosphere, oceans and soil; the totality constituting a feedback or cybernetic system which seeks an optimal physical and chemical environment for life on this planet', he was invoking organismal status for the entire planet. This analogy implies that the earth was 'born' and will someday 'die', as individual organisms do; that it will suffer major failure if one of its key parts is killed or disabled, and may even 'die' if certain components are missing (e.g. the loss of certain species); that it has a central organization (God or Gaia?) analogous to the genetic code of an individual; and that it is purposive (that is, the biota is involved in processes to create ideal conditions for itself, as opposed to individual species adapting to the conditions produced by physical and geochemical/geological processes, to the composition of communities being altered by such changes, and to the changes that these communities produce in their environment as a by-product of their life activities). The error of the uncritical transfer of concepts from one level of biological integration to another has become a problem in ecology. Wikipedia presents an analysis of the arguments for and against the trans-organizational-level leap of faith implicit in awarding organismal status to the planet (http://en.wikipedia.org/wiki/Gaia_hypothesis).

This chapter is intended to demonstrate some of the difficulties that follow from the uncritical use of terminology from lower levels of biological organization and integration, or from social sciences, and their application at the ecosystem level without an adequate understanding of their applicability to ecosystems. It also asserts that because the policy implications of various interpretations of these concepts and terms generally cannot be explored empirically, and because theoretical analysis that fails to account for ecosystem processes and complexity is not useful, there is an urgent need for process-based, scenario and value trade-off analysis tools at the ecosystem management level with which to undertake assessments of the potential implications of these concepts and various definitions of environmental terms.

Gone are the days when the job of a forester was mainly to produce as much wood volume as cheaply as possible. Such a strategy made short-term sense during wartime, when wood was a strategic material and public attention was not focused on non-strategic resources. Much of European forestry over the past few centuries, including central, southern and northern (Fenno-Scandinavian) Europe, has reflected a long history of conflict and wars, both terrestrial and maritime, and the importance of wood during this period as a military material for armies and navies. Such a timber-oriented strategy makes no sense with a global population that has increased by about ten times over the past 500 years, and nearly seven times over the past 200 years; nor during an unprecedented period without global military conflict (many local conflicts notwithstanding) when much of society, at least in wealthy countries, has been able to focus on non-timber forest values. With great increases in health, wealth, nutrition and living standards in many countries, an increasing proportion of the world's population expects forests to be managed for a much wider set of resources.

As limits to growth come ever closer (Meadows et al, 2004), and the global human footprint causes widespread changes to oceanic, aquatic and terrestrial ecosystems and the atmosphere, the public in many countries is demanding change (Rockström et al, 2009). This is supported by the increasing recognition of the overwhelming cost of replacing ecosystem services that are being impaired locally and globally. The mandate for foresters today is to manage forests to sustain these services, while at the same time maintaining the supply of both conventional and non-timber forest products and values.

The difficulty faced by forestry is not in recognizing and accepting the need for change; it is in the identification of the changes that will sustain the new set of desired values, while at the same time respecting the ecology of our diverse forests. The public, informed (but also sometimes misinformed) by environmental groups (Kimmins, 1993a, 1999a), has supported a variety of beliefs about ecosystems in general and forest ecosystems in particular that sometimes are in conflict with the mainstream scientific understanding of these systems (Botkin, 1990). Unfortunately, these belief systems, which are often applicable for certain values in some forest ecosystems but are incorrect and even dangerous for many others, are also supported by some biologists and other scientists. We believe that this results from their training and experiences having been mainly at levels of biological organization below that of the ecosystem. As a consequence, some of these scientists share with some members of the public and environmental groups an incomplete understanding of ecosystems, or at least of forest ecosystems.

These belief systems are politically powerful, and it is important that the changes in forest policy and practice that result from the public pressure they create be framed in an understanding of the differences between the metaphors and assertions of such belief systems, and the current science-based understanding of forest ecosystems (Botkin, 1990; Attiwill, 1994; Perera et al, 2004; Kimmins, 1997b, 2004a). It is also important that this science be incorporated into forestry decision-support systems to facilitate partnerships of government and industry foresters, other resource specialists, civil society organizations and scientists: partnerships that can explore the range of possible outcomes for multiple values of changes in forest policy and management that reflect current public opinion.

Aldo Leopold, the modern North American grandfather of thought on conservation and environmental ethics, noted in his essay 'The land ethic':

A thing is right when it tends to preserve the integrity, stability, and beauty of the biotic community. It is wrong when it tends otherwise.' (Leopold, 1949, p240)

This has been widely interpreted to mean that if there is any change in the species list, the *integrity* of the ecosystem has been lost: that without one of the species present today the ecosystem will collapse, just as an old-fashioned Swiss watch will not work if a single one of its many cogs is removed. Unlike forest ecosystems, there is no redundancy in Swiss watches. Unmanaged 'natural' forest ecosystems are continually changing their species list as the ecosystems pass first through the phases of stand dynamics, and then through the several seral stages of succession, and this does not lead to biotic or ecosystem collapse. Clearly, there is sufficient functional redundancy (see references in Rockström et al, 2009) in most forest ecosystems so that the temporary or even permanent absence of a particular species may have little effect on overall ecosystem function. In tropical forests in which there are highly evolved relationships between individual species, such as one species of insect being responsible for pollinating one species of tree, the loss of that insect may mean the failure of that tree to reproduce. However, in the much younger (on an evolutionary timescale) temperate and northern forests such specificity is rare, and the absence of any one animal, plant or microbial species will rarely cause the loss of another species or ecosystem function. Also, in the species-rich tropical forest, the loss of a single tree species may have little effect on the overall forest ecosystem structure and function. Of course, there are examples of communities in which one species plays a key role in maintaining a particular instantaneous level of some measure of biodiversity and community structure in particular ecosystems (Levin, 1999a), or in sustaining a particular seral stage, but the loss of that species and the consequent change in seral stage and biotic community is not a threat to the integrity and functioning of the ecosystem.

In the same vein, *stability* in this Leopold quote has widely been interpreted to mean constancy or lack of change, despite the overwhelming evidence that change at the stand level is both inevitable and essential in most forests for the long-term maintenance of ecosystem productivity and landscape-level biodiversity. Likewise, *beauty* in this quote has been interpreted by some to mean that if it looks ugly, untidy and 'wasteful' it is wrong: that aesthetic beauty is a good measure of sustainability and stewardship. This essentially European philosophy has little place in many of the forests in the world that still have many of the (frequently 'untidy') characteristics of unmanaged forests. It does not respect the disturbance ecology of forests. However, despite this ecological perspective on beauty, if aesthetic beauty is considered an important forest value, foresters should try to develop management strategies that are ecologically sustaining of multiple values but also aesthetically acceptable.

We are an emotional species that depends heavily on our visual senses. This can lead to confusion between the aesthetics of an ecosystem and its ecological condition (Kimmins, 2000b; Sheppard and Harshaw, 2000). It raises the following questions: are beautiful stands and landscapes always sustainable, are sustainable stands and landscapes always beautiful, and is small-scale, low-severity disturbance always more appropriate than larger-scale, more severe disturbance? (Kimmins, 1999a). Which of these levels of disturbance is better for ecosystem stability, 'health', resilience and integrity? The following discussion should help to illuminate some aspects of this question, but the obvious answer is: it depends! It depends on the many things that are involved in the complexity of our forests.

Interestingly, on the very next page of 'The land ethic', Leopold wrote: 'The evolution of a land ethic is an intellectual as well as emotional process. Conservation is paved with good intentions which prove to be futile, or even dangerous, because they are devoid of critical understanding either of the land, or of economic land-use.'

The first of these two quotes has been widely reproduced in the environmental debate, but often without an adequate understanding of what Leopold intended. This second quote has rarely been reproduced, but it shows that he believed that decisions in conservation should be made in the head as well as the heart.

Elsewhere in his writing, it is clear that Leopold understood the dynamic and ever-changing character of ecosystems, and that 'ecological integrity' refers to the maintenance of ecosystem processes, not the preservation of particular species lists or unchanged ecological conditions. By 'stability' he was referring to non-declining patterns of change at the stand level, and a shifting mosaic of local change of fairly constant overall character, if evaluated at a sufficiently large landscape spatial scale. By 'beauty', he was referring to the ecological beauty of ecosystem complexity, diversity, function, resilience and – when considered at the appropriate spatial and temporal scales – continuity of all five fundamental attributes of ecosystems, rather than aesthetic beauty.

This chapter explores some of these concepts, and in so doing identifies some of the performance capabilities we believe foresters need in ecosystem management decision-support systems. These should be capable of assessing the policy and management implications of these concepts.

Ecosystem stability

There are many concepts in ecology about which there is as yet no unanimity, and the concept of stability is notable among them. Grimm et al (1992, p144) note the problems associated with defining and characterizing 'stability':

> *The debate about stability in ecological theory is marked by a frightful confusion of terms and concepts. Stability concepts can only be applied in clearly defined ecological situations. The features of an ecological situation determine the domain of validity of statements about stability.*

There are numerous complications. Does 'stability' refer to gene diversity, genetic make-up, population size, persistence of a species locally or regionally, community composition and structure, the five key characteristics of ecosystems (structure, function, complexity, interaction of the parts, and change in all of these over time), or to some other biotic or ecosystem variable? 'Stability' is often thought of as constancy and lack of change, but because all of these biotic categories and ecosystems are continually changing, it would seem that in this interpretation there is no stability in nature, in which case why would maintaining 'stability' be a management and conservation objective?

Ives and Carpenter (2007, p58) captured this complexity when they said:

> *Understanding the relationship between diversity and stability requires knowledge of how species interact with each other and how each is affected by the environment. The relationship is also complex, because the concept of stability is multifaceted; different types of stability describing different properties of ecosystems lead to multiple diversity–stability relationships. A growing number of empirical studies demonstrate positive diversity–stability relationships. These studies, however, have emphasized only a few types of stability, and they rarely uncover the mechanisms responsible for stability. Because anthropogenic changes often affect stability and diversity simultaneously, diversity–stability relationships cannot be understood outside the context of the environmental drivers affecting both. This shifts attention away from diversity–stability relationships toward the multiple factors, including diversity, that dictate the stability of ecosystems.*

Recognition of the dynamic nature of all things biological and ecological has led to a variety of definitions of ecosystem stability (Orians, 1975):

- Constancy: the lack of change in some parameter of the ecosystem.
- Persistence: the period of time over which ecosystem parameters are constant.
- Inertia: the resistance of the ecosystem or one or more of its structures and processes to change induced by disturbance. The inertial aspect of stability is what Holling (1973) referred to as 'resilience'. Inertial resilience is one of two definitions of resilience;
- Elasticity: the speed with which an ecosystem or one of its components or processes returns to its pre-disturbance condition following disturbance. This is also referred to as elastic stability, the second measure of resilience.
- Amplitude: the degree to which an ecosystem or one of its components or processes can be changed and still return rapidly to its initial condition.
- Cyclic stability: the ability of an ecosystem to change through a series of conditions (seral stages) that bring it back to its initial condition. This is successional or dynamic stability.
- Trajectory stability: the tendency of an ecosystem to return to a single final condition after different disturbances have altered it to a variety of new conditions. This is the stability implied by successional convergence as exemplified by the monoclimax successional theory.

Applying these definitions to two different forest types in British Columbia (western Canada) illustrates the complexity of the stability concept.

A mature, early seral lodgepole pine ecosystem in the central interior of British Columbia

This is a fire-origin ecosystem condition that is very susceptible to fire; it has very low inertial stability to such a disturbance. However, because of its canopy seed-bank in serotinous cones, the site will rapidly be colonized by another lodgepole pine stand; it has very high elastic stability. As an early fire-origin seral stage, the pre-fire pine-dominated ecosystem condition is re-established as soon as the trees reach the same age as the original stand. Similarly, a mountain pine beetle infestation will kill mature and old-growth lodgepole pine stands, which therefore have low inertial stability with respect to this insect, whereas a young pine stand will not be attacked; it has greater inertial stability. If the young pine stand is burned in an extensive stand-replacing fire before it can develop a new canopy seed-bank, it will have low elastic stability. On the other hand, if the fire leaves scattered patches of unburned trees, the elastic stability will be intermediate. Lodgepole pine stands also have high inertial stability in the face of drought. Because fire or insect-killed older lodgepole pine stands are rapidly recolonized by pine, such pine stands have high constancy and persistence stability as long as stand-replacing fire and insect disturbances continue to occur. Such stands have low persistence stability in the absence of such disturbance because they will be replaced by the community of the subsequent seral stage. Such later seral stands can have various levels of inertial stability, depending on species composition and the type and severity of disturbance, but will have lower elastic stability following high-severity stand-replacing disturbance than the pine stand, since the area has to progress through a longer period of successional development to return to the pre-disturbance condition. Clearly, there is no simple stability relationship here.

A climax stand of western red cedar and western hemlock (both thin-barked, fire-susceptible species) in coastal British Columbia

This stand has very low inertial stability in the face of fire, but high inertial stability in relationship to insect pests (although, rarely, hemlock can be killed by defoliators). If the hemlock is killed, the red cedar remains, so the ecosystem processes are less changed than if all trees were killed; the overall ecosystem processes of the mixed stand have moderate inertial stability. However, both these species are sensitive to severe drought, and the mixed species stand has low stability in the face of such moisture stress. If fire removes the entire stand, its forest floor and much of its coarse woody debris a new succession will develop, but the re-establishment of pre-disturbance conditions may require a lengthy period of ecosystem development, suggesting low elastic stability of such climax old-growth stands. Hemlock/red cedar stands are susceptible to wind damage but tend to regenerate rapidly back to hemlock/red cedar, so the species composition of this community has high persistence and constancy stability in the face of wind, but low stability of mature forest structure The structure of an old-growth stand has low elastic stability, since the processes of stand dynamics and succession must operate for a long time to re-establish the pre-disturbance stand structure. Again, the stability concept is complex is such a stand.

Use of the term 'stability'

Both inertial and elastic stability will depend on the type of ecosystem, the seral stage it is in, the type, severity and spatial scale of disturbance, and the individual species adaptations to different types of disturbance agent, agreeing with the assessments of Ives and Carpenter (2007) and Grimm et al (1992).

Much of the use of the term 'stability' appears to refer to the parameters of the biotic community, but this biotic application of the concept is often uncritically applied to the entire ecosystem. The most stable (unchanging) aspects of an ecosystem are its geology, topography, soil, aspect, elevation, slope, climate (increasingly debatable in the face of climate change) and the characteristic natural disturbance regimes. Using the metaphor of 'ecological theatre' (Figure 2.1), these variables define the ecological 'stage', which in turn defines the biotic potential of the ecosystem, which is generally more stable (constancy and persistence, inertial resilience) than the biotic realization of this potential. This biotic potential defines which series of biotic communities can occupy and be replaced in the ecosystem over time following a disturbance: i.e. what successional sequences and pathways are possible. In this metaphor, such a sequence is the ecological 'play', each seral stage being analogous to the individual 'acts'. The ecological play defines which ecological 'actors' (i.e. species) will be present on the 'stage' at any particular time. The ecological 'stage' has the greatest inertial stability but the least elastic stability of the ecosystem components. The ecological 'play', in conjunction with periodic disturbance that initiates a new ecological 'play', may exhibit cyclical stability and may also exhibit trajectory stability. The ecological 'actors' generally have low persistence and constancy, depending on how many seral stages they appear in and their maximum longevity. They have low inertial stability since many of them are replaced naturally during succession, but this low individual species stability is part of the overall successional and ecosystem stability. If there is a mosaic of seral stages across the landscape, local extirpations will be followed by re-invasions from adjacent local populations as habitat conditions are re-established. In fact, local extirpations are often a component of the long-term stability of the landscape-scale meta-population (e.g. Huffaker, 1958).

Much of the literature on stability refers to aquatic or intertidal ecosystems, or to herbaceous plant communities that have short life cycles. Much of it deals with simple theoretical models that fail to represent explicitly the key ecosystem determinants of the object of the stability analysis. Such models were developed from simple mathematical models of simplified population or community systems, and as a consequence the assessment of the relationship between stability and various measures of biological diversity reveals a wide range, from positive through neutral to negative (Ives and Carpenter, 2007). Perhaps because of the long timescales of forest stands and successional pathways, and the natural propensity of forest ecosystems to change naturally, there has been little theoretical or empirical examination of the stability issue in forest ecosystems, studies such as Huston and Smith (1987), McIntire et al (2005) and Blanco (2010) notwithstanding. Perhaps it also reflects the relative lack of process-based, ecosystem-level models with which to explore this issue, or a failure to use such models where they are available. The patterns of change that define 'stability' must be assessed at both stand and landscape levels and over timescales that match the successional and disturbance ecology of the ecosystems in question.

Stand-level change: stand dynamics – a population or community-level phenomenon

A population or community of trees and other plant life forms invades and colonizes an area from which the previous tree population/community has been removed by

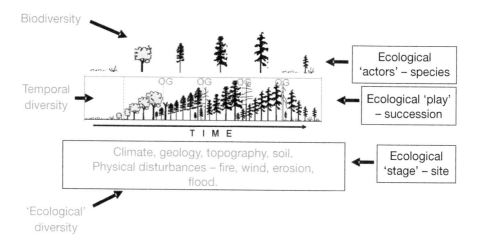

Figure 2.1 *Diagrammatic representation of the concept of 'ecological theatre'*

Note: Only one or a few plant species ('actors') are shown for each 'act' (seral stage and its biotic community) of the ecological 'play' (the successional sequence of communities/seral stages), whereas in reality there will be many species 'on stage' in each 'act'. This diagram implies that the 'script' or 'storyline' for the 'ecological play' is constant, and will be repeated exactly following ecosystem disturbance. In reality, it can vary according to different types, severities, spatial scales and timings of disturbances, differences in ecosystem character and condition, and the resultant variation in the processes of ecosystem development. 'Biodiversity' refers to the diversity of genes, species and biotic communities. The physical diversity of the landscape and physical disturbances constitutes 'ecological diversity'. This is the environmental framework within which biodiversity develops; it is not in itself 'biodiversity', contrary to the many definitions that claim that it is. 'Temporal diversity' is the change in biotic and local physical conditions over time. Note that 'old growth' (OG) – a phase of stand dynamics – develops at the end of each 'act' (seral stage) of the 'ecological play'.

Source: modified from Kimmins et al (2005)

disturbance or tree mortality. This 'stand initiation' phase of stand dynamics leads to canopy closure and the dense, highly competitive stage called 'stem exclusion', also referred to as the 'nudum' phase (because the dense closed canopy, deep shade and abundant leaf and branch litterfall eliminates nearly all ground vegetation, including bryophytes). Trees dying in this phase from competition or disease are small with narrow crowns, so their death lets little light through the canopy (Pretzsch 2009). As the remaining trees become larger with bigger canopies, gaps are left when they die, and if the tallest, largest crowned trees die or are wind damaged, significant and more persistent canopy gaps may form. This facilitates invasion by shade-tolerant shrubs and herbs, and the reduction in tree leaf litterfall may permit bryophyte development if the understorey is not too dense. Seedlings of shade-tolerant tree species may also invade the stand at this stage, if seed is available. This understorey 'reinitiation' phase leads into what is called the 'old-growth' phase, in which dying trees are so large that they leave significant and permanent canopy gaps that are large enough to permit recruitment into

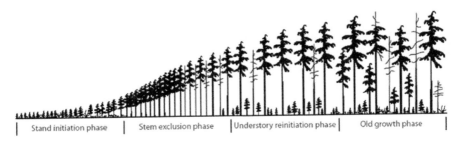

| Stand initiation phase | Stem exclusion phase | Understory reinitiation phase | Old growth phase |

Figure 2.2 *The four main phases of stand dynamics: a population model*

Source: after Oliver and Larson (1990)

the canopy of more shade-tolerant trees of the next seral stage, or of the same species if the canopy species are shade-tolerant (Figure 2.2). This series of phases occurs for each tree population in a seral stage and for each seral stage, from the first tree-dominated population/community of a succession through to the climax community.

Accompanying the different phases of stand dynamics, there will be a change in the abundance and species composition of the animals and microbes, as well as the herbs and shrubs (Pretzsch 2009). Early in the stand initiation phase following a stand-replacing disturbance, there will be increased soil moisture and an 'assart period' of increased nutrient availability. The herb and shrub composition during this period can be quite different from the characteristic indicator species for the site; there will be species more characteristic of moister and richer sites, but these will not persist, being replaced by species that are competitive under the long-term moisture and nutrient characteristics of the site. The high diversity and abundance of animals during this phase is in marked contrast to the relative impoverishment of visible biodiversity (animals and plants) during the stem exclusion phase. Young trees dominate to the exclusion of most other plants during this phase, and there is little habitat for most animals, apart from legacies of any snags (standing dead trees) or coarse woody debris from the previous stand. However, animal and plant species diversity and plant structural diversity increase again as the stand enters the understorey reinitiation phase, and may reach their highest levels in the old-growth phase, especially if this is seral old growth and not climax old growth. Late successional or climax old growth can have markedly reduced levels of some measures of biodiversity, but may have high levels of some other measures.

Snags and coarse woody debris often have a U-shaped pattern of abundance over time during the stand cycle, with legacies from the previous stand declining through the first two phases, new snags being too small to sustain the biomass of these structures. There is an increase in the size and persistence of snags and coarse woody debris during the last two phases as increasingly large trees die, creating new legacies to be passed on to the next stand initiation and stem exclusion stages (either post-disturbance or in the subsequent seral stage).

Much has been written about the change in tree, shrub, herb and bryophyte communities over the four phases of the stand cycle and the associated animal communities. Much less is known about the dynamics of the microbial community over this stand cycle. The nudum, stem exclusion phase has been called 'a biological desert' by some environmentalists, because it has an even-aged and often monoculture tree population – even in 'natural', unmanaged stands – and little animal habitat. However, with the large turnover of dead organic matter there can be a rich soil microbial and micro/meso-faunal community. The 'biological desert' designation reflects a failure to understand the full complexity and dynamics of the ecosystem.

Stand-level change: succession

If a seral stage is dominated by a single tree species, stand dynamics is a population process; in contrast, succession is a community- and ecosystem-level phenomenon. Stand dynamics as a component of succession can refer to either stands in which the stand initiation phase involves several tree species that vary in shade tolerance and longevity (the initial floristic composition successional model of Egler, 1954); or stands where shade-tolerant species invade and colonize the stand during the understorey initiation or old-growth phases of each successive seral stage (Egler's relay floristic model).

Where the tree species entering the stand during the understorey reinitiation or old-growth phases are more shade-tolerant, later-seral species, the initial population will be replaced during the old-growth phase by the tree species of the next seral stage. This will only happen if there is a supply of seed (or other reproductive units) for these new and more shade-tolerant species, without which the previous population may regenerate beneath its own canopy, but only when the death of older trees creates large enough canopy gaps to provide sufficient light and below-ground soil resources. In some forests in British Columbia, canopy gaps in the old-growth phase of any seral stage may be invaded by such dense understorey vegetation that tree regeneration is limited to stumps and large decomposing logs (nurse logs). If there is an insufficient abundance of such regeneration, gaps may not fill in with trees, and the stand may break up into shrub woodland. Only if there is sufficient ecosystem disturbance to disrupt this competitively dominant shrub/herb layer will closed forest reoccupy the site. Thus, closed forest is not the climax community in the long-term absence of disturbance in many northern forests characterized by vigorous, vegetatively reproducing shrubs; it is ericaceous woodland dominated by rhizomatous shrubby vegetation. Such shrub woodland or shrub 'ser-climaxes' can occur at the end of any seral stage if there is sufficient abundance and dominance of the shrubs and insufficient establishment of tree populations of the next seral stage (Kimmins, 1996b). Frequently, it is disturbance that facilitates the establishment of the next tree-dominated seral stage, or reverts the ecosystem to a tree-dominated stage from shrub-dominated open woodland.

As plant succession proceeds, there will be a change in animal and microbial species, just as there is as the phases of stand dynamics proceed. As suggested earlier, successional change is somewhat like the different acts in a play: some actors are present on stage in every act (the generalist species), while others (specialist species) may be present for only one act (seral stage) of the ecological play.

Stability at the stand level: what does it mean?

Considering the changes in biota, forest floor and some mineral soil horizons associated with both stand dynamics and succession, what does 'stability at the stand level' mean? The only concept that seems consistent with these realities is that of non-declining patterns of change. But non-declining patterns of what? Individual species come and go. Light and nutrient levels rise and fall, as does stand biomass. The abundance of snags and coarse wood debris of various sizes and conditions fluctuates in concert with their production, courtesy of competition and disturbance, and their loss through decomposition.

If natural disturbance keeps the ecosystem in a particular seral stage over many repeated stand dynamics cycles ('one-act plays'; see discussion of the metaphor of 'ecological theatre' above), stability may be the repeated attainment of a particular stand biomass and ecosystem condition over many cycles of disturbance. Such a definition requires a minimum of three cycles, because a trend cannot be described by two points. This is why decision-support tools that cannot examine ecosystem processes and conditions realistically over at least three cycles of disturbance are unable to assess stand-level stability and sustainability, their value for other applications notwithstanding.

However, in some ecosystems, repeated stand cycles of the same species may be associated with the development of soil conditions that are undesirable for that species (e.g. red alder in coastal British Columbia in Canada and Washington State in the USA), or increases in diseases (e.g. root rots) or insect pests. Long-term non-declining change in some forest types may require an alternation of seral communities, in much the same way that rotation farming maintained soil fertility and reduced pests and diseases before the days of fossil-fuel-based agricultural chemicals. But if a series of different seral stages is required for successional stability, it becomes difficult to define what 'stability' is, other than in terms of successional sequences (the 'ecological play').

The complexity of this issue requires the use of ecosystem process-based models that can simulate alternative management strategies to assess, for multiple values, which combination of management practices will result in the long-term repetition of a given set of ecosystem conditions. Only through ecosystem-level simulation can we assess stability and sustainability over the necessary timescales, and evaluate it for all the different ecosystem components and key processes in the face of a diversity of 'natural' and management-related disturbance. A conceptual framework for such scenario analysis is that of 'ecological rotations', which are explained diagrammatically in Figure 2.3. What this shows is that sustainability defined at the stand level for repeated stand cycles can potentially be achieved by adjusting three variables: frequency of disturbance, severity of disturbance, and rates of stand development and ecosystem 'recovery'. However, this is a very simplistic model, as it does not deal with soil, insect and disease issues associated with repeated stand cycles of certain species in particular ecosystem types. It also fails to represent successional sequences, although Kimmins (2004b) presents an extension of this ecological rotation model to a multi-seral-stage scenario. As noted above, a closer approach to the complexity of real life involves the use of scenario analysis in ecosystem management models of adequate complexity.

Landscape-level change: shifting mosaics

If you think that defining stability at the stand level is complex, definition at the landscape scale can be even more complex. In the simplest case, landscape stability is a shifting mosaic of non-declining change at the stand level. But what if the spatial pattern of natural disturbance-driven change produces an ever-changing mosaic of shapes, sizes

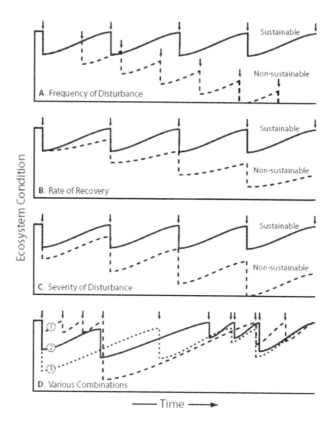

Figure 2.3 *Concept of ecological rotations*

Note: the post-disturbance restoration of ecosystem conditions and values depends on the severity and frequency of disturbance and the rate of ecosystem recovery. Non-sustainability can result from a frequency that is too high for a particular severity and recovery rate (A); a rate of recovery that is too slow for a particular frequency and severity of disturbance (B); and a severity of disturbance that is too great for a particular frequency and rate of recovery (C). A, B and C depict a sustained repetition of the levels of the three variables, something that is generally unlikely to occur. D shows a more realistic scenario in which the severity and frequency of disturbance will vary over time. It shows that a wide range of disturbance severity and frequency can be sustainable if the other of these two variables is changed appropriately. Management alteration of the rate of ecosystem recovery adds another dimension of complexity and choice (after Kimmins, 2004c). This simple stand-cycle model does not address successional stability adequately. An extension to successional sequences can be found in Kimmins, 2004b.

and arrangements of disturbance patches on the landscape? What if the landscape one is concerned about is too small to include all the seral stages and age classes of the ecosystems involved, with the result that the overall character of the mosaic is continually changing? While a landscape of sufficient size may have a stable overall character (the shifting mosaic), a smaller area that is only a portion of this larger landscape may appear to be unstable. There may never be a repeat of any particular pattern in many unmanaged landscapes, and attempts to produce a fixed landscape pattern with a fixed proportion of stand dynamics phases and seral stages through management generally will not reproduce natural patterns. Does this mean that such 'fixed' landscape patterns represent non-sustainable landscape management?

The reality is that there is no 'correct' landscape pattern or composition. For any particular wildlife species and for any other particular forest value, one may be able to define a desirable landscape pattern of seral stages and ecosystem conditions. However, increasingly we are learning that a focus on single values or species does a disservice to other values and species (e.g. Spies and Duncan, 2009). The best approach is usually to try to represent all 'acts' of the 'ecological play' in the landscape (i.e. examples of all seral stages), and all age classes of stands within a particular seral stage (i.e. all phases of stand dynamics), which requires that evaluations and management must be conducted at sufficiently large landscape scales. The current paradigm is to emulate natural forest disturbance (ENFD) patterns to within the range of natural variation (RONV). However, depending on the size of the landscape being assessed and the forest region, this could range from trees being harvested from entire valleys or local landscapes and a single age class of young forest established over the entire area, to almost no stand-replacing disturbance anywhere. Also, from which historical period do you take your RONV model? The little ice age? The warm period before or after that period? The patterns imposed by First Nations' use of fire? The pattern after a century of fire control? How do we know the range of natural variation for times for which we lack pollen records, tree ring chronologies, First Nations' traditional knowledge, and other lines of historical evidence? These questions must be addressed before we can be confident that we are managing landscapes 'sustainably', and that the landscape is 'stable'. Some examples of different sustainability approaches to forest management at the landscape level can be found in Sayer and Maginnis (2007).

Defining stability at the landscape scale must consider the effects of climate change. Climates have always been changing, even within the lives of trees, especially long-lived trees. As a consequence, species ranges have changed; trees have invaded alpine meadows during periods of warm winters with low snow packs. Trees have invaded low-elevation grassland during wetter periods. Hamann and Wang (2006) present a view of future climatic regions in British Columbia under various climate-change scenarios, and use statistical correlations to suggest how this might affect tree-species distributions. However, a re-assessment of their earlier predictions using a more complex and realistic bioclimatic model suggests that their initial forecasts for ecozone movement are probably excessive.

The need to include strategies of adaptation to, and mitigation of, effects of climate change is widely recognized. Although there have been many estimates of the effects of climate change on forests, we cannot yet predict with confidence the long-term consequences of climate warming (Redmond, 2007). The major effect may be increases in fires and insect epidemics, and possibly some forest disease issues (Bergeron and Flannigan, 1995). There will undoubtedly be direct effects on seed production, regeneration and tree physiology, and climatic zones may move significant distances in regions with flat topography (Hamann and Wang, 2006), although the error associated with the predictions for these conditions is also high. Recent research has documented the migration of tree species in the USA (Woodall et al, 2009) and increased rates of tree mortality in the Pacific north-west (Van Mantgem et al, 2009).

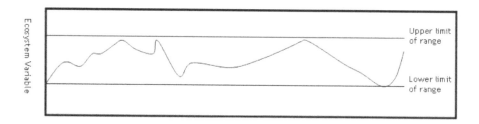

Figure 2.4 *Variation in a given ecosystem attribute (e.g. percentage of landscape in recently disturbed, mid-seral or late-seral conditions) over time*

Source: adapted from Morgan et al (1994)

While bioclimatic models have become a widely used tool for assessing the potential responses of species ranges to climate change (Beaumont et al, 2005), some researchers have criticized this 'bioclimatic envelope' approach because bioclimatic models do not represent biotic interactions, evolutionary change (adaptability) and species-dispersal strategies and limitations (Pearson and Dawson, 2003). It has been suggested that without accounting for interspecific competition, migration rates, seedling production, invasibility or disturbances, climate envelope models lack practical utility (Davis et al, 1998; Grace et al, 2002; Thuiller et al, 2008). For example, Bergeron et al (2004) showed that the limits between mixed-wood and coniferous forest in north-eastern North America, which apparently match climatic boundaries, are actually the result of wild fires. The climatic conditions of present species distributions are also not necessarily a valid proxy for possible future tree distributions, because forests, especially in the northern hemisphere, have not yet reached equilibrium after the last glaciation (Bergeron et al, 2004). The complexity of climate-change effects on forests is so great that it will require a linkage between climate-change models and process-based ecosystem models that incorporate climate effects.

In more mountainous topography, the effects of climate in determining forest composition are strongly modified by aspect, slope, slope position, and soil moisture and fertility. Biogeoclimatic zones are not areas of homogeneous vegetation that will move as a body when climates shift; they are a mosaic of different successional sequences of biotic communities ('ecological plays'), the overall character of which reflects the mosaic of different site types (spatial patterns of different 'ecological stages') as determined by soil, geology, slope, slope position, and soil moisture and plant nutritional conditions. Climate change is likely to have more subtle effects on the composition of this mosaic than has been suggested, especially in physically diverse areas. Also, our long-lived species have survived through major climate shifts over the past millennium. Effects on trees may have more to do with seed production and recruitment of seedlings than with the mortality of mature trees, resulting in considerable time lags in any changes in tree species distributions, unless the forest is regenerated by planting, or unless changed insect, fire and disease regimes become the major agents of climate-change effects.

Lo (2009) suggested that the migration of current ecosystems northwards or upwards as whole units is unlikely because of site- and species-specific responses to different climatic and non-climatic features. What could be climatically positive for one species (increased summer precipitation in the montane spruce zone of British Columbia for lodgepole pine) could also benefit other species (e.g. hybrid spruce), but be irrelevant for yet others (e.g. Douglas-fir). The response to changing climate is species-specific, and may vary between local populations and provenances of the same species due to within-population genetic diversity (Hurtley, 1991; Wilmking et al, 2004). In addition, many species can grow well in environments warmer than their current ranges, but are prevented from doing so through mechanisms of competition with faster-growing species, not because of poor adaption to climate (Hurtley, 1991). Different species will migrate at different speeds, and many current tree populations will remain in their present ranges, making it difficult for southern species to successfully establish themselves, unless the present populations are eliminated via other disturbances. As climates change, climatic zones may cross geological and soils boundaries, changing the relationships between climate and trees. As a result, new biological communities will be created in a process similar to post-glaciation colonization, which in some areas is still under way. Until we understand more about climate-change effects and their potential variation in different parts of the country, coherent policy will be difficult to develop.

In concluding this section on ecosystem stability, it is difficult to support any definitions or concepts of stability that do not incorporate change, since this is a fundamental characteristic of ecosystems. As with the concept of sustainability, the concept of 'no change' cannot exist at the global level in the face of the continued growth of the human population, its per capita resource and energy consumption and its ecological footprint (Rees, 1992). Stability and sustainability at the landscape scale are similarly related to escalating global warming, whatever its causes, and all the other impacts that humans have on the environment. Similarly, the concept of 'no change' is inapplicable at stand and local or regional landscape scales.

What we are faced with is continuing changes in the earth's ecosystems. What we are challenged with is how to define stability and sustainability in the face of these changes, and how to regulate and manage human–environment interactions in a way that is consistent with whatever definitions we agree on.

Ecosystem resilience

The term 'resilience' was introduced by Holling (1973) as a component of his exploration of measures of stability in ecological systems. The original concept revolved around the question: how much disturbance can an ecosystem endure without fundamental change in its character? Since then, ecologists have developed resilience theory in an effort to explain the 'surprising' and non-linear dynamics of complex adaptive systems (see a collection of the most influential papers on this topic in Gunderson et al, 2009). This interpretation of the term 'resilience' reflects the system's 'inertial stability', but as noted above there is a second meaning: 'elastic stability', or the system's speed of recovery after disturbance (its 'bounce-back rate'). Neither meaning is easy to define for real forests, as opposed to theoretical populations, communities or ecosystems, without detailed information about the type of forest ecosystem, the age, composition and seral stage of the biotic community, the biological legacies from the past, and the spatial scale at which the evaluation is made (e.g. a single leaf, a branch, an individual tree, a group of trees, a species or entire biotic communities over 100ha, 10,000ha or even larger scales). It also depends on the type of disturbance, such as fire (crown, ground or surface, and the severity of each), disease, insects or pollution; and the climate regime may influence the ability of the biota to resist (i.e. tolerate with little change) these disturbances. Thick-barked Douglas-fir, ponderosa pine and western larch have high inertial resilience to some categories of fire, but not to others. They are resilient to bark beetles under some climatic conditions, but not others. Thin-barked species (such as shade-tolerant climax or old-growth species) are not resilient to any fire, and may or may not be susceptible to diseases and insects.

Change in forest ecosystems caused by succession and stand dynamics is inevitable, so for forests it is not clear what is meant by the question: how much disturbance can an ecosystem endure without fundamental change in its character? This suggests that a resilient ecosystem will not change in the face of disturbance, yet disturbance is often a key component of natural successional change, and for many forests is a key factor in sustaining productivity and several or many measures of biological diversity. It would appear that this concept is more appropriate for populations than for ecosystems, and/or for aquatic ecosystems rather than for forest ecosystems.

The inertial stability interpretation of ecosystem resilience appears to relate more to the effects of disturbances such as acid rain, climate change and aquatic eutrophication than to disturbance regimes in forest ecosystems. 'Elastic stability' is a more useful concept for forest ecosystems, in most of which disturbance is a natural and necessary factor for 'normal' ecosystem function. However, the concept is not relevant where disturbance merely accelerates stand development or successional change.

As suggested in Figure 2.3, stability and sustainability at the stand level as a non-declining pattern of change cannot be defined by a single variable; it is a function of the combined action of the degree of retrogressive ecosystem change caused by a disturbance (related to the inertial stability aspect of resilience), the frequency of disturbance, and the rate at which the ecosystem returns to pre-disturbance conditions (related to the speed of ecosystem recovery, and thus to 'elastic stability'). 'Elastic stability' is determined by a whole suite of factors that relate to the speed of succession, including serial stage, soil, climate, availability of seed and the determinants of production ecology. However, 'elastic stability' only refers to one possible outcome of disturbance – successional retrogression – whereas disturbance may in fact advance succession (Kimmins, 2004b).

Stand-level resilience may thus be either the ability of the ecosystem to resist change in the face of disturbance, or the speed with which the ecosystem returns to initial conditions following disturbance. These different concepts of ecosystem resilience lead to different management strategies to sustain ecosystem resilience, and to the considerable confusion over this term and concept. The concept of resilience works poorly when disturbance accelerates rather than retards succession.

'Landscape-level resilience' refers to variables such as pattern, scale, patch-size frequency, fragmentation and connectivity. The inertial resilience of such characteristics may be related to the relatively unchanging physical factors of the ecosystems that often influence the type and severity of disturbance: soil, topography, aspect, slope, etc. The inertial resilience of landscape pattern will thus tend to be higher in physically diverse landscapes (such as mountains) than in flat topographic regions. If the above characteristics are altered by disturbance in landscapes lacking strong physical diversity, elastic stability may be very low. However, disturbance probability and severity are sometimes related to characteristics of the forest plant community, so the biota may contribute to inertial stability in some forest landscapes with low physical diversity, at least under moderate- to low-severity disturbance.

The whole issue of resilience, whatever the interpretation of this term, is confounded until we know what ecosystem variables are being considered; different variables have different levels of resilience and respond differently to different types, severities and frequencies of disturbance. We need to know over what timescale resilience is being considered, and over what spatial scales. The inertial stability of the landscape is different from that of an individual small stand. Different seral stages have different resilience properties. To talk about ecosystem resilience in the absence of such detail is to have a conversation that leads nowhere.

If forest ecosystem complexity is not addressed, it is unlikely that we will get satisfactory forest policy and practice. Clearly, we cannot manage for ecosystem stability and resilience unless we understand what they are, and clearly they are complex issues. An exploration of these concepts cannot be undertaken empirically because of time and spatial constraints. What is needed is process-based, ecosystem-level models at stand and landscape scales that can represent the key processes and resultant key ecosystem features.

Ecosystem health

It was perhaps inevitable that as the public became increasingly concerned about the environment, it would use human health as a metaphor for ecosystem condition. Overwhelmed by visual images of deforested, eroding and/or drought-ravaged landscapes, and the accompanying scenes of human starvation and illness, the concept of 'dying ecosystems' became firmly entrenched in the minds of the public. Scenes of polluted air and poisoned rivers and lakes, together with pictures of dead and dying plants and animals, made the idea of 'sick ecosystems' easy for the public to embrace. As environmentalists and the media increasingly portrayed 'sick' and 'dying' ecosystems as being worthy of the same concern as sick and dying people, the question of forest and ecosystem health became an important issue in the evolving debate about the environment.

It is widely accepted that human health requires 'ecosystem health' (Rockström et al, 2009), though exactly what this means is rarely defined. This homology between individual organism health and ecosystem health reflects a widespread use (outside of the community of ecosystem-level ecologists) of the organismal concept of ecosystems, despite there being no science in support of this. Under this philosophy, ecosystems can die; in reality, this is something that does not happen unless all living organisms are removed (e.g. a thermonuclear blast zone, a toxic mine-tailings pond, or a massive landslide down to the bare rock), and even then there may be a microbial or cryptogrammic community interacting with the physical environment: an ecosystem. This philosophy considers that ecosystems can become 'sick', just as individual organisms can.

The problem with the idea of 'ecosystem health' is that is has embraced a combination of language and ideas from natural (i.e. biophysical), social and human health sciences, and has taken concepts from the individual organism, population and community levels from outside and within ecology and applied them directly to ecosystems. This is not a metaphor or philosophy that can be a useful foundation for understanding forest ecosystems and managing them sustainably, because ecosystems are at a different level of organization and have characteristics that are different from the other levels of organization and from human systems. Some definitions of 'ecosystem health' on the internet are outlined below:

- 'Ecological health', 'ecological integrity' and 'ecological damage' are used to refer to symptoms of an ecosystem's pending loss of carrying capacity, its ability to perform nature's services or a pending ecocide due to cumulative causes (such as pollution). The term 'health' is intended to evoke human environmental health concerns, which are often closely related (but as a part of medicine, not ecology). See http://en.wikipedia.org/wiki/Ecosystem_health for the current public discussion on this topic.
- 'Ecosystem health' is defined as a systematic approach to the preventative, diagnostic and programmatic aspects of the management of human interaction with ecosystems, and to understanding the relationships between ecosystem health and human health. See www.eartheconomics.org/ecolecon/ee_definitions.html.
- 'Ecosystem health' is a systematic approach to the preventative, diagnostic and prognostic aspects of ecosystem management, and to the understanding of relationships between ecosystem health and human health. It seeks to understand and optimize the intrinsic capacity of an ecosystem for self-renewal while meeting reasonable human goals. It encompasses the role of societal values, attitudes and goals in shaping our conception of health at human and ecosystem scales. See www.schulich.uwo.ca/ecosystemhealth/education/glossary.htm.

Note the intimate association between human health and ecosystem health, but a ubiquitous failure to concisely explain what ecosystem health is and why we should use human health as a model for some ecosystem condition. Forest legislation in Canada (e.g. the Forest Practices Code of British Columbia (Government of British Columbia, 1994)) and the USA (references in Kolb et al, 1994) requires that maintaining forest health is a fundamental responsibility of forestry. European governments, concerned about the impact of acid rain and air pollution on forests, have invested large sums of money in research to provide technical explanations for, and solutions to, examples of forest decline that are attributed to 'ecosystem sickness' (Innes, 1993). In many countries, maintaining ecosystem health has become a legal requirement and a major preoccupation in the debate over the environment (e.g. Joseph et al, 1991; Costanza et al, 1992; Monning and Byler, 1992; Society of American Foresters, 1993; USDA Forest Service, 1993; Covington and DeBano, 1994; Everett, 1994; Sampson and Adams, 1994). But the question remains as to what it is.

An analysis of the complexities of ecosystem health requires ecosystem models that incorporate major ecosystem processes and can represent a range of social as well as biophysical values and conditions. A more detailed discussion of ecosystem health is presented in Kimmins (1996b, 2004d, pp547–550) and in the other references cited above.

Forest ecosystem integrity

Integrity is the state of being complete, unbroken and unimpaired: a perfect condition, wholeness, entireness (*Websters New 20th Century Dictionary*, 1978). Images of eroding hillsides, rivers choked with logging debris, river beds filled with sand and silt from landslides, and rocky hillsides from which organic matter and soil have been removed by fire or over-grazing are the antipodes of our feelings about the 'integrity' of an ecosystem or landscape. The visual impact of a recently clearcut forest often elicits the same reaction.

With growing public concern about the environment, many people have come to think that any disturbance that creates visual ugliness, which disturbs the 'perfect condition' of a beautiful old-growth forest or that appears to have violated the 'completeness' of the forest ecosystem, has damaged ecosystem integrity (Sheppard and Harshaw, 2000). In contrast to clearcut areas or partially harvested areas with soil disturbance, partially harvested forests with little or no soil disturbance appear to the general public to be complete, whole and unbroken. By comparison, the former (especially clearcuts) appear to some people to be broken, violated and incomplete, as though their integrity has been impaired. A frequent consequence of these responses to visual stimuli is a rejection of ecosystem disturbance on the basis that it destroys or damages ecosystem integrity: therefore, ecosystems should be protected from disturbance, their 'delicate integrity' conserved, and harvesting trees should create as little physical and visual disturbance as possible and minimize any changes in the ecosystem.

As was the case with ecosystem health, there is now legislation in several countries that requires 'maintenance of ecological integrity' as well as maintenance of ecosystem health and biodiversity (references in Woodley et al, 1993). Such legislation requires a clear and accurate understanding of what ecosystem integrity is before it can achieve its intended objectives (Kay, 1993). This concept is complex and is generally over-simplified in the debate about how forest ecosystems are managed; it is so complex that its characteristics cannot be expressed through a simple indicator or descriptor (De Leo and Levin, 1997). 'Integrity' cannot be defined by snapshot post-disturbance evaluations, nor can it be based on steady-state or equilibrium concepts, which are rarely useful in the context of forest ecosystems.

There is a curious redundancy in the idea of ecosystem integrity. An ecosystem is defined as an ecological system that exhibits the criteria of structure, function, complexity, interconnectedness and change over time. By definition, it is a complex, structural, functional system in which the parts and processes are highly linked and continually changing. By definition, it has integrity as a dynamic system, and if an ecological system is a functioning ecosystem, it has integrity. The following definition is based on these thoughts. It views ecosystems as continually changing, with the integrity of one ecosystem condition being replaced by the integrity of the new condition.

A particular forest landscape unit (at either the stand or landscape scale) will retain its integrity as an ecosystem as long as its structure, function, complexity, interactions, and pattern and rate of change over time are consistent with the historical range in these variables, and can maintain production ecology, stand dynamics and succession. Its integrity will remain intact despite changes to the structure, function and composition of the biotic community as disturbance and succession occur. An ecosystem will also have integrity if it strays outside of that historic range, but it will be the integrity of the new ecosystem condition, or seral pathway. The integrity of the original condition will have been lost, but replaced by that of the new, as always occurs during ecological succession in forests unaffected by humans.

Much of the debate over ecosystem integrity and clearcutting results from the confusion between ecosystem integrity and the integrity of a particular seral stage. Both natural and human-caused disturbances and natural processes of stand dynamics and succession result in changes in ecosystems, which 'destroy' the integrity of the pre-disturbance condition but replace it with the integrity of the subsequent ecosystem condition. Clearcutting 'destroys' the integrity of the old-growth condition; it may 'destroy' parts or occasionally all of the biotic community of a climax/late seral stage of North American west coast old-growth forests (or any other late-seral or climax forest). Wildfire or other stand-replacing natural disturbance similarly 'destroys' the integrity of the pre-disturbance ecosystem condition as it creates an earlier seral stage. Unless poor road location, construction, maintenance and/or 'debuilding', or fire or wind damage cause slope instability and soil erosion, or loss of root strength causes slope failure, or insufficient soil disturbance permits herbs and/or ericaceous or other shrubs to dominate the site after fire, wind, insects, diseases or logging, the prompt regeneration that normally occurs in these west coast forests following disturbance ensures a rapid return to forest conditions. The integrity of the forest ecosystem has not been damaged, one seral stage has simply been replaced by another. Thus, the integrity of a particular seral stage is a very different question from the integrity of the ecosystem of which that seral stage is merely one of several alternative conditions.

The debate over ecosystem integrity often refers to the differences between the ecosystem consequences of 'natural' vs anthropogenic disturbance (see the discussion of 'natural' by Peterken, 1996). While these two categories of disturbance result in different post-disturbance ecosystem conditions and can lead to different successional pathways, the wide variation in scale, frequency, type and severity of both 'natural' and human-caused disturbance undermines the utility of this over-simplified distinction. What is needed is careful evaluation of the effects on specific ecosystem components and processes of particular types of disturbance, irrespective of the origins of the disturbance (for a discussion of disturbance, see Attiwill, 1994; Frelich, 2002; Perera et al, 2004; Kimmins, 2004c). Conclusions should be site- and ecozone-specific, and should not be generalized across different types of forest ecosystem and different types of disturbance.

An exploration of the complexity of the concept of ecosystem integrity requires the use of process-based ecosystem management models of sufficient complexity to represent the key processes of the ecosystem. A further discussion of ecosystem integrity is presented in Kimmins (2004a, pp550–553), and De Leo and Levin (1997). They note that the term 'ecosystem integrity' is as much a social (relating to the value of ecosystems to humans) as an ecological concept. This confusion between social and ecological sciences becomes a problem if it is used as the basis for ecosystem-specific policy and management. Unfortunately, De Leo and Levin perpetuated the classic and oft-repeated error of generalizing about the differences between 'natural' and anthropogenic disturbance of forests based on a study in a tropical rainforest, without extending the comparison to the literature on the disturbance ecology of disturbance-driven temperate and northern forests.

Are ecosystems 'complex adaptive systems'?

Another term that has crept into the discussion of forestry and the ecosystem attributes discussed above is 'adaptive ecosystems'. This concept has become widely used amongst some of the leading thinkers in theoretical ecology, but we assert that the usage of 'adaptive' in the context of forest ecosystems is inappropriate. The term was advanced by Levin (1999b) and developed by Holling and Gunderson (2002) in association with their 'panarchy' concept. It is also used in a recent book on silviculture (Puettmann et al, 2009).

'Adaptation' has been defined as: 'the adjustment or changes in behaviour, physiology and structure of an organism to become more suited to an environment. "Adaptive" is the capability for adaptation' (www.biology-online.org/dictionary). It is a term normally applied to an organism (through acclimation) or to a species (through genetic selection). To suggest that ecosystems are adaptive is to give them organismal or species status, a view of ecosystems that is not supported by current ecological sciences. We suggest that such terminology be excluded from discussions of forest ecosystems.

Individual organisms may adapt through behaviour and physiology. Populations may become adapted by changes in the genetic diversity of the population through selection for genetically controlled characteristics that are better adapted to a changed environment or biotic interactions. A community may become adapted to changing environmental conditions by changes in the species composition of the community, losing species that are not adapted to the new circumstances, and gaining new species that are; but this is the result of the individual, genetically controlled tolerances of the species involved, and not a purposive action by the biotic community acting as a unit. In contrast, the climate, topography, geology, soil, hydrology, and physical disturbances such as wind, fire, flood and landslide – the physical components of the ecosystem – do not 'adapt'. They may become altered, but usually on long to very long timescales. Thus, the biotic component of the ecosystem may undergo adaptation on an individual organism or species basis, but the ecosystem as a functional, structural, complex, interacting and changing physical and biotic system does not. Changes in forest ecosystems in response to changed physical or biotic factors are driven by a complex of successional processes, and do not represent an adaptation of the system as a whole designed to increase the survival or competitiveness of the ecosystem, both of these terms referring to the biota and not the ecosystem. In short, all biotic communities capable of undergoing succession are thus adaptive through individual organism and species responses; the physical components of the ecosystem, and thus the ecosystem as an ecological unit, are not.

'Adaptive ecosystems' is simply a misuse of the term 'adaptive'. The biotic component can be adapted to current conditions, and multi-species communities may remain unaltered in the face of some types and severities of disturbance events, more so than low species-diversity communities (i.e., the former may have greater inertial stability and could thus be considered more 'adaptive'). However, as noted above, multi-species communities can have lower inertial stability in response to some other types of disturbance than low species-diversity stands, and often have lower elastic stability. Thus, in designating a biotic community as 'adaptive', one must define this in terms of what types of stability it exhibits and what types of disturbances it experiences. There do not appear to be any generalities that are useful.

Management can be adaptive because it is a human activity that is purposive. Ecosystems are not known to be purposive, and therefore are not adaptive. Once again, the confusion between social and cultural attributes, or attributes from lower levels of biological organization, cloud the debate about how to manage forest ecosystems sustainably.

To be useful, decision-support tools in forestry should be capable of representing biotic acclimatization and adaptation, but they do not need to represent 'ecosystem adaptation' since this is not a reality. As with 'resilience', we feel that the use of 'adaptive ecosystems' reflects a biocentric rather than an ecosystem-centric background and perspective.

Old growth

'Old growth' (OG) forests have become an icon of the environmental movement (Spies and Duncan, 2009), but rarely in the public debate is the term linked to a definition that is specific to the ecosystem in question. The OG condition can occur as the final phase of any of the forested seral stages of a successional sequence (Figure 2.1), and is not restricted to the final or 'climax' seral stage (sometimes referred to as 'ancient' forests). Seral OG is a relatively temporary condition, whereas climax community OG is often more permanent but not always (Kimmins, 2003).

Definitions are required to identify the location and extent of OG, and to differentiate it from not-OG. These definitions will vary according to the seral stage they are found in, the type of ecosystem, the previous disturbance history and the time over which this condition has been developing. There is great variation in the structure (both vertical and horizontal), tree age, tree size, species composition and various other attributes of stands that are described as OG (e.g. Trofymow et al, 2003). This creates problems for identification, mapping and conservation planning for this ecosystem condition. Definitions have been advanced for some types of forest (e.g. Wells et al, 1998), for British Columbia, but what the public thinks OG is remains vague.

Forests develop and then lose OG characteristics as they age and pass through successive successional stages; the degree to which they satisfy various definitions of OG varies over time. This problem led to the adoption of the concept of old growth index (OGI), which measures the 'old-growthiness' of a stand. Most ecologically based definitions of OG involve multiple ecosystem characteristics, and any particular stand may have high OGI scores for some characteristics but low scores for others. So, is such a stand OG or not? By scoring each OG attribute out of ten, adding the scores and dividing by the maximum possible score (i.e. 'perfect' OG), one gets the OGI. One can then set conservation targets based on what levels of OGI combined with scores of individual OG attributes of concern are worthy of conservation. A difficulty with this approach is that some desired OG values may be found in stands that score low on other OG values. Nevertheless, from a policy and management point of view the OGI concept is very helpful.

While the OG condition is frequently associated with high stand structural diversity, stands of old trees that the public thinks of as OG can be single aged or have only two age cohorts; sometimes such stands can be monocultures; sometimes they can lack large standing dead trees and large decomposing logs on the ground.

One of the confusions over OG is the different values people associate with it. Biodiversity conservation is one, and the public, led by past statements from some OG scientists, generally believes that OG is high in biodiversity. It can be, but it is not necessarily so; younger stands or earlier phases of seral stages may support higher species diversity. Periods of transition between seral stages may have greater diversity of some groups of organisms than late seral-stage forests. Such statements sometimes result from comparisons between unmanaged stands with high OGI values and intensively managed, short rotation, monoculture timber plantations that are repeatedly harvested at the end of the 'stem-exclusion' phase of stand development (Figure 2.2). There is little argument that the former offers higher levels of many biodiversity values than the latter, but stands managed to sustain 'biological legacies' with retention of patches of old trees and appropriate densities can equal or exceed some unmanaged OG for many measures of biodiversity.

The assertion that high structural diversity is important for ecosystem functioning is also questionable. Many natural or managed monoculture, early- to mid-seral stands can have very high ecosystem productivity, rates of nutrient cycling and nutrient retention, often exceeding that of very old stands. True, energy flow in such stands is mainly via trees, whereas in open OG stands the herbs, shrubs and bryophytes account for a much larger proportion of energy flow and biogeochemistry. But in terms of overall ecosystem function, low-OGI stands generally exceed high-OGI stands in overall ecosystem function. On the other hand, if one defines ecosystem function as providing habitat for species that require access to high-OGI stands for at least part of their life cycle, then this statement may be supportable. But not in terms of the basic ecosystem functions of energy capture, storage, flow and biogeochemistry.

Many people think that all OG should be conserved, without considering the longevity and sustainability of the condition. Seral OG is temporary, and an OG strategy should include conservation of younger stands with moderate OGI scores that will replace seral stands with currently high OGI scores as they transition to the next seral stage with initially declining OGI values. In contrast, in the absence of climate change, some high-OGI late-successional (climax) stands may have such longevity that 'permanent' conservation is appropriate, unless (in the long-term absence of disturbance) they lose the integrity of closed OG and transition into shrub woodland that does not have a high OGI score (Kimmins, 1996b). These may have either lower or higher vertebrate species diversity, but often have higher bryophyte diversity.

Numerous researchers have noted that a fixation on OG does not serve conservation well; biodiversity is more of a landscape than a stand issue, and to maintain all native species one needs a shifting mosaic of different ages and seral stages across the landscape, including sufficient areas with high OGI to satisfy the needs of species associated with OG (Spies, 2009; Forsman in Noon, 2009; Carey, 2009). The idea that forests with high structural complexity are beneficial for biodiversity emphasizes the fact that second-growth forest, managed to sustain snags, large decaying logs and structural diversity through thinning and intermediate timber harvests, can be as useful as OG for wildlife. In coastal British Columbia and adjacent Pacific north-west US states, early seral species such as red alder will provide large decaying logs and snags for cavity-nesting species much faster than intermediate seral species like Douglas-fir or late seral species like western hemlock or western red cedar. By fixing nitrogen, alder accelerates the growth of these other species. Consequently, periodic (e.g. every second or third rotation) higher-level stand disturbance – e.g. clearcutting with soil disturbance – with variable retention, alternating rotations of different length, and a local landscape mosaic with a diversity of forest ages and seral stages (i.e. all 'acts' of the 'ecological play'), including areas of unmanaged high-OGI stands, will sustain overall landscape biodiversity better than entire landscapes of OG.

Exploration of the OG issue requires ecosystem management models that can represent stand dynamics, succession, and the development of the structural and wildlife habitat attributes that define OG. The rapidly increasing understanding of wildlife habitat needs suggests a focus on the specific habitat elements needed by native species, and how these develop over time in specific seral stages of specific forest ecosystem types, rather that a mindless commitment to conservation of OG without understanding exactly what this is and how it relates to ecosystem attributes and their temporal dynamics.

Dangers of the misuse of terminology: feeding inappropriate belief systems

Language involves the association of specific meaning(s) with specific sounds (words), and we cannot communicate effectively unless we agree on the meaning of specific sounds. The same sounds have different meaning(s), and a single meaning will have different sounds in different languages. A significant problem in discussions of the type of ecosystem attributes addressed in this chapter is that different people have adopted different meanings for particular words, leading to a breakdown in communications: an environmental Tower of Babel. This greatly complicates the discussion of ecosystem-related forestry issues. All engaged in debates over these issues should agree at the outset on the meanings of these ecological terms. Without such agreement the debates remain confused and ineffective, as has been noted by many authors.

The ultimate threat to the 'normal' functioning of forest ecosystems is human population growth, increased per capita impact (ecological footprint) on the environment, and accelerated climate change caused by human activities. These are global concerns. Considering 'stability' and 'resilience' at the global scale is a different matter from the stand and landscape scales that have been considered here, involving different and much larger-scale processes driven by as-yet incompletely understood mechanisms. Global environmental impacts of human activity are of enormous concern, partly because of the social and political consequences that result, and because poverty, social conflict and war are amongst the greatest threats to conservation and forests worldwide (IUCN, 2008).

'Stability' and 'resilience' at forest scales have different meanings in different parts of the world. What may be an accurate interpretation for an Amazon rainforest is likely to be questionable or inappropriate for north temperate and boreal forests. This problem of uncritical geographical extrapolation reached its zenith when Canada was accused of being the 'Brazil of the north', a label that revealed more about the lack of understanding of the ecological, social and political differences between the two countries than of real environmental problems. Again, we need to agree on the language we are using if communications are to result in improvements in sustainability and stewardship.

Take-home message

The discussion in this chapter is presented to illustrate the complexity and frequent lack of understanding about forest ecosystem issues, and problems related to the misuse and misapplication of terms and concepts from lower levels of biological organization and from human systems. It would be dangerous for forest policy and practices to be based on belief systems that lack an adequate understanding of ecosystem structure, function, dynamics and other attributes, the most notable of which is change. There are many examples of public pressure through the political process resulting in ineffective and sometimes counter-productive policy and practice. These have not ensured that the values desired by society will be sustained, something that Aldo Leopold warned about in 'The land ethic'.

How, then, can we avoid the problems associated with policy based on poorly informed belief systems? Science frequently has not proven helpful, since much of it is reductionist and conducted at levels of biological organization, spatial scales and temporal scales that are so far removed from the real issues as to result in 'jigsaw puzzle' policy that does not work (Kimmins et al, 2007). Experience alone is not adequate because it is of the past or present, whereas ethical forestry involves consideration of long-term future consequences of which we cannot have experience. Theoretical approaches to foretelling possible forest futures tend to be grossly over-simplified, and are often at the wrong level of biological organization. What is needed, we believe, is a generation of decision-support tools that permit credible, multi-value, scenario and value trade-off analyses at multiple spatial scales. Ecosystem management models that combine experience, traditional knowledge and a scientific understanding of individual ecosystem components and processes, integrated up to the complexity, time and spatial scales of real forest stand and landscape issues, are an important part of moving ahead towards the sustainable ecosystem management that we aspire to.

Tools that will provide these capabilities should incorporate the best available understanding of ecosystem processes, calibrated for the local ecosystems in question. They should be linked to social values, and must be able to scale up from the stand to the landscape scale. To do this while avoiding excessive complexity will involve meta-modelling: linking model sub-units into decision-support systems of appropriate complexity and scale.

Additional material

Readers can access the complementary website (www.forestry.ubc.ca/ecomodels) to access the full versions of the following additional material:

- Slide show on ecological succession presented by J. P. Kimmins, at the Nanjing Forestry University, China, 2006.
- Slide show on modelling forest ecosystem management across scales, presented by J. P. Kimmins, at the Fifth NAFEW, Ottawa, ON, 2005.

Chapter 3

Hybrid Simulation in the Context of Other Classes of Forest Models, and the Development of the FORECAST Family of Hybrid Simulation Models

Introduction

Forest models can be classified in many different ways: by level of biological integration (physiological, autecological, population, community or ecosystem); empirical or theoretical (data driven or equation driven); spatially explicit or non-spatial; spatial scale (individual tree, stand or landscape); see Chertov et al, 1999a; Mäkelä et al, 2000; Monserud, 2003; Landsberg, 2003a and Brugnach et al, 2008 for other typologies. Kimmins (1988) suggested a triage classification: historical bioassay (HB), process simulation (PS), and hybrid simulation (HS; combined HB and PS). This chapter examines this three-level classification in the context of the evolution of forest modelling, and introduces the approach to hybrid simulation modelling developed in the FORECAST-HORIZON family of models. Ecosystem-level, mechanistic hybrid simulation is thought to be the best approach to the development of predictive forest ecological or resource management tools.

A three-level classification of models

Historical bioassay (HB) models

The early days of forest ecology, and of general ecology in the context of forested stands and landscapes, were dominated by descriptions of the present and past states of ecosystems and their population or community components: essentially, experience-based ecology. The use of forest models as an aid to management planning and regulation similarly started with experience-based tools. This descriptive, inductive, experience-based stage of forest ecology and management is a necessary starting point, and is efficient for modelling ecosystems whose future changes over time are anticipated to be similar to historical rates and patterns of change: i.e. for futures that are anticipated to be similar to the past over comparable time periods. It is necessary as it provides knowledge with which to develop the conceptual models that are the progenitor of computer models, and data with which to parameterize (calibrate) the latter. It is efficient because the historical record of tree growth and stand development (for models restricted to prediction of tree growth/stand development) implicitly incorporates all the physiological, population, community and ecosystem processes that have been involved. This obviates the need to represent all these determinants explicitly in the HB model (an important practical consideration), but only if one is willing to assume unchanging environmental (climate, soil and natural disturbance) and management regimes.

Models based on experience and descriptions of the past are both reliable and an efficient basis on which to forecast future forest states for unchanging conditions. However, as suggested in Chapter 1, such rear-view mirror driving is likely to produce unexpected and undesirable results ('surprises') in ecosystems in which significant changes in either natural or management-related ecosystem disturbance regimes are anticipated, or significant environmental alteration (e.g. human-accelerated climate change) is predicted. 'Surprises' in forest modelling are more to do with modelling at the wrong level of biological integration (Figure 1.2) or using models that lack explicit representation of key processes than with real unpredictable events, although these can occur. Similarly, 'emergent properties' of forests are generally the outcome of conceptualizing a forest model at a level of integration below that of the real system being considered. Most 'surprises' and 'emergent properties' of forests, as seen from a population or community perspective, are in fact predictable characteristics of forest ecosystems.

Forest models based entirely on experience or descriptions of historical growth and stand development are referred to in our classification as HB models. The approach assumes the integration in the historical dataset of all the key ecosystem, community, population and ecophysiological processes that determine the growth of trees and other plants, stand development and successional change. It uses this bioassay of site potential under past conditions to forecast future states of the stand or landscape. Simulation of change over time in either stand structure and biomass, successional sequences or landscape patterns is based on empirically derived transition probabilities. We consider this the most reliable and efficient basis for predicting the future where all key soil, climate, natural disturbances and biotic interactions are expected to remain essentially the same. Where this is not the case, such models have little useful predictive value for conservation and sustainable resource management for multiple values, except possibly over timescales much shorter than those of the key issues in ecology and forest management.

Process simulation (PS) models

The inflexibility of HB models was one of the motivating forces for the development of modern forest science. Failure of HB growth and yield models in northern Germany in the early 1800s (Rennie, 1955; Assmann, 1970) led to the conclusion that forecasting future stand growth and timber yields should be based on an understanding of the key processes that determine growth and yield (see Chapter 1). Despite this nearly 150-year-old conclusion, only recently has pressure increased to utilize in our forest models our rapidly developing understanding of the key structures and processes at the physiological, population, community and ecosystem levels. Recent suggestions to use more understanding in our model building include Korzukhin et al, 1996; Mäkelä et al, 2000; Mäkelä, 2003 and Landsberg, 2003a.

PS models are mechanistic models that describe the major components/compartments in the system being modelled (state variables) and the key rates of transfer (system variables) of energy, water, nutrients and biomass between these in individual trees or a forest stand. The full complexity of the components and processes that determine the present character and possible future states of forest ecosystems has generally rendered it impossible to go beyond the development of conceptual models of this complexity. The development of PS computer models of forest ecosystems has always involved a simplification or omission of some or many components and processes. Forest computer models are frequently limited to those processes that can be parameterized from existing empirical data or from a theoretical knowledge of the processes. Most process models of forests are either higher-level, generalized models that lack site-specific detail (which can nevertheless be useful for addressing questions over large areas in which averages are assumed to be useful), or are models of one of physiology, population processes or community dynamics in which details of processes at other levels are averaged or simply omitted. This greatly limits the efficacy of such models as decision-support and value trade-off tools with which to assess alternative site-specific forest management and policy issues and management practices, especially in the face of predicted climate change and changing management practices and disturbance regimes, and the public's demand for the sustainability of multiple values.

As the power of computers increases, as our understanding of the complexity of forest ecosystems improves, and as recognition grows that reliable prediction can only come from ecosystem-level models that incorporate explicit representations of key processes that are anticipated to change in the future, the utility of PS models will continue to increase. There already exists a number of such models for relatively simple forests in which site does not vary greatly over considerable distances, and they are increasingly used as predictive models for the specific forests for which they are parameterized (e.g. Battaglia et al, 2004).

The existence of such models is a tribute to the foresight of researchers, research managers and granting agencies who conducted or supported the targeted team research needed to calibrate such models. The lack of such models in other areas often reflects a 'jigsaw puzzle' approach to science and the research it is based on.

Hybrid simulation (HS) models

There is growing interest in developing and using a combined approach to forest modelling in which the power of each component approach is used to compensate for the weaknesses in the other approach (Landsberg, 2003b; Mäkelä, 2009; Tomé and Soares, 2009; Weiskittel et al, 2009; other papers in Dykstra and Monserud, 2009). The empirical reliability and believability of HB models is used as the foundation for an HS model, while the inflexibility of the HB model in the face of environmental, management and social (values) change is addressed in the PS component. The flexibility of PS in addressing change is used for forecasting possible forest futures, while the difficulty of calibrating ecosystem-level PS models is reduced through a combination of the output from an HB model and our understanding of ecosystem and other processes, data from a rapidly expanding research literature, or from targeted research.

Calibration of much of the PS component of an integrated HS model can be achieved by combining the historical patterns of tree or other plant growth and variables such as tissue nutrient content with our current understanding of the physiological and population processes that have been involved in determining the historical record. Combined with a series of simple field observations/measurements and information from the extensive databanks that are now available in many parts of the world, we can draw logical conclusions about the rates at which a variety of key processes must have occurred to have resulted in the observed historical pattern. For example, field observations of variables such as light levels below the plant canopy at various levels of canopy biomass (or simply using a Beer's Law equation calibrated for the forest in question), the duration of leaf retention, the accumulation of biomass components, the reduction of stand density (trees per hectare) and the height of the bottom of the live canopy (obtained from chronosequences of stands or long-term field experiments) can yield credible estimates of net photosynthesis, biomass turnover, litterfall for various biomass components, nutrient uptake, internal cycling and nutrient return in litterfall. Competition-related mortality in stems and branches (mainly for light) can be inferred, and estimates of density-independent mortality from other causes can be obtained from historical records, field experience and/or expert opinion. Log dimensions can be determined from taper equations, snag survivorship predicted on the basis of the simulated mortality of trees of known size, and decomposition data obtained from field observations or literature studies of decomposition. Coarse woody debris inventories can be obtained from simulated tree/snag fall and decomposition rates. A more complete explanation of the internal calibration of the PS component of HS models from its HB component is provided in Kimmins et al (1999) and in Chapter 5.

Brief history of the development of these three categories

Lacking a body of understanding of the processes that determine tree growth, stand development and change, early foresters had no choice other than using experience to guide their management decisions. As formal tools were developed to assist forest management, it was therefore inevitable that such tools also be based on the extrapolation of the past performance of a forest. There was little change in this approach until the advent of computers, the development of systems ecology, and sufficient research-based understanding of ecological processes to permit the development of process-based models: models based on understanding.

The development and use of process models was initially an academic/research activity conducted by ecologists trying to understand ecosystems and their sub-levels. It dates from the late 1960s, with major advances accompanying the International Biological Program (IBP; 1964–1974), which attempted to coordinate large-scale ecological and environmental studies to examine the biological basis for ecosystem productivity and human welfare. Organized following the successful International Geophysical Year of 1957–1958, the IBP was an experiment in applying the methods of 'big science' to ecosystem ecology and environmental issues. However, unlike other more successful applications of the big science model of scientific research, the IBP was said to lack a clear socially and scientifically pressing goal. Many biologists, especially those operating at lower levels of biological organization, were critical of the IBP, which they considered a waste of research funding since, in their opinion, it had poorly defined objectives and addressed relatively unimportant problems. They also criticized the targeting of research to describe all key components and processes of ecosystem models, an activity involving research teams working from a common conceptual model that limited the freedom of individual scientists to choose their own research projects and research directions.

Despite these criticisms, the IBP was responsible for major advances in the integration of disciplinary understand up to the level of ecosystems. For the first time it permitted ecosystem-level ecologists to explore and compare the ecological character of five major biomes. It also was the genesis of ecosystem-level forest modelling. Many of the IBP process models were complex, required large research teams and substantial funding to calibrate, and resulted in system-specific models. These attributes limited the practical application of these models outside of basic ecosystem studies. They were of little value for day-to-day decision making and policy development in forestry.

A crucial stage in the development of hybrid ecosystem models was the development at Brookhaven National Laboratories on Long Island of the JABOWA model by Botkin et al (1972; see also Botkin, 1993a, 1993b). The JABOWA model stimulated the development of a large family of so-called 'gap models', and must be considered the most successful lineage of ecological forest models to date (Figure 3.1, Table 3.1). However, not all of the offspring of JABOWA remained ecosystem-level or truly HS models, and none of them include simulation of early tree growth or competition from herbs and shrubs following stand-replacing disturbance, so they may not be considered true ecosystem-level models.

Another early forest model was the tree and stand simulator (TASS) of Mitchell (1969, 1975). While this is an even-aged, spatial, individual tree stand population model, it has elements of the hybrid approach in that it modifies historical data on tree and stand development according to an index of crown competition. A later version of this pioneering stand growth simulator is still in use in British Columbia as a growth and yield model, and a multiple cohort model is under development (BC Ministry of Forests, 1999; Di Lucca, 1999; www.for.gov.bc.ca/hre/gymodels/index.htm).

The FORECAST-HORIZON family of hybrid simulation models

Thirty-five years of HS model development at the University of British Columbia

The origins of the FORECAST project was the Arab oil embargo of 1973 when the Organization of Arab Petroleum-Exporting Countries (OAPEC, a sub-group of the Organization of Petroleum-Exporting Counties, OPEC) proclaimed an oil embargo in response to the USA's decision to re-supply the Israeli military during the Yom Kippur war. Lasting only until 1974, the embargo had profound effects on western nations due to a fourfold increase in the price of oil. The economic shock that resulted led to the formation of the International Energy Agency (IEA) to coordinate measures in times of oil supply emergencies. As energy markets have changed, so has the IEA. Its mandate has broadened to incorporate the 'three Es' of balanced energy policy: energy security,

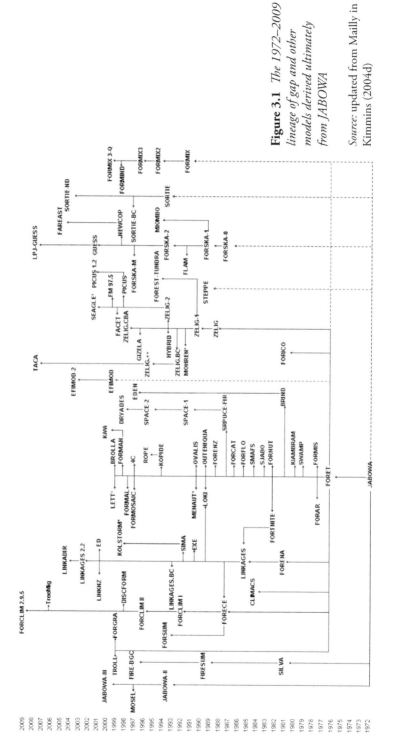

Figure 3.1 *The 1972–2009 lineage of gap and other models derived ultimately from JABOWA*

Source: updated from Mailly in Kimmins (2004d)

Table 3.1 *The main geographic locations and forest types, authors, and key literature citations for the family of gap models presented in Figure 3.1*

Model	Forest type	Location	Reference
4C	Mixed conifer/ hardwood	Central Europe	Bugmann et al (1997)
BRIND	Montane eucalypt	Australia	Shugart and Noble (1981)
BROLLA	Mediterranean forest	Spain	Pausas (1999)
CLIMACS	Pacific NW conifers	Oregon	Dale and Hemstrom (1984)
DISFORM	Sub-alpine forests	Swtizerland	Löffler and Lischke (2001)
ED	Tropical rainforest	Amazon basin, Brazil	Moorcroft et al (2001)
EDEN	Montane eucalypt	Australia	Pausas et al (1997)
EFIMOD	European boreal conifers	North Europe	Chertov et al (1999b)
EFIMOD 2	Boreal forests	North Europe	Komarov et al (2003)
EXE	Temperate hardwood	Eastern North America	Martin (1992)
FACET	Pacific NW conifers	Oregon	Acevedo et al (2001)
FAREAST	Boreal conifer forests	Northern China, eastern Russia	Xiandong and Shugart (2005)
FIRE-BGC	Coniferous mountain forests	Rocky Mountains	Keane et al (1996)
FIRESUM	Montane conifer	Montana	Keane et al (1989)
FLAM	Spruce/pine forests	Central Sweden	Fulton (1991)
FM 97.5	Conifer forest	Sierra Nevada, California	Miller and Urban (1999)
FORAR	Pine, oak/pine	Arkansas	Mielke et al (1978)
FORCAT	Hardwood forests	Tennessee	Waldrop et al (1986)
FORCLIM	European Alps	Switzerland	Bugmann (1991)
FORCLIM II	European Alps	Switzerland	Bugmann (1996)
FORCLIM 2.9.5	European Alps	Switzerland	Didion et al (2009)
FORECE	Mixed conifer/ hardwood	Southern central Europe	Kienast and Kuhn (1989)
FORENA	Deciduous forest	Eastern North America	Solomon (1986); Solomon et al (1981); (Solomon and Shugart, 1984)
FORENZ	Mixed forests	New Zealand	Develice (1988)
FORET	Appalachian hardwood	Eastern Tennessee	Shugart and West (1977)
FOREST– TUNDRA	Forest–tundra transition	North-eastern Canada	Sirois et al (1994)
FORFLO	Floodplains	Southern Carolina	Pearlstine et al (1985)
FORGRA	Scot pine forest	The Netherlands	Jorritsma et al (1999)
FORICO	Rainforest	Puerto Rico	Doyle (1981)
FORMAN	Caribbean mangroves	Coastal Florida	Chen and Twilley (1998)
FORMIND	Tropical rainforest	Sabah, Malaysia	Köhler and Huth (1998)
FORMIS	Floodplains	Mississippi	Tharp (1978)
FORMIX	Tropical rainforest	Sabah, Malaysia	Bossel and Krieger (1991)
FORMIX2	Tropical rainforest	Sabah, Malaysia	Bossel and Krieger (1994)
FORMIX3	Tropical rainforest	Sabah, Malaysia	Huth et al (1998)
FORMIX 3-Q	Tropical rainforest	Sabah, Malaysia	Ditzer et al (2000)
FORMOSAIC	Tropical rainforest	Malaysia	Liu and Ashton (1998)
FORNUT	Appalachian hardwood	North Carolina	Weinstein et al (1982)
FORSKA	Boreal conifer	Sweden	Leemans and Prentice (1987)
FORSUM	Sub-alpine forests	Switzerland	Kräuchi and Kienast (1995)
FORTNITE	Northern hardwood	New Hampshire	Aber and Melillo (1982)
GIZELA	Pacific NW conifers	British Columbia	Robinson (1996)
HYBRID	Pine forest	Montana	Friend et al (1993)
JABOWA	Northern hardwood	New Hampshire	Botkin et al (1972)

Model	Forest type	Location	Reference
JABOWA-II	Deciduous forests	Eastern North America	Botkin (1993b)
KIAMBRAM	Sub-tropical rainforest	Australia	Shugart et al (1980)
KELLOMAKI*	Boreal forests	Finland	Kellomäki and Väisänen (1991)
KIWI	Mangrove forests	Northern Brazil	Berger and Hildenbrandt (2000)
KOLSTRÖM*	Boreal forest	Fennoscandia	Kolström (1998)
KOPIDE	Korean pine/broadleaves	North-eastern China	Shao et al (1994)
LETT*	Birch forests	Eastern North America	Lett et al (1999)
LINKAGES	Deciduous forests	Eastern USA	Pastor and Post (1986)
LINKAGES.BC	Pacific NW conifer	North-eastern Vancouver Island	Keenan (1993)
LINKANDIR	Northern temperate hardwoods	Adirondacks, Quebec	Lafon (2004)
LINKNZ	Temperate hardwoods	New Zealand	Hall and Hollinger (2000)
LOKI	Boreal conifer	Alaska, Canada	Bonan (1989)
LPJ-GUESS	European mixed forest	Central Europe	Smith et al (2001)
MENAUT*	Tropical rainforest	Côte d'Ivoire	Menaut et al (1990)
MIOMBO	Tropical dry woodlands	Central Africa	Desanker and Prentice (1994)
MOHREN*	Dry sand soils	The Netherlands	Mohren et al (1991)
MOSEL	Hardwood forest	Eastern North America	Malanson (1996)
NEWCOP	Mixed conifers	North-eastern China	Yan and Zhao (1996)
OUTENIQUA	Sub-tropical rainforest	South Africa	Van Daalen and Shugart (1989)
OVALIS	Appalachian hardwood	Virginia	Harringon and Shugart (1990)
PICUS	Beech/spruce forests	Austrian Alps	Lexer and Hönninger (1998)
ROPE	Korean pine/broadleaves	North-eastern China	Shao et al (1994)
SEAGLE*	Riparian forests	Maryland	Seagle and Liang (2001)
SILVA	Montane conifer	California	Axelrod and Kercher (1981); Kercher and Axelrod (1984)
SIMA	Boreal forests	Finland	Kellomäki et al (1992)
SJABO	Conifer forests	Estonia	Tonu (1983)
SMAFS	Acadian forest	Nova Scotia	El Bayoumi et al (1984)
SORTIE	Hardwood forest	Eastern North America	Pacala et al (1993)
SORTIE-BC	Mixed forests	British Columbia	Kobe and Coates (1997)
SORTIE-ND	Mixed hardwood/conifer	Quebec	Beaudet et al (2002)
SPACE	Appalachian cove hardwood	North Carolina	Busing (1991)
SPRUCE-FIR	Appalachian spruce/fir	North Carolina	Busing and Clebsch (1987)
STEPPE	Semi-arid grassland	Colorado	Coffin and Lauenroth (1989)
SWAMP	Floodplains	Arkansas	Phipps (1979)
TACA	Douglas-fir forests	British Columbia	Nitschke and Innes (2008)
TreeMig	Central European forests	Central Europe	Lischke et al (2006)
TROLL	Tropical rainforest	French Guiana	Chave (1999)
ZELIG	Appalachian hardwood	Tennessee	Smith and Urban (1988)
ZELIG.BC	Pacific NW forests	British Columbia	Burton and Cumming (1991)

Note: * = unnamed model, ascribed here to the name of its author economic development and environmental protection. Current work focuses on climate-change policies, market reform, energy technology collaboration and outreach to the rest of the world, especially major consumers and producers of energy like China, India, Russia and the OPEC countries.

Canada's contribution to the IEA was the ENFOR programme operated by the Department of Energy, Mines and Resources. Standing for Energy from the Forest, the research involved was undertaken by the Canadian Forestry Service and initially focused on an inventory of forest biomass, studies of forest biomass utilization as biofuels, and some of the environmental impacts of biofuels harvesting. The initial focus has broadened and now there is major emphasis on the total carbon budget of Canadian forests in the context of greenhouse gas emissions and climate change (Kurz and Apps, 1999, 2006; Kurz et al, 2008) and on non-conventional forest bioenergy such as ethanol production (Pimentel and Patzek, 2005; Lin and Tanaka, 2006; González et al, 2009).

One sub-programme of the original ENFOR project was intended to develop a model to predict the energy efficiency and biophysical sustainability of short rotation, intensive culture (SRIC) bioenergy plantations. What is the ratio of the fossil fuel energy required to manage these plantations vs the bioenergy delivered? And for how long can such plantations be harvested before soil fertility is reduced and the crops need fertilization (reducing the energy benefit–cost ratio)? The Forest Ecology group in the Faculty of Forestry at the University of British Columbia (UBC) was contracted in 1976 to produce such a SRIC bioenergy analysis and decision-support tool. The resultant FORCYTE (Forest nutrient cycling and yield trend evaluator: Kimmins and Scoullar, 1979) model was developed over a ten-year period, extending from the original SRIC context to the whole question of the sustainability of several biophysical and social values in even-aged forest stands under a variety of management strategies. At the end of the contract the model was delivered to the Canadian Forestry Service, and the UBC modelling group switched to the development of a more advanced, multi-value HS model, FORECAST (Forestry and environmental change assessment tool). From the outset, this family of models addressed issues of ecosystem sustainability and both forest management and forest policy issues. However, to address such topics in a scientifically credible manner, it was decided that the models had to be based on an explicit representation of the key ecosystem structures and processes that determine the outcome of environmental change and forest management. Consequently, these models are basic ecological models designed to address pressing and complex practical issues.

Sequence of events that led to the development of the different models in the FORECAST family

The Arab oil crisis and interest in SRIC bioenergy plantations were the progenitors of FORCYTE. Over the period of model development, the oil crisis passed and peak oil prices declined, and so the mandate of the model was expanded from its initial narrow focus. Public awareness of forestry issues had developed rapidly over this period, and the need had switched from a focus on timber and bioenergy sustainability to the sustainability of a wide range of forest values, and the trade-offs between these accompanying different management strategies.

In response to this change in the modelling needs for forest policy and management, and also for concerned public groups, the UBC team embarked on an open-ended development of the FORECAST model – a non-spatial stand-level model – and extensions of this core model to a spatial model of local forest landscapes, and to a spatial individual tree model for application in complex forest stands and agroforestry. The transition from non-spatial to spatial at various spatial scales involved the development of an engine for spatial modelling: HORIZON. This spatial framework was the foundation for the landscape model (LLEMS), the complex stand model (FORCEE) and for an educational version of the landscape model (Possible Forest Futures). This last was developed from a simpler educational model (FORTOON) developed from FORCYTE. The motivation for the two educational gaming tools was the need to make the basic models more accessible for school, college, university and public workshop applications. FORTOON was originally developed under contract to Canada's National Film Board to promote the use of a science-based forest management educational game in high schools.

Major categories of hybrid simulation models we have developed

Stand-level model: FORCYTE

Because the terms of reference of the original contract that initiated the 33 years of model development was to examine the sustainability and energy efficiency of SRIC bioenergy plantations in the context of soil fertility, FORCYTE was designed as a non-spatial, light- and nutrient-limited model. Because herbs and shrubs can be both competitors for and conservers of nutrients, the model incorporated both trees and minor vegetation: herbs, shrubs, and, for certain applications, the ability to represent bryophytes. As an HS model, the growth of trees and lesser vegetation was represented as a series of Chapman-Richards equations (the HB component of the hybrid model), a separate equation being developed from available empirical data (chronosequence data, data from long-term monitoring of field plots, or simply from a well-established empirical plant growth model) for each main biomass component of each species being represented. The equations were based on data that described the growth of each species in monoculture, since the data were assumed to represent a bioassay of the potential of each species on the site. Data were required from a series of sites that varied from high (fertile) to low (infertile) productivity. The model did not address the differences in moisture status between fertile and infertile sites since it was assumed that the effects of moisture are represented in the fertility levels. This assumption is clearly untenable for extremely dry or wet sites or climates, but SRIC management was not considered for such sites or climates, and the management was assumed to have no effect on soil moisture status. Consequently, the lack of explicit moisture representation was deemed acceptable (Kimmins et al, 1990; Kimmins, 1993b).

The FORCYTE model allowed for the use of simplified representations of the biomass components of plants if detailed data were not available, although it was recognized that the greater the simplification, the less reliable the model output was likely to be. For example, trees could be represented simply as roots, stems and leaves; shrubs and herbs as leaves and roots. Division of stems into bark, sapwood and heartwood could be omitted, although this detail is required for an accurate representation of nutrient inventories, uptake demand, nutrient cycling and decomposition. Some representation of roots was required, since the simulation of nutrient cycling and nutrient limitation of growth required estimates of soil occupancy by fine roots. In the early stages of FORCYTE development leaves were required because they are a major component of litterfall and nutrient cycling, and a major component of shade and light competition. As the inadequacies of representing growth by a Chapman-Richards growth equation were recognized, the model structure was altered to drive growth from estimates of the photosynthetic efficiency of leaves, so leaves gained an additional importance. Details of how the photosynthetic efficiency of leaves is estimated are given in Chapter 5.

A key element of the FORCYTE model was the preparation of an 'initial state' for all the soil and plant variables represented in the model. This involved running the model set-up to represent the known or estimated history of natural and management-related disturbance of the site, and capturing the output values of all variables in a file (the 'starting state' file) at the end of this simulation. This set of values then becomes the starting point for all future simulations for that ecosystem with that assumed history. Early experience with the model revealed that its predictions were very sensitive to the legacies of soil resources, organic matter and existing herb, shrub and tree populations at the start of a simulation, something lacking in many models, which limits their ability to simulate aspects of ecosystem sustainability and resilience (elastic stability).

FORCYTE went through 11 stages of development, the last of which was a well-developed stand-level, non-spatial, forest ecosystem management simulator that could also address natural disturbance events such as fire and windthrow (Kimmins, 1993b). Most major silvicultural practices could be represented, and the model evaluation carried out by Sachs and Trofymow (1991) demonstrated good results for the FORCYTE approach in Douglas-fir plantations on the Canadian west coast. However, much remained to be improved, and the question of temperature and moisture effects needed to be added before the important issue of climate change could be addressed. These developments, and several improvements and other additions, awaited the development of the next model, FORECAST.

Because the concept of sustainability is rooted as much in social as in biophysical sciences (World Commission on Environment and Development, 1987), and because the origins of FORCYTE were concerned with both aspects, this model included subroutines that calculated the social values of employment and stand-level economics, and the social/environmental variables relating to carbon balance, fossil fuel use and energy benefit–cost ratios of management. While some modellers disagreed with our development of multi-value, multi-criteria models, we felt that this is precisely the type of decision-support tool needed today as we face the demands of managing forests for multiple values. Because the social and environmental values included in FORCYTE and its derivative models are derived from forecasts of ecosystem variables produced by these ecosystem management decision-support tools, there is every reason to use them in multi-value scenario and trade-off analyses.

Stand-level model: FORECAST

Building on the HS approach developed for FORCYTE, FORECAST represents an extension to correct known deficiencies, add capabilities, and above all to enable the model to address issues of soil moisture that had limited its application in very dry forests, and to examine the possible consequences of possible future climates. Improvements included adding representations of snags (survivorship and decay of standing dead trees), canopy response to increased canopy space, age-related tree vigour (controlling tree response to thinning), improved representation of understorey and of fire and windthrow effects, and the addition of soil moisture, water balance and drainage, and wildlife habitat suitability sub-models. Details of FORECAST are presented in Chapter 4.

HORIZON: a spatial engine for the extrapolation of the non-spatial FORECAST model

In order to extend the non-spatial stand-level FORECAST model to a spatial landscape-level model, a framework – HORIZON – was developed that permitted the use of FORECAST to simulate each of the spatial sub-units of a landscape. Initial states are prepared for each sub-unit, each of which can be subjected to independent management and natural disturbance events defined by the user. There is no interaction between sub-units, however, which limits the minimal size of sub-units that can be simulated in a credible manner. HORIZON became the foundation for the LLEMS landscape model and for the watershed management educational game Possible Forest Futures (see below and Chapter 7). Extensions of HORIZON were first reported by Kimmins et al (1999). We do not describe HORIZON in a separate chapter, but its application in LLEMS, FORCEE and PFF are described in Chapters 4, 5 and 6.

LLEMS: a local landscape spatial extension of FORECAST, built on HORIZON

The lower spatial limit on HORIZON and the lack of polygon interactions limited the spatial resolution of simulations. Because of the desire to use FORECAST to examine small patch and variable retention harvesting (Franklin et al, 1997, 2002; Mitchell and Beese, 2002) at the landscape scale, LLEMS (local landscape ecosystem management simulator) was developed, in which polygon interaction in terms of shading and seed dispersal are simulated. The resolution is 10 × 10m, permitting essentially individual tree simulation in a mature stand, although this is not the intended use of the model. LLEMS has an interactive interface (CALP-Forester) that permits the user to invoke different harvesting patterns and methods on-screen with a mouse, and to prepare 3D images and movies of temporal as well as spatial patterns of change in forest structure, composition and function over time. LLEMS is described in Chapter 6.

FORCEE: a stand-level, individual tree extension of FORECAST, based on HORIZON

Although LLEMS has a minimum spatial resolution of 10 × 10m, it is not a true individual tree model, being driven by FORECAST. In order to simulate complex stands involving intimate mixtures of trees of different species and sizes/ages, we extended the modelling approach used in FORECAST to a true individual tree model: FORCEE. The motivation for this development was the analysis of alternative agroforestry strategies, and the simulation of management in complex uneven-age and mixed-species stands (mixed woods). FORCEE is described in Chapter 5.

FORTOON: a simple educational and gaming tool based on HORIZON and FORECAST

Ecosystem management models are generally developed for use by forest managers and policy makers, and available to public groups concerned about forests. However, they can also make an important contribution to education about the complexities of forest ecosystems and their management. At the request of the National Film Board of Canada we developed FORTOON (forestry cartoon) for a grade 11–12 (age 16–18) resources course, and for use at college and other introductory post-secondary institutions. It also proved to be helpful at various forest management training workshops. The FORTOON game can be downloaded from our website (www.forestry.ubc.ca/ecomodels).

FORTOON challenges players to manage eight areas for six different values: the social values of employment, wealth creation and timber supply, and the ecosystem values of wildlife habitat, soil fertility and general environmental quality (a surrogate for a balance of forest age classes, including mature/OG stands). The player selects a management strategy for each of the eight areas, is warned when any single values is being depleted or in short supply, and is provided with optional pictorial, tabular and graphical presentations of the results of the selected scenarios for the entire management area, or for each of the eight areas individually. There is an interactive game in which the player flies a helicopter with a water bucket to put out a forest fire. If successful, the management area is saved, but if the player fails to put the fire out, all values for that area are lost until the forest regrows. The objective of the game is to achieve the top score for each of the six values, and for the overall management. The top scorer gets their name on a virtual granite wall of fame, and it remains there until another player achieves a higher score for any one value or the overall 'Best Forester'. To assist beginners there are 'classrooms' for ecology, forestry and wildlife. There is an English and a French version, and the text of much of the game can easily be converted to another language, if desired, by editing user-accessible text files. More details are given in Chapter 7.

Possible Forest Futures (PFF): a more advanced, watershed-level educational and gaming tool based on HORIZON and FORECAST

FORTOON is a very simplistic, introductory scenario analysis and value trade-off educational gaming tool. Its limitations led to the development of a more advanced, watershed landscape game in which a wider range of scenarios could be examined in a spatial landscape framework. PFF is introduced in Chapter 7.

Take-home message

The environment is changing, as it always has, but the rate of change appears to be increasing due to population growth, increasing per capita ecological footprints, and anthropogenically induced climate change. Society's values and expectations with respect to the environment in general, and to forests in particular, are also changing. With the growth in the power and utilization of the internet, there is increased dissemination of the products of science that are relevant to these issues, but also of belief-system-based opinions about them. This, coupled with an increased scepticism about science and the rise of postmodernism (e.g. www.as.ua.edu/ant/Faculty/murphy/436/pomo. htm) has increased the complexity of the debate about issues of forest stewardship and sustainability.

Faced with considerable uncertainty over what constitutes sustainability and the many other environmental/ecological concepts introduced in Chapter 2, policy makers and forest managers require scenario and value trade-off analysis tools that will permit the exploration of the policy and management implications of the diverse perspectives on forestry issues. Experience alone is no longer sufficient. Single-value decision-support tools are no longer useful for many issues. Process-based tools on their own offer many advantages, but also drawbacks. We conclude that suites of models are needed that combine experience, traditional knowledge and contemporary science-based understanding of both 'big picture' questions, and also the details that define them: combined top–down and bottom–up decision-support tools. We believe that HS models as defined in this book offer the greatest chance for progress on these issues.

The next chapter starts a detailed look at the family of HS models we have developed, leading to a discussion of their testing and application, and their combination with other models in meta-models and communication tools.

Additional material

Readers can access the complementary website (www.forestry.ubc.ca/ecomodels) to access the full versions of the following additional material:

- a poster illustrating hybrid ecosystem-level forest models as tools for forest management and research, presented by J. A. Blanco at the SISCO winter workshop, Penticton, BC, 2006;
- a slideshow on ecosystem management modelling, presented by J. P. Kimmins, UBC, Vancouver, BC, 2005; and
- a slideshow on complexity in decision-support tools, presented by J. P. Kimmins at the IUFRO World Congress in Quebec City, QC, 2006.

Chapter 4

Forestry in Transition:
The Need for Individual Tree Models

Introduction

Until relatively recently, forest management practices in Canada used clearfelling as the principal means of timber extraction, followed soon after by manual restocking with the aim of regenerating an even-aged monoculture stand. The rationale for this harvest and planting regime was largely economic: minimize harvesting costs and the interval between subsequent harvests. The result was usually stands with a spatial structure considerably less complex than the pre-harvest condition. Furthermore, with a principal focus on growth and yield, forest managers could utilize relatively simple stand-level empirical models for predicting the outcome of their management actions (see Messier et al, 2003 for review of empirical models and their application). Recent decades have witnessed a paradigm shift in how forests are viewed and managed. Rather than being solely a production system for fibre, there is now widespread appreciation and consideration for the broad range of goods and services that forests provide. This ecosystem-based approach (see Bourgeois, 2008) encompasses a set of guiding principles based upon the concept of sustainable forest management (SFM; Kneeshaw et al, 2000). A typical feature of SFM is the retention of complex stand structural and functional attributes through the application of variable retention (VR) harvesting. VR systems range from single-tree selection (highest retention) to harvesting regimes that remove a significant proportion of the total merchantable stems (Sougavinski and Doyon, 2002).

The application of SFM principles as embodied in VR harvesting introduces greater complexity into the management cycle. Dispersed retention with low removal may mean that sub-canopy light levels remain relatively low; larger openings, in contrast, have higher light levels but competition from pioneer species can be severe (Sullivan et al, 2006; Lindgren et al, 2007). Selection of appropriate species for replanting therefore requires consideration of their light requirements and competitive ability. As they develop, VR systems promote the creation of structurally more complex stands because structural elements are retained after the final harvest and because of the additional tree cohort that develops from planting (Franklin et al, 1997). This complexity has two important implications.

First, few empirically based growth and yield models are suitable for VR systems. This is because their representation of forest structure is overly simplistic and they are calibrated for harvesting and silvicultural systems (clearfelling and even-aged stand management) that differ fundamentally from VR. It has been argued, therefore, that process-based models are more likely to meet the information requirements for implementing SFM, a view advocated by Korzukhin et al (1996) and Messier et al (2003). Typically, a process model of tree growth simulates rates of change in several variables that represent the main processes associated with carbon capture, allocation and loss (Godfrey, 1983; see below). The second implication of VR forestry is that generalized stand-level attributes (as reflected in many stand-level models; see Chapter 5) are no longer sufficient as the basis for management decisions in VR stands. This is because the configuration and density of the retained trees following VR harvesting is an important determinant of resource distribution. To successfully represent resource heterogeneity requires models that simulate resource capture and growth by individual trees, and which include explicit representation of their spatial configuration (Le Roux et al, 2001).

Another potential application of an individual tree model is in agroforestry. An agroforestry system is designed to realize the benefits from combining trees and shrubs with crops and/or livestock on the same plot of land. In this respect, these systems have structural attributes and a complexity similar to VR forestry. Traditional agroforestry systems were developed through a trial-and-error process from which knowledge was accumulated as to which techniques were successful (see Kimmins et al, 2008b for examples). Intergenerational knowledge transfer was therefore critical to the success of the system, and effective techniques were often integrated within the social and cultural fabric of the population through customs, codes of practice and formal religious beliefs. This approach is impractical for modern agroforestry systems, in part due to the protracted length of time necessary to evaluate and substantiate suitable options, and because climate change might result in future growing conditions that differ considerably from historical trends.

To be useful, agroforestry models must represent the competition between tree, shrub and/or herbaceous crop species for the three essential resources – light, water and nutrients – in a coupled vegetation–soil system (Mobbs et al, 1998). In this respect, models that represent only water use or light interception have limited utility (Wang and Jarvis, 1990). Important interactions also occur among the different components of a system in terms of space occupation (vertical and horizontal stratification, both above and below ground) and resource allocation (light, water, nutrients) (Anderson and Sinclair, 1993). Hence, as with VR forestry, a useful agroforestry model has to be capable of incorporating key ecosystem drivers within the context of evaluating management options such as species selection, planting density, thinning regimes, grazing and harvest, in order to properly define the most efficient management strategy (Balandier et al, 2003).

The ideal individual tree model

The fundamental unit of information in an individual tree model is the tree itself. Given that a single tree is also the basic unit of a forest stand, these models should in principle be capable of simulating development at scales that vary from the individual tree to

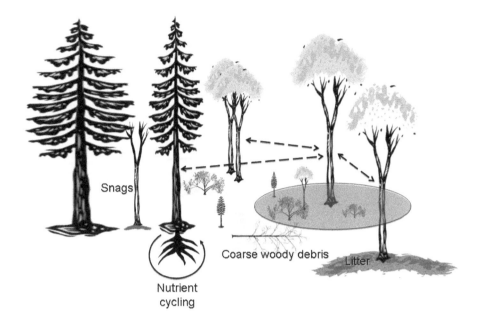

Figure 4.1 *Elements influencing one individual tree in the forest*

stands comprised of one or more species, age cohorts and different spatial configurations (Figure 4.1). Because individual tree models must account for the feedback between stand structure and individual growth, their representation of ecosystem processes needs to be more comprehensive than that which is sufficient for a typical stand-level model (Figure 4.2; Pretzsch et al, 2008).

An individual tree model should take account of the principal physiological processes involved in carbon metabolism (photosynthate production, respiration, reserve dynamics, allocation of assimilates and growth), at least to some degree (Le Roux et al, 2001). A given model design will typically emphasize a subset of these processes using a combination of process-based and statistical formulations, depending on model objectives and how well a particular process is understood.

Physiological process

The relationship between plant function and the biotic and abiotic environments (Figure 4.2) is commonly considered in terms of the flow of energy through one or more biogeochemical cycles (e.g. carbon, nitrogen) and the hydrological cycle (Osborne, 2004). Particular emphasis is placed on the physiological processes involved in the exchange of essential resources between the plant and its environment. Carbon is acquired by fixing atmospheric CO_2 through photosynthesis (Lawlor, 2000), and lost as a by-product of the energy generated from aerobic respiration (Amthor, 2000). In land plants, nutrients such as nitrogen or phosphorus are acquired primarily from the soil matrix through root absorption, or in aquatic plants such as algae directly from the surrounding water.

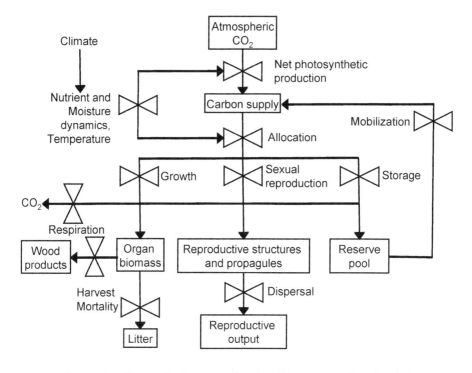

Figure 4.2 *Key tree-level processes that should be represented in the ideal individual tree model*

Note: boxes refer to pools; triangles are processes.

Nutrients are lost through the shedding of plant parts, usually after they reach senescence (Aerts and Chapin, 2000). Land plants obtain water mostly through their roots (leaf absorption can occur but is of secondary importance), and water is lost by transpiration from leaf surfaces through the stomata (Lawlor, 2000).

Modelling forest growth in terms of carbon balance involves calculating the net assimilation rate of carbon and its distribution (Figure 4.2; Mäkelä et al, 2000), but the influence of environmental factors should also be represented. The latter can include direct effects such as light and temperature regimes, nutrient availability and competition, or indirect effects via the impact of soil water content on, for example, stomatal conductance. Some tree-growth models simulate net carbon acquisition only (West, 1993; Kellomäki and Strandman, 1995) and do not include an explicit representation of respiratory losses. The annual net gain of above-ground organic matter is estimated by applying a conversion efficiency factor to calculations of total radiative gain. Calculations of photosynthesis are then regressed against growth (Brunner and Nigh, 2000; Courbaud, 2000). Despite its simplicity, this approach can successfully represent qualitative differences in productivity across sites and between individual trees (Mäkelä et al, 2000).

Carbon allocation within a plant results from the linkages among organs that function as a carbon source (mainly leaves) with those that are the main carbon sinks (stem sapwood, branches and roots, and reproductive structures). A number of carbon models have been developed in recent decades as tools for analysing the rules governing how sources and sinks interact within a plant to drive vegetative growth and reproductive allocation (see Génard et al, 2008). Although there is considerable knowledge of individual processes within plants, regulation of these source–sink relationships is not well understood (and thus can be difficult to represent within a model) due to the complexity in feedback mechanisms, interactions among different processes and the spatial distribution of the various plant organs (Wardlaw, 1990; Le Roux et al, 2001). There are two main approaches to this issue (Génard et al, 2008).

The first, a top–down approach, can be seen as an extension of classical ecological theory. In this approach, individual tree models provide a generalized representation of one or more responses to environmental variation that may or may not be constrained by population and community processes (see DeAngelis and Mooij, 2005 for examples). These models can accommodate greater complexity than traditional analytic approaches and they may yield general principles of carbon capture and allocation that are applicable to a wide range of plant species (West et al, 1997, 1999; Enquist, 2002). The second is a bottom–up approach in which knowledge of source–sink relations is translated into a set of rules that guide resource-allocation decisions within each individual. In this case, populations and biotic communities display properties that result from the adaptive response of individuals (DeAngelis and Mooij, 2005). In reality, both approaches are necessary to understanding carbon dynamics but neither is sufficient. The top–down approach sacrifices detail about selection pressures that shape allocation within individuals, while a difficulty with the bottom–up approach is that allocation decisions are not independent of population-level processes: frequency dependence, for example. Hence, the optimal allocation decision on the part of individuals may depend on what other individuals in the population and community are doing, and the relative costs and benefits of alternative strategies, a question best answered using game theory (see Vuorisalo and Mutikainen, 1999, for example).

Effects of architecture on tree growth

Plant structure (number, shape and size of organs) is intimately linked to physiological function. The number and size of leaves in a plant canopy, for example, are key factors in determining total water consumption (Osborne, 2004), and canopy architecture is thus tightly coupled with the physiology of water transport and water loss through transpiration (Woodward, 1987). A similar coupling of structure and function has been demonstrated in models that relate above-ground processes and carbon allocation to root development and morphology (Osborne and Beerling, 2002; Woodward and Osborne, 2000).

Trends in carbon allocation for each organ type (e.g. leaves, roots, stems) across species are usually consistent within a given environment (Friedlingstein et al, 1999) and this relationship has often been employed in models of carbon partitioning (Woodward and Osborne, 2000). Typically, carbon is invested in organs that function to acquire the most limiting resource. Hence, in nutrient-limited environments, for example, carbon is allocated preferentially to fine root growth at the expense of stem and leaf growth, resulting in higher root:shoot ratios (Gleeson, 1993).

Recent modelling studies have used a recursive approach to linking physiological processes and morphological development (de Reffye et al, 1997; Perttunen et al, 1998; Prusinkiewicz and Rolland-Lagan, 2006). Growth, for example, is represented by a function that describes biomass acquisition, and is allocated to different plant organs according to their relative demand for limiting resources. Biomass acquired in the next time step is then driven in part from allocation decisions made previously, and so on (Yan et al, 2004; Drouet and Pages, 2007). These approaches consider architectural units as independent elements connected through topological rules. As a consequence, a complete description of plant structure has to be computed at every growth stage, making the calculations time-consuming (Fourcaud et al, 2008).

Management activities can have a significant impact on nutrient dynamics (either positively or negatively), while competition for soil moisture from understorey vegetation can have a strong negative impact on tree productivity (Kimmins, 2004d). Despite its importance in resource capture, however, root growth is usually not represented in any great detail in models of plant growth (Fourcaud et al, 2008). Although the root compartment is sometimes divided into fine and coarse roots, individual roots are seldom represented. Two exceptions are the model TREGRO (Weinstein et al, 1992; Weinstein and Yanai, 1994), which uses soil layers and associated root biomass to simulate nutrient uptake; and ECOPHYS, which uses a 3D representation of the root system (Le Roux et al, 2001). A more accurate representation of the root system is important in individual tree models because resource distribution within the soil matrix is highly heterogeneous in both space and time. Similarly, the absorptive capacity, order and spatial position of a root, as well as soil volume and layer occupied, determine its ability to absorb water and nutrients (Jourdan and Rey, 1997; Pages et al, 2004; Wu et al, 2007).

Spatial resolution

Individual tree models can be classified as distance-dependent or distance-independent, depending on whether or not they include information on the spatial arrangement of individual trees (Munro, 1974). Including distance dependence adds to model complexity and computational load. An important question then is whether simulating the interactions among trees in a population is worth these additional costs (Busing and Mailly, 2004). When Pacala and Deutschman (1995) eliminated horizontal spatial heterogeneity from their stand-level model SORTIE, the simulated forest had unrealistically low biomass and diversity. In subsequent work, however, Deutschman et al (1999) tested versions of the model that differed in the spatial complexity of the light regime. In this case, community dynamics did not differ substantially between model versions.

A fine spatial resolution is required if the objective is to simulate tree architectural dynamics within the context of the surrounding population. On the other hand, a coarse spatial resolution (and thus a crude representation of tree architecture) is adequate if the objective is to simulate an average growth response at the plot level or higher (Le Roux et al, 2001). At the intermediate scale are models that simulate tree dynamics in heterogeneous stands, or forest growth models that focus on the heterogeneity of individual trees within a stand (Le Roux et al, 2001). In this case, modellers generally represent an individual tree as an ensemble of growth units, or, more often, as clusters of growth units, such as leafy shoots or branches. This representation can capture essential features of the competition between trees within stands without using a complex, organ-based approach. High-resolution models are often difficult to parameterize, whereas lower-resolution models may be adequate for achieving management objectives and are much easier to parameterize, calibrate and test. The important technical and conceptual trade-offs associated with spatial resolution therefore are the maximum area that can be simulated by the model (set by practical limits of computing resources), and the degree to which model objectives need to incorporate mechanistic detail and spatial interaction (Mladenoff, 2004).

One application of a high-resolution model can be to provide parameter values for input into a lower-resolution model (Luan et al, 1996; Berninger and Nikinmaa, 1997; Le Roux et al, 2001). This is the approach employed to develop a meta-model (see Chapter 10). For instance, SICA is a high-resolution model in which photosynthesis and transpiration are calculated contemporaneously. The detailed data on these two variables were summarized into yearly values and then used as inputs into SIMFORG, an annual time-step process model that distributes photoproduction and nutrient uptake between plant organs (Berninger and Nikinmaa, 1997). Sinoquet and Le Roux (2000) developed a detailed mechanistic model that computed the instantaneous photosynthetic rate for multiple growth units within an individual tree. The model showed that daily light-use efficiency did not vary significantly despite changes in the location of a foliage unit and the associated light regime. This modelling exercise provided a confirmation that a constant value for light-use efficiency can be used to compute carbon gain at a range of spatial scales, from individual leaves to sections of the crown (Le Roux et al, 2001).

Individual tree models as management tools

A recent challenge in tree growth modelling is the trend in forest management to promote silvicultural methods that more closely resemble natural patterns of recruitment and stand development (Mäkelä, 2003). This includes techniques for promoting natural regeneration, leading to a wider mixture of species of variable age, conducting selective harvest to encourage a greater size distribution among surviving trees as well as increasing spatial heterogeneity, and retaining decaying wood material to support seedling recruitment and as habitat. Furthermore, there is evidence (e.g. Anten and Hirose, 1998; Wang et al, 1998) that response to heterogeneity in resource availability at the population or community level is more than simply a linear combination of individual-based responses. Hence, there is a need for individual tree models with the capacity to accommodate these complexities. A number of empirical or semi-empirical models have been available for several decades (Mitchell, 1975; Franc and Picard, 1997), but their application is restricted to management scenarios for which there is sufficient empirical data for calibration, and they are unable to account for changing climatic and soil conditions. The suitability of process models to address these issues is well recognized and these models are now seeing broad application within the context of natural resource management (Kolström, 1993; Bartelink, 2000; Kokkila et al, 2002; Pretzsch et al, 2006, 2008).

Process-based individual tree models are particularly well suited to commercial forestry activities. In northern temperate forests, for example, broadleaf trees have until recently, and in some cases are still, routinely eliminated from conifer plantations to satisfy reforestation requirements and increase conifer productivity (Lieffers et al, 1996; Comeau et al, 2000). Although the latter is indeed enhanced following complete hardwood removal (Simard et al, 2001), there are concerns around forest resilience and diversity (Simard et al, 2001; Aitken et al, 2002; see Chapter 2 for a further discussion of these concepts). Removal of broadleaf species reduces the overall diversity of stand types on the landscape by converting mixed conifer/broadleaf stands into multi-species conifer or conifer monoculture stands. At the stand level, lower species diversity may reduce any inertial resilience produced by the overlap of ecological function among species of different functional groups (Peterson et al, 1998). These issues are amenable to evaluation with the appropriate model.

The complexities of mixed-wood management

Mixed woods constitute a major forest type across temperate North America and are an important multi-value resource. From an ecological perspective, defining a mixed wood is problematic because at a landscape level it can constitute a mosaic of pure conifer and pure hardwood stands or any combination in between, and over this continuum there is a wide range of spatial graininess, from intimate tree-by-tree mixtures to an intermixed landscape mosaic of monospecific patches of various sizes. Gap models are one of the few types of individual tree-based forest models that can simulate a mixed-wood forest (Bartelink, 2000, Figure 3.1). These models have often employed a distance-independent tree-level approach, where mixed-wood dynamics are simulated within one or more small regeneration patches (Botkin et al, 1972, Bugmann, 2001). JABOWA, for example, is a typical stand-level gap model, applicable to small and medium-sized canopy gaps (the neighbourhood plot size is usually set at 10 × 10m; Botkin, 1993a), but it contains no spatial representation of individual trees (though see Busing, 1991; Mailly et al, 2000). This represents a significant limitation for mixed-wood forests of the boreal region, where conifers have narrow crowns and their spatial relationship is an important determinant of gap size (Oliver and Larson, 1990). In some models, the spatial interaction between trees has been simulated indirectly via its effect upon resource depletion (e.g. Wu et al, 1985; Pacala et al, 1993).

Early gap models assumed that light was transmitted vertically through the forest canopy (Botkin et al, 1972). In more recent models, the daily and annual geometry of light transmission has been included, and it can exhibit a profound effect upon vegetation dynamics (Canham et al, 1994), particularly at high latitudes where sun angles are low. This 3D approach to modelling light transmission thus represents an important addition to individual tree models (Canham et al, 1994; Groot, 2004). Examples include SORTIE (Pacala et al, 1996), FORSKA (Leemans and Prentice, 1989) ZELIG (Larocque et al, 2006), and FORCEE (see below).

To date, most gap models have been used to explore theoretical and research-based questions of forest development – such as successional sequences in the absence of disturbance – rather than issues related directly to management practice. To properly manage mixed woods requires a thorough understanding of competitive relations among species, and when competition thresholds can be used to identify management activities that promote beneficial interactions and avoid undesirable outcomes (Burton, 1996). There is a relatively small number of empirical models that have been developed specifically for mixed-wood management (Hasenauer, 1994). One example is SORTIE, an empirical light-driven model (Canham et al, 1999). In SORTIE, the crowns of individual trees (from seedling to canopy tree size) are represented as cylinders, with the radius of the crown (i.e. the cylinder) estimated as an empirical function of tree diameter at breast height. The model has been designed such that all of the functions needed to specify the crown dimension can be parameterized from empirical data (Canham et al, 1994). In British Columbia, a variant of the model has been used to simulate partial cutting prescriptions in temperate deciduous, boreal and temperate coniferous mixed-species forests (Coates et al, 2003). Model simulations indicated how the species, and the amount and spatial pattern of canopy tree removal had a major influence on understorey light dynamics.

TASS is a spatially explicit individual tree model (Di Lucca, 1999). It was developed to generate growth and yield estimates for even-aged, monoculture stands in British Columbia. The model has been used to represent multiple tree species in a stand, but it does not represent their interactions explicitly. Work is under way to develop an uneven-aged version and to improve its capability for mixed-wood management (see Di Lucca et al, 2009, for a description of the planned extension of TASS to complex cutblocks). The model is driven by the height and crown expansion of individual trees. Crowns add a shell of foliage each year, and a given crown can expand or contract asymmetrically in response to internal growth processes, physical restrictions imposed by the crowns of competitors, environmental factors (applied as OAFs: operational adjustment factors, representing environmental effects that are recognized as important for tree growth but not represented in TASS) and silvicultural practices. The volume increment produced by the foliage is then distributed over the bole on an annual time-step.

MOSES (Hasenauer, 1994) and SILVA 2.2 (Pretzsch and Kahn, 1995; Pretzsch, 2001) simulate multiple species and ages. Current annual height and diameter increment are calculated in accordance with a dynamic growth reduction function representing abiotic conditions, competition and management (Pretzsch et al, 2006).

PROGNAUS, a derivative of the PROGNOSIS model (Monserud and Sterba, 1996; Sterba and Monserud, 1997), uses a distance-independent approach to predict individual tree growth from inventory data. Growth and yield are calculated from tree size, site factors, stand density and level of competition. PROGNOSIS[BC] is another variant of PROGNOSIS, adapted for use in British Columbia from the North Idaho version of the forest vegetation simulator (FVS) model (Wykoff et al, 1982; Dixon, 2002). PROGNOSIS[BC] has been used to forecast future stand conditions for mixed-species and/or multi-aged (complex) stands in various ecological zones located in south-east and central British Columbia (Marshall et al, 2008).

BWIN (Nagel, 1997) is a distance-independent model that calculates tree productivity from stand input variables. Height increment is a function of potential height growth modified by competition, age and thinning regimes, while diameter increment depends on crown surface and competition. The mixed-wood growth model (MGM; Huang and Titus, 1994) is a deterministic, distance-independent, individual tree-based, stand growth simulation model. The general modelling approach is similar to that of FVS.

One important limitation of empirically based models is that they are applicable only to the range of the data with which they were calibrated (Bartelink, 2000; Kimmins, 2004d). In reality, the number of potential combinations of species, management regimes, and site-dependent interactions is so high in mixed woods that empirical data are likely to always represent a constraint on deriving appropriate management practices. Models will therefore be required that have the ability to represent spatial relationships and the processes underlying ecosystem function (Bartelink, 2000). Ideally, these models should utilize empirically derived relationships as a guideline for process simulation: an approach termed 'hybrid simulation' (Kimmins, 2004d; see Chapters 2, 3 and 5).

Agroforestry: an emerging paradigm

In conventional timber management forestry, minimizing competition from non-target species is an important management objective. In addition, trees are planted at a spacing close enough to achieve maximum site occupancy and induce self-pruning to improve stem form and produce high-quality timber (Cabanettes et al, 1999). Agroforestry systems are managed differently in that trees are typically grown at a wider spacing, usually in close association with other non-tree crops and/or animals. This intermixture of species introduces considerable complexity into the interactions among community members. The addition of legume crops, for example, can improve soil nitrogen and thereby enhance tree productivity; these same plants, however, may also be strong competitors for water, thereby depressing tree growth. Similarly, the droppings from grazing animals can provide a ready source of nitrogen to the plant community, but these same animals can trample young trees, browse new growth, and damage the bark on older trees when stems are used as rubbing posts (MacDicken and Vergara, 1990; Cabanettes et al, 1999).

The ideal agroforestry model should be capable of simulating the complex interactions that occur among the components of an agroecosystem, including spatial relationships (vertical and horizontal stratification, both above and below ground) and resource allocation (light, water, nutrients) (Anderson and Sinclair, 1993), within the context of evaluating options for species selection, planting density, thinning regimes, grazing and harvest (Balandier et al, 2003). A variety of models has been developed that satisfy these criteria to varying degrees (see Ellis et al, 2004, for a comprehensive review). The HyPAR model (Mobbs et al, 1998) is a combination of two models: HYBRID, a forest canopy model (Friend et al, 1997), and PARCH (Fry and Lungu, 1996), a crop growth model developed for application in the dry tropics. HyPAR incorporates biophysical processes to calculate light interception through a disaggregated canopy of individual trees in known positions, along with water and nutrient competition, daily carbon allocation, hydrology, and the impact of various management scenarios (Mobbs et al, 1998). Lawson et al (1995) used this framework to evaluate crop productivity under a range of environmental conditions by:

- coupling HYBRID and PARCH;
- coupling an individual tree model (MAESTRO; Wang and Jarvis, 1990) with PARCH; and
- incorporating a combined model of evaporation and radiation interception by neighbouring species (ERIN).

APSIM (agricultural production systems simulator; McCown et al, 1996) is a modular, component-based design that allows individual models to communicate via a common engine. At present, several individual crop and pasture modules are available within APSIM, including a forest production module (Huth et al, 2001).

Traditional non-modular or self-containing models have also been developed for agroforestry systems. WaNuLCAS (water, nutrient and light capture in agroforestry systems; van Noordwijk and Lusiana, 1999) is designed to model tree–soil–crop interactions for a wide range of agroforestry practices. The model simulates litterfall production, soil fertility, shading effects of trees on crop yields, and plant competition for resources. WaNuLCAS can accommodate a range of management options, including variation in plot size, tree spacing and choice of tree species, cropping cycles, pruning, organic and inorganic fertilizer inputs and crop residue removal. SCUAF (soil changes under agroforestry; Young and Muraya, 1990) is a nutrient cycling model that predicts annual changes in soil conditions (e.g. soil erosion, carbon, nitrogen, phosphorus and organic matter) and its effect on plant growth and harvest. FALLOW (forest, agroforest, low-value landscape or wasteland; van Noordwijk, 2002) is a model that scales up the assessment of land-use systems by evaluating the impact on soil fertility, crop productivity, biodiversity and carbon stocks of shifting cultivation or crop–fallow rotations at the landscape scale. Finally, ALWAYS is a management-oriented biophysical model (Balandier et al, 2003) used to simulate a silvo-pastoral system (a broadleaved tree/pasture/sheep association) and its management. The model was calibrated for two sites with contrasting soil and climate. Despite its relatively simple structure, model output was consistent with empirical data.

Although a broad range of agroforestry models has been developed, in many cases their practical application was restricted to the specific systems for which they were created (Ellis et al, 2004). Furthermore, the majority of models are designed more for research than for management (as reflected in their high degree of complexity and poor development of any management interface), and very few have made their way into the hands of decision makers and landowners. The result is a lack of model application in extension and planning activities (Huth et al, 2005). Ellis et al (2004) have argued that agroforestry decision-support tools must be flexible enough to accommodate a broad range of potential issues at a variety of spatial scales, and in a manner that responds to users' needs and resources.

FORCEE: a comprehensive, spatially explicit, individual tree management and agroforestry model

One shortcoming in many stand-level forest productivity models is that their projections of growth and yield are made without explicit consideration of the underlying biological processes that drive productivity. For example, the MGM (Huang and Titus, 1994) and TWIGS models (Miner et al, 1988) are based on empirical (non-mechanistic) relationships between height and diameter, which are then used to predict the future growth of individual stems. This approach is reasonable for predicting growth over relatively short time periods under relatively constant conditions. If management objectives and/or environmental conditions change, however, past observations of diameter and height growth may not be reliable as the basis for predicting the future growth response (Seely et al, 2008). Hence, as Ellis et al (2004) have pointed out, reliable, realistic and useful decision-support tools should be capable of representing real world processes and predicting outcomes for scenarios that may not always have historical precedents. The FORCEE (forest ecosystem complexity evaluator) model was designed in an effort to address these requirements.

FORCEE is a spatially explicit, individual tree, ecosystem management model. The model is built using the framework developed for the non-spatial, stand-level model FORECAST (see Chapter 5). In this respect, FORCEE grows trees and understorey plants from simulated photosynthesis, nutrient cycling, nutritional regulation of growth, and competition for light and nutrients; moisture dynamics is currently under development. The simulation rules are derived, in part, from data that describe historical patterns of tree and plant growth on sites of differing nutritional quality. Rates of some key processes are then estimated using a back-casting procedure, i.e. from a calculation of what the rates must have been to generate the observed historical patterns of tree and plant growth. Other processes are calibrated with local empirical data or literature values.

FORCEE is designed to simulate development in complex stands: the complexity that derives from the vertical and horizontal structure and stratification of vegetation (including the understorey community), variable species composition, multi-age tree canopies, and the spatial and temporal heterogeneity in forest floor and soil characteristics. The model operates at a 10 × 10cm spatial resolution. It employs a series of visualization screens to directly illustrate the outcomes of the simulation routines, and is equipped with a well-developed management interface similar to the FORECAST model (see Chapter 5). Hence, examples of management activities that can be simulated with FORCEE include site preparation, planting and regeneration options, weed and stocking control, fertilization, pruning, and variable harvesting dates and utilization levels. Management practices can be spatially specific, or imposed on all members of a specific group (a particular age cohort, size class or species, for example). Its representation of ecosystem components is as follows.

Foliage

The total light available to a given cell of foliage is simulated using a hemispherical view model (Brunner and Nigh, 2000). Each 10 × 10cm foliage cell can intercept light and will cast shade (see Figures 4.3–4.5). A foliage–light algorithm is used to determine the amount of net primary production (NPP) in relation to light interception. Allocation of NPP to foliage itself varies, depending on the amount of light captured by a given foliage cell, foliar age, and the carbon cost of supporting stem segments and roots. As a result, foliage will grow differentially into areas with the best light conditions, though limits on branch length constrain the distance from the stem at which new foliage cells can be produced (Figure 4.4). Poorly performing foliage cells are more likely to die. Following death (either seasonally or because of poor light conditions), leaves drop onto the forest floor and accumulate as litterfall. Over time, the canopy for each tree becomes hollow as the cells around and above them shade the foliage of inside cells (Figure 4.4). In the northern hemisphere, less foliage will grow on the north side of a tree because of poorer light conditions relative to sun angle; lower foliage also dies sooner on the north side. Depending on their foliage placement, adjacent trees can shade each other, and each tree casts its own shade towards the forest floor (Figure 4.5), displaced according to sun angle. The latter is important as a source of shading for the understorey.

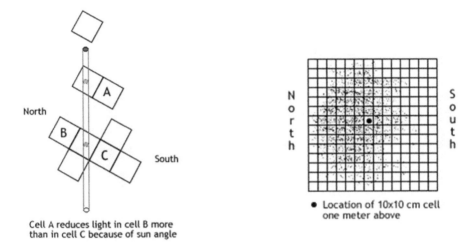

Figure 4.3 *Influence of light on foliage growth*

Note: Left: hemispherical view approximation used to reduce light to lower layers. Right: Pattern of shading cast by one foliage cell on the layer below from hemispherical view analysis.

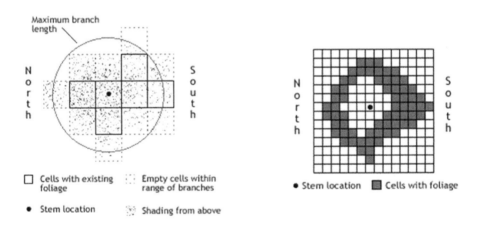

Figure 4.4 *Schematic representation of foliage growth in FORCEE*

Note: Left: growth of foliage into new cells. Right: foliage death or a failure to replace litterfall is greater in cells with less light, causing a hollowing in the total foliage disc.

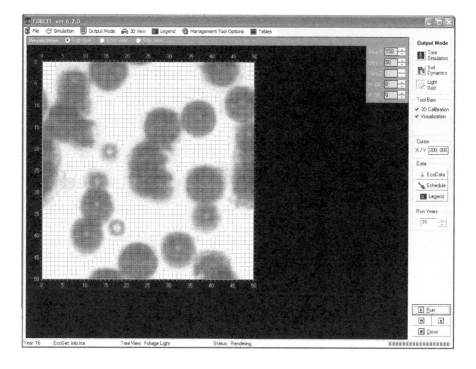

Figure 4.5 *FORCEE maps of light levels on the ground*

Stems

Stems are represented by series of nested cylinders (Figure 4.6), with each cylinder comprised of up to three concentric rings. Moving from the inside to the outside of the stem, each ring defines heartwood, sapwood and bark, respectively (Figure 4.6). A given cylinder can expand to a maximum length of 1m. Lower cylinders are of greater diameter because they have had more time to accumulate biomass than the younger cylinders above. Each cylinder can diverge from the vertical depending upon light conditions; this provides a means for representing stem lean. A stem may lean in any direction, but it can also recover from lean as it elongates (Figure 4.6a). A visual depiction of stem growth is shown in Figure 4.7.

If appropriate, tree stems can divide, as occurs, for example, during coppicing. Tree roots originate from the bottom-most stem, spreading through the soil layers to access and uptake nutrients.

Topography and soil

Soil is represented by a 3D grid of 10 × 10 × 10cm pixels (Figure 4.8). Each pixel (or cell) can contain litter, humus and mineral soil, or a pixel can represent an impermeable layer, such as bedrock. Bedrock surface and mineral soil surface are mapped as landform grids. As litter falls from a foliage cell and lands on a forest floor pixel, it accumulates in that pixel. When/if that (top-most) pixel is fully occupied, litter is added to a pixel above. Litter decomposes and transitions over time to humus. Litter decomposes at a variable

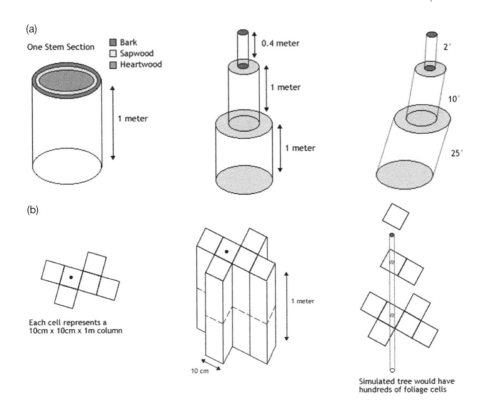

Figure 4.6 *Representation of tree stems in FORCEE and connection between foliage and stems*

Note: Top left: stems are cylinders composed of up to three layers: heartwood, sapwood and bark, respectively. Top centre: example of a 2.4m-tall stem with exaggerated taper. Top right: Example of bent stem with lean angle that is undergoing correction as the stem grows into an area of higher light. Bottom left: flat grid depiction of foliage cells located at stem section point (black dot). Bottom centre: foliage cells have 1m depth. Bottom right: Example of 2.4m-tall stem with exaggerated foliar cell size.

rate depending on species, litter type and age, whereas humus decomposes at a rate that is constant but much slower than litter. Both the litter and humus decomposition rates are user-defined inputs, and are site-specific. Decomposition reduces the occupied volume of a given pixel. When this occurs in a lower pixel, material drops down from the pixel immediately above and ensures that lower pixels are fully occupied, if possible. Each pixel is defined by its available nutrients from decomposition, and its litterfall and humus content. Litterfall accumulates differentially on the forest floor, with larger amounts occurring around the base of a given tree (Figures 4.8 and 4.9). FORCEE can simulate prevailing wind conditions, and this can influence the build-up of litter on the forest floor. Released nutrients diffuse vertically down between pixels, but a given tree can access available nutrients only from the pixels occupied by its roots.

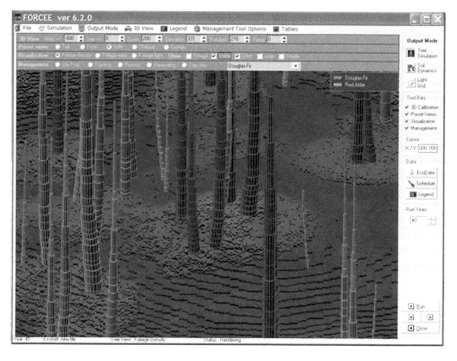

Figure 4.7 *A visual representation of stem growth in FORCEE*

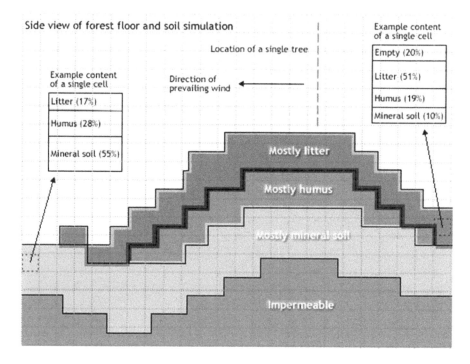

Figure 4.8 *Representation of humus and litter in soil cells*

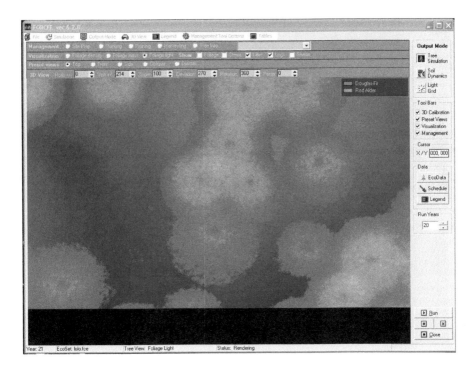

Figure 4.9 *Litter maps in FORCEE illustrating its accumulation around the base of the trees*

Root system development and fine root access to available nutrients

Each cell contains a record of any root mass that develops within that cell from one or more trees. When seeds sprout, their fine roots are located in a single surface cell and this then represents the seedbed: the upper surface of the top soil layer. Root mass can increase annually within a cell, and roots can also expand into adjacent cells. Roots grow in proportion to nutrient availability within the cells occupied by a given tree in a given year, though growth can be limited by a species-specific maximum root elongation rate and rooting depth.

Tree roots occupying the same cell access nutrients according to the relative mass of their roots and their relative demand for nutrients. Simulated photosynthetic rates and growth allocation to different organs (each with their own nutrient requirements) determine total nutrient demand. A tree invests more root growth in cells with higher available nutrients, and they expand into adjacent cells when the latter contain relatively higher nutrient levels. Trees thus tend to grow more roots into cells with greater decomposition outputs and less competition. Cells with higher nutrient levels will generate greater root mass overall.

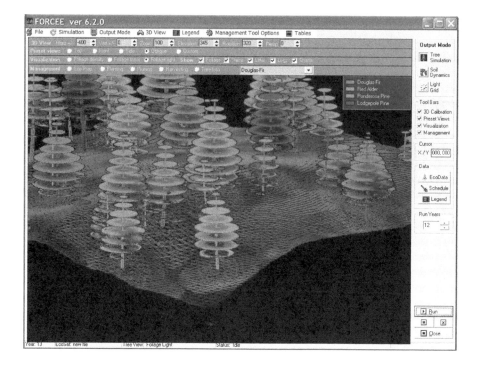

Figure 4.10 *A visual representation of a mixed species stand in FORCEE*

Note: A colour version of this figure is shown in the plate section as Plate 1. Each colour denotes a different tree species.

Understorey

Stand establishment following disturbance can be significantly hindered by competition from a well-established understorey for nutrients and moisture (Harvey et al, 1995; Man et al, 2008). FORCEE addresses this issue by simulating understorey dynamics, either by species or as a series of functional groups. The latter approach is employed much more often, simply because (with very few exceptions) there are insufficient data to calibrate the model for individual species. Typical categories for functional groups include: large, medium and small shrubs; grasses; herbs; nitrogen-fixing species; and degree of shade tolerance. If desired, ground-dwelling bryophytes can be represented. Light and nutrient capture are simulated in the same way for the understorey as with the tree population. There is relatively less detail, however, on the architecture and specific growth form of the understorey population.

Take-home message

Forestry is in transition as management practices move away from a reliance on clearcutting and single-species planting in favour of more complex systems such as variable retention, agroforestry and mixed-species stands. As a consequence, the application of traditional forest productivity models that represent stands simply as an aggregate of trees is no longer sufficient. This has necessitated the development of a new set of modelling tools with the capability to simulate the growth and development of individual trees, either in single or mixed-species stands (Figure 4.10). Because these individual tree models must account for the feedback between stand structure and individual growth, their representation of ecosystem processes needs to be more comprehensive than that which is sufficient for a typical stand-level model. The idealized tree model should include a representation of the physiological processes involved in carbon metabolism, the effects of architecture on tree growth, and the spatial configuration of each individual within the population. An example of this approach is FORCEE: a spatially explicit, individual tree, ecosystem management model built using the framework developed for the non-spatial, stand-level model FORECAST.

Additional material

Readers can access the complementary website (www.forestry.ubc.ca/ecomodels) to download the following material:

- Paper with a basic description of FORCEE: Welham C., Blanco J.A., Kimmins J.P. (2008) 'FORCEE helps to manage mixed woods', Link, vol 10(2), pp12
- FORCEE Screenshots (Slide show in PDF file)

Chapter 5

Stand-Level Hybrid Models as Tools to Support Ecosystem-Based Management

Introduction

Forest resource management in the past few decades has undergone a gradual transition from the basic goal of sustaining levels of resource (dominantly timber) extraction to the more comprehensive objective of sustaining the forest ecosystems that provide these resources (e.g. Thomas, 1995; Rauscher et al, 2000). This holistic approach to forest management, generally referred to as ecosystem-based management, requires resource managers to consider the merchantable components of forests within the broader context of ecosystem structure and function. Moreover, ecosystem-based management demands that forest managers address the multiple objectives associated with the wide range of ecosystem services provided by forests at multiple spatial and temporal scales. These objectives generally include, but are not limited to: a sustainable and economically viable harvest of forest products, the maintenance of biodiversity and wildlife habitat, the preservation of forest area for recreation opportunities and visual resources, the preservation of forested areas as watersheds, and the maintenance of the roles forests play in global cycles of carbon, oxygen and atmospheric gas exchange. To meet these diverse goals, forest managers have developed a wide variety of management techniques and silvicultural systems, the nature and complexity of which have expanded dramatically over the past 20 years. The days of even-aged, single-species plantation management are steadily declining. Instead, the challenges of multi-objective forest management at the stand and landscape levels require a wide variety of management systems that are typically developed and selected based on both ecosystem characteristics (e.g. species composition, nutrient cycling, productivity and natural disturbance regimes) and management objectives. Some examples of these include: mixed-wood management (intimate mixtures of deciduous and conifer species), variable retention harvesting (partial harvesting leading to the development of uneven-aged stands with complex structural characteristics), and harvest systems designed to emulate natural disturbance regimes with respect to the frequency and intensity of disturbance created by harvesting.

As the needs, goals and methods of forest management shift, the types of modelling tools required to support the application of new and relatively untested management systems have also changed. Several authors (e.g. Korzukhin et al, 1996; Johnsen et al, 2001; Landsberg, 2003b) have suggested that empirical or statistical models developed from historical observations of forest growth with little representation of ecological processes have limited application as decision-support tools to guide the application of ecosystem-based management. The limitations of such models are even more pronounced when the long-term impacts of climate change are taken into consideration. As described in Chapter 3, process-based forest models offer a much greater level of flexibility for applications involving the simulation of complex forest management systems and changing environmental conditions for which we have little long-term field experience. However, the difficulty in calibrating and verifying highly detailed process-based models has significantly limited their application as decision-support tools for forest resource management. In Chapter 3 we described the development of hybrid models as a means to capture the flexibility and understanding of process-based simulation, while maintaining the believability inherent in historical bioassay empirical models. While such hybrid models, particularly at the stand level, have been advocated for use in supporting the development and application of ecosystem-based forest management systems (e.g. Mäkelä et al, 2000; Landsberg, 2003b; Kimmins et al, 2007), there is a great deal of variability in the structure and functionality of hybrid stand-level models reported in the literature. The variability of these so-called hybrid models is largely a reflection of the wide range of management issues and systems that are often addressed in ecosystem-based forest management. Further, this variability in modelling methods and structures among stand-level hybrid models is helpful in meeting the diverse needs of multi-objective forest management.

In this chapter we begin by exploring the different types of stand-level hybrid models commonly employed in forest management, and group them into three general types based on their construction and functionality. We provide some examples of each type and their general utility. The remainder of the chapter is focused on providing a detailed description of the stand-level model FORECAST, developed at the University of British Columbia (UBC). A summary of its origins is provided in Chapter 3. The description provided in this chapter will include a detailed discussion of the hybrid simulation approach employed in FORECAST, including data requirements, key simulation algorithms and equations, and model structure. We also provide an overview of a series of case studies in which the model has been evaluated against long-term field data for different forest types and management systems. The chapter also includes a description of several projects for which FORECAST has been used to examine issues associated with the application of alternative silviculture systems to meet a range of forest management objectives, including ecosystem carbon storage, the recruitment and maintenance of stand structural attributes important for wildlife habitat, and the projection of merchantable volume accumulation. Finally, the chapter includes an overview of the development of the FORECAST Climate model, which includes an explicit representation of the effects of climate on forest growth, organic matter decomposition rates, water stress and water competition between tree and minor vegetation species.

Classification of stand-level hybrid models

The fundamental approach employed by hybrid forest growth models is to utilize synthesized knowledge of the biophysical processes underlying forest growth and development in such a way that allows for an incorporation of standard forest mensuration inputs and outputs. This link to mensuration inputs and outputs is often achieved through statistical rather than mechanistic relationships (Mäkelä et al, 2000). Several authors have attempted to classify stand-level forest growth models (e.g. Korzukhin et al, 1996; Landsberg, 2003b) using different attributes and schemes for separation. Mäkelä (2009) presented a particularly useful scheme to distinguish between three main hybrid modelling approaches, with a focus on the method of hybridization and a recognition of the core simulation approach utilized in the model. We have reproduced her classification scheme with a few minor changes for the purposes of this chapter. While specific models may contain elements of each group, the scheme is useful for identifying different approaches to constructing hybrid models. For a more detailed review of stand-level models designed to support forest management, the reader is referred to Landsberg (2003b), Messier et al (2003) and Van Oijen et al (2004).

Hybridized empirical models

These are models that utilize process elements as sub-models to provide additional inputs to empirical growth models. The inputs are often expressed as modifiers of statistically derived growth functions. Examples of process elements include simulations of intercepted radiation, gross or net photosynthetic production, soil water status, climate relationships, and physical constraints such as growing space.

Models within this group are typically constructed from statistical relationships between height and diameter growth and site or environmental condition variables. This approach generally requires large datasets to derive functional response relationships, and models of this type tend to be fairly site-specific, as relationships based on correlations don't always capture process interactions well. One of the recognized strengths of the approach is a representation of competition between stems within a stand. Examples of models that fit within this group include SORTIE-BC (Canham et al, 1999), TASS (www.for.gov.bc.ca/hre/gymodels/tass), and FVS-BGC (Milner et al, 2003).

Hybridized process models

These are models in which potential growth rates, biomass allocation patterns and mortality functions are derived from empirical data and modified using a complex system of physiologically based functions or sub-models. Process rate parameters are often derived from field-based measurements of process results (e.g. biomass accumulation, height growth, stand density).

The efficacy of the approach depends on the degree to which the structure of the growth and development processes within the model captures the key forcing factors regulating growth rates (e.g. light availability, nutrient availability and moisture availability) and other aspects of stand development. Examples of this type of model include the FORECAST model (Kimmins et al, 1999) and many of the gap models derived from the work of Botkin et al (1972) and Botkin (1993b). Mäkelä (2009) suggests that models developed with this approach have a good potential to capture the combined strengths of biometrics and process modellers.

Reduced-form process models

These represent a distinct subset of hybridized process models derived from detailed mechanistic models through an aggregation of growth processes and parameters into core components (carbon acquisition and allocation) to increase portability and ease of use. Empirical elements relate, for example, to the allocation of carbon within and between trees, but also to the process of parameterization and calibration of the model as a whole.

One of the benefits of reduced-form process models is that their mechanistic foundations provide insight into the impact of environmental forcing factors (particularly climate variables) on the relative productivity of different tree species. Model parameters can be estimated using a variety of different approaches ranging from Monte Carlo techniques (e.g. Mäkelä, 1988) to Bayesian synthesis (e.g. Green et al, 1999; Van Oijen et al, 2005). These models typically rely heavily on empirical sub-models to translate productivity and carbon allocation into mensurational output, including height and diameter increments important for growth and yield applications. Further, the stand-level focus of the growth functions limits the application of these models for evaluations of within-stand competition and other issues associated with stand structure. Examples of reduced-form models include 3-PG (Landsberg and Waring, 1997), Pipestem (Valentine et al, 1997) and G'DAY (Comins and McMurtrie, 1993).

Description of the hybrid modelling approach employed in FORECAST

Model overview

FORECAST is a management-oriented, stand-level forest growth and ecosystem dynamics simulator. The model was designed to accommodate a wide variety of harvesting and silvicultural systems and natural disturbance events (e.g. fire, wind, insect epidemics) in order to compare and contrast their effect on forest productivity, stand dynamics and a series of biophysical indicators of non-timber values. Projection of stand growth and ecosystem dynamics is based upon a representation of the rates of key ecological processes regulating the availability of, and competition for, light and nutrient resources. An alternative version of the model, called FORECAST Climate, includes moisture and climate feedbacks on growth rates and decomposition (see below). A complete description of the growth equations and simulation algorithms employed in FORECAST, as well as a detailed description of data requirements, can be found in Kimmins et al (1999).

Hybrid modelling approach employed in FORECAST

As outlined in the classification scheme above, FORECAST is best described as a hybridized process model. The model is structured such that the rates of ecological processes are calculated from a combination of historical bioassay data (biomass accumulation in component pools, stand density, etc.) and calculated measures of

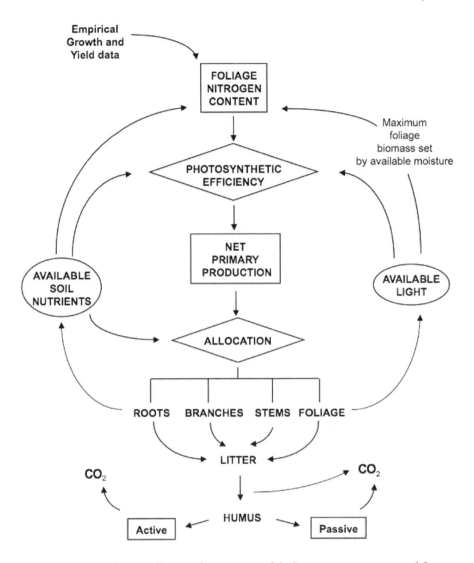

Figure 5.1 *A schematic diagram showing some of the key ecosystem processes and flows represented in FORECAST*

specific ecosystem variables (decomposition rates, foliar N efficiency and nutrient uptake demand, for example). This is achieved by relating biologically active biomass components (foliage and small roots) to calculations of nutrient uptake, the capture of light energy, and net primary production. Using this internal calibration (or hybrid) approach, the model generates a suite of growth properties for each tree and understorey plant species that is to be represented. These growth properties are subsequently used to model growth as a function of resource availability and competition (see Figure 5.1). They include (but are not limited to):

- photosynthetic efficiency per unit of foliage biomass and its nitrogen content, based on relationships between foliage nitrogen, simulated self-shading, and net primary productivity after accounting for litterfall and mortality;
- nutrient uptake requirements, based on rates of biomass accumulation and measures of nutrient concentrations in different biomass components for sites with a range of site nutrient quality, but accounting for internal cycling within plants; and
- light-related measures of tree and branch mortality, derived from stand density and height of the bottom of the live canopy input data in combination with simulated light profiles (light levels at which foliage and tree mortality occur are estimated for each species).

Representation of individual stems, snags and logs

FORECAST performs many calculations at the stand level, but it also includes a sub-model that disaggregates stand-level productivity into the growth of individual stems. The sub-model is designed to simulate a divergence in stem size as trees age as a consequence of small genetic differences between individuals, spatial variations in microclimate and soil resources, and competitive interactions not represented in the model (Figure 5.2). Top height and diameter at breast height (DBH) are calculated for each stem and used in a taper function to calculate total and individual gross and merchantable volumes.

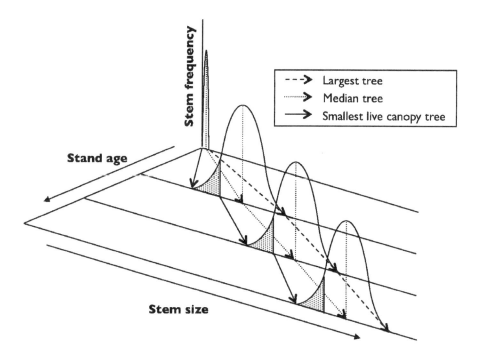

Figure 5.2 *A diagram of FORECAST's representation of the distribution of individual stems in a even-aged cohort into stem biomass size classes through time*

Source: after Kimmins et al (1999)

Snags and logs are created in the model from natural stand self-thinning and from different types of user-defined disturbance events such as insect/disease-induced mortality, windthrow and non-commercial thinning. Snag fall rates and log decomposition are simulated using species-specific and tree-size-specific decay parameters derived from literature reviews and expert opinion.

Decomposition and nutrient cycling

Decomposition and dead organic matter dynamics are simulated in FORECAST using a method in which specific biomass components are transferred, at the time of litterfall, to one of a series of independent litter types. An example of decomposition rates used for the main litter types represented in the model is shown for a sub-boreal forest type in Figure 5.3. The mass loss data required to drive the decomposition simulation in FORECAST are generally derived from the results of extensive field incubation experiments conducted across BC and elsewhere (Prescott et al, 2000; Camiré et al, 2002; Trofymow et al, 2002). Residual litter mass and associated nutrient content is transferred to active (85 per cent) and passive (15 per cent) humus pools at the end of the litter decomposition period (when mass remaining is approximately 15–20 per cent of original litter mass). Mean residence times for active and passive humus types are estimated from regional long-term climate data based on relationships reported in the

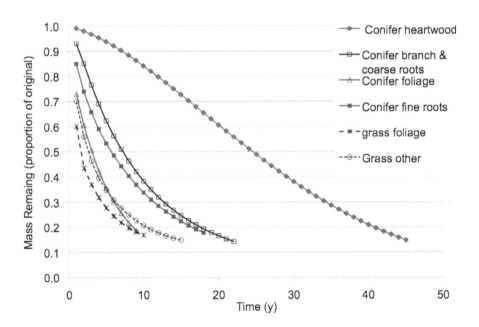

Figure 5.3 *An illustration of the decomposition rates for the main litter types represented in FORECAST*

Note: mass remaining at the end of the litter decomposition period (15–20 per cent) is transferred to humus pools, which have constant mass loss rates.

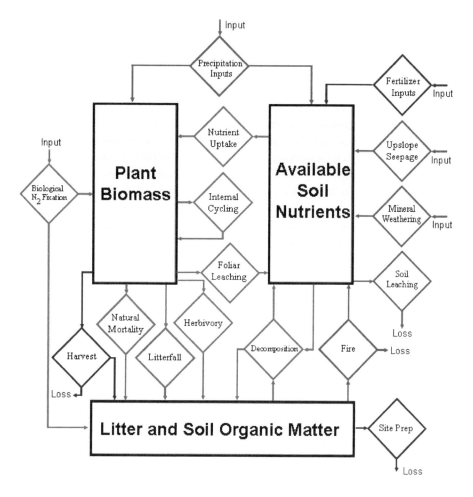

Figure 5.4 *A schematic diagram illustrating the representation of the mass balance approach to nutrient cycling employed in FORECAST*

Note: the three main pools of nutrients and the simulated transfer processes that regulate the movement of nutrients between pools and into/out of the ecosystem are shown.

Source: after Seely et al (2008).

literature. Using this approach, residence times for ecosystem types for similar climate regimes typically range from 25 to 50 years for active humus, and 400 to 750 years for passive humus.

The simulation of nutrient cycling in FORECAST is based on a mass balance approach designed to track the total pool of nutrients within the forest ecosystem and the import or export of nutrients to and from the ecosystem (Figure 5.4). Nutrient storage, and release from decomposing litter and soil organic matter into the plant-available pool, represents the largest source of nutrients for uptake by plants. Plant nutrient demand for new growth is satisfied from this pool and by simulated translocation from older tissues. Growth of plants (trees, herbs, shrubs and bryophytes) is limited by the availability of nutrients if the supply from uptake and translocation is less than that required to support the expected annual biomass increment. Actual growth rate is restricted to an increment level supported by the simulated uptake plus that obtained from translocation.

Site quality change

Most models require the user to declare the site quality for which a simulation is being conducted, and assume that this site quality remains constant for the duration of the run. In reality, nutritional site quality is a dynamic ecosystem characteristic that varies over the stand cycle (the various phases of stand dynamics within a seral stage, or simply the time between successive stand-replacing disturbances) and between different seral stages, and it is affected by both management-induced and natural ecosystem disturbances. Site moisture quality can also vary under the influence of climate variation, ecosystem disturbance and different biotic conditions, but much less so than nutritional site quality. It is more a function of soil texture, depth, aspect, slope and slope position, all of which are essentially constant over the timescales being considered.

In contrast, FORECAST simulates site quality change over time. It does so by determining the temporal patterns of site nutrient availability (and in FORECAST Climate, site moisture availability) that probably supported the historical patterns of plant growth and biomass accumulation. These patterns then become the definition of site quality. In a simulation, the model compares the simulated site quality values with the definition values, and if differences are found, the model adjusts the perceived (for individual plant and tree species) site quality accordingly. These adjustments translate into shifts in variables controlling levels of potential growth rates, carbon allocation patterns, and associated nutrient requirements for each species represented in the model. This enables the model to simulate both site quality improvement and decline over time. More detail of the simulation of site quality change can be found in Kimmins et al (1999).

Model calibration and application

FORECAST has four stages in its application (after Blanco et al, 2007):

1   data assembly and input verification;
2   establishing the ecosystem condition for the beginning of a simulation run (by simulating the known or assumed history of the site);
3   defining a management and/or natural disturbance regime; and
4   simulating this regime and analysing model output.

The first two stages represent model calibration. Calibration data are assembled that describe the accumulation of biomass (above- and below-ground components) in trees and understorey vegetation for three chronosequences of stands, each one developed under homogeneous site conditions, representing three different site nutritional qualities. Tree biomass and stand self-thinning rate data can be obtained from the height, DBH and stand density output of locally calibrated traditional growth and yield models, or from long-term growth and yield plot measurements, in conjunction with species-specific component biomass allometric equations. To calibrate the nutritional aspects of the model, data describing the concentration of nutrients in the various biomass components being represented – from local measurements or literature values – are required. FORECAST also requires data on the degree of shading produced by different quantities of foliage (from field measurements or from a locally calibrated Beer-Lambert equation) and the photosynthetic response of foliage to different light levels; both shade- and light-adapted foliage can be represented in the model. A comparable but simpler set of data for understorey vegetation must be provided if the user wishes to represent this ecosystem component. Lastly, data describing the rates of decomposition of various litter types and soil organic matter are required for the model to simulate nutrient cycling (see above). The complete list of data requirements for FORECAST is provided in Kimmins et al (1999).

The second aspect of calibration involves running the model in set-up mode to establish a starting ecosystem condition, which is required for the simulation of nutrient cycling and nutrient feedback to forest growth (see Seely et al, 1999). The detailed representation of many different litter types and soil organic matter conditions makes it impractical to measure initial litter and soil conditions directly in the field; consequently, the model is used to generate starting conditions for all litter and soil organic matter compartments. This initial ecosystem state is established by running the ecosystem model without nutrient feedback to force the vegetation to grow as it has historically, and to allow the ecosystem to accumulate soil organic matter, forest floor litter and coarse woody debris as expected under a defined historical disturbance regime. The output of this run is saved as a starting condition file for future use.

Once a starting condition file has been created and selected, a set of management activities and/or natural disturbance events is defined and scheduled to occur during the ecosystem simulation period. The range of management activities and natural disturbances that may be simulated in FORECAST are listed in Table 5.1. After a set of management activities has been defined, soil nutrient feedback is switched on, and the ecosystem program is run to simulate future ecosystem development under the defined management/natural disturbance scenarios. The output of each simulation run is saved into a series of text files and a graphics file for analysis of model results.

Overview of FORECAST evaluation studies

The ongoing transition to ecosystem-based forest management, and eventually to full ecosystem management, requires resource managers to project the probable outcomes of alternative management options within the context of multiple resource values. This has elevated the level of complexity in forest management and necessitated the development of decision-support tools that allow for greater flexibility in representing management and environmental conditions. Before any model can be used with confidence, however, it is necessary to establish its validity by comparing its forecasts against available field measures of forest growth and development.

The capability of the FORECAST model to reproduce observed long-term trends in forest growth and development has been evaluated in a number of projects through direct comparisons of model output to field measurements. The following section provides a

Table 5.1 *Silvicultural systems, management activities and natural disturbance events that can be simulated by FORECAST*

Silvicultural systems	Management Activities	Natural Disturbance Events[1]
Even-aged (clearcuts)	Site preparation	Wildfire (variable intensity)
Intensive plantations	Planting	Disease-related mortality
Mixed wood	Regeneration[3]	Insect-related mortality
Shelterwood or dispersed retention harvesting[2]	Weed control	Windthrow[4]
Nurse crops	Stocking control	Insect defoliation
Variable rotation lengths	Pruning	Wildlife browsing
Variable utilization levels	Fertilization	
Agroforestry	Biosolids application	
Uneven-aged (multiple cohorts)[5]	Intermediate harvests	
	Final harvest	
	Slash burns	

[1]Natural disturbance events are predetermined, not stochastic. Snags or logs are created depending on the disturbance.
[2]The model assumes a uniform distribution of individuals.
[3]Regeneration timing and numbers must be specified by user.
[4]Numbers and size classes of individual trees to be affected must be provided by user.
[5]The model simulates multiple cohorts of even-aged trees regenerated after episodic disturbance events.

summary of two model evaluation projects published in peer-reviewed journals, and a third in review. These include the following studies: Blanco et al (2007); Seely et al (2008); and Gerzon and Seely, in review. Additional evaluation studies of FORECAST performance in tropical and subtropical plantations can be found in Bi et al (2007) and Blanco and González (2010).

Fertilization and thinning effects in coastal Douglas-fir

The ability of FORECAST to reproduce a 29-year record of stand response to factorial thinning and fertilization treatments was tested in a Douglas-fir (*Pseudotsuga menziesii*) plantation (Blanco et al, 2007). Model output was compared against field measurements of height, diameter, stem density, above-ground component biomass and litterfall rates, and estimates of nutrient uptake, foliar biomass efficiency and understorey vegetation biomass.

Description of field trial

The study site was located in a Douglas-fir plantation of approximately 5000 stems ha^{-1} at Shawnigan Lake, on southern Vancouver Island, BC, Canada. The plantation had been established on the site through planting and natural regeneration in 1946 following a stand-replacing fire in an existing Douglas-fir stand. Sixty 0.04ha plots were established, each surrounded by a 15m treated buffer. Four replicate plots were assigned to each of the major treatments, two in each of 1971 and 1972 (when the trees were 25–26 years old). The main treatments consisted of three levels of thinning (T0, no thinning; T1, removal of one-third of the basal area; and T2, removal of two-thirds of the basal area) and three levels of fertilization (F0, no fertilization; F1, fertilization with urea at 224kg N ha^{-1}; and F2, fertilization with urea at 448kg N ha^{-1}). Data describing 24 years of tree growth and stand development at this study site were summarized by McWilliams and Thérien (1997), and additional field data were collected until 2003 by the Pacific Forestry Centre, Victoria, BC, resulting in a record of 29 years of tree growth.

Model application and evaluation

Model performance was evaluated firstly using a regional calibration dataset (Phase 1), and secondly with site-specific data collected from the control plots (Phase 2). An initial condition file was prepared for the Phase 1 and 2 runs by simulating the stand history of this plantation reported by Crown and Brett (1975). The quantities of soil organic matter and litter created by these set-up runs were similar (within 13 per cent) to those measured in the field at the time of stand establishment at Shawnigan Lake (Crown and Brett, 1975). FORECAST was subsequently run to simulate each of the different thinning and fertilization treatment combinations described above. Runs were repeated for both the regional and site-specific calibration datasets.

Since biomass values were not directly measured as part of the field measurement, component biomass (stemwood, bark, branch, and foliage) were estimated using published allometric biomass equations for the control plots on the site (Barclay et al, 1986). These estimates were compared against model output. A series of statistical tests for goodness of fit were conducted to evaluate the capability of the model to reproduce the observed and estimated field measurement of stand growth. Only the R^2 and the modelling efficiency (Vanclay and Skovsgaard, 1997) statistics are reported here. The modelling efficiency (ME) statistic is calculated as follows:

$$ME = 1 - \frac{\sum D_i^2}{\sum (observed_i - \overline{predicted})^2}$$

where D_i = $observed_i$ – $predicted_i$. This statistic provides a simple index of performance on a relative scale, where ME = 1 indicates a perfect fit, ME = 0 indicates that the model is no better than a simple average; and negative values for ME indicate that the model is showing opposite trends to observed data. Graphical comparisons of model output to field data were used to evaluate temporal trends in model performance.

Summary of results and conclusions

When calibrated with regional data, the results from graphical comparisons, three measures of goodness of fit, and equivalence testing demonstrated that FORECAST can produce predictions of moderate to good accuracy, depending on the variable (Table 5.2). A comparison of some of the ecological process rate values predicted by FORECAST to those estimated for the control plots in the Shawnigan Lake site are shown in Table 5.3. Model performance was generally better when compared to field measurements (e.g. top height, DBH and stem density) as opposed to outputs derived from allometric and volume equations. An example of temporal patterns in simulated growth response of Douglas-fir to the different thinning and fertilization treatments is shown for average stem DBH and total stem density in Figures 5.5 and 5.6, respectively.

Use of site-specific data to calibrate the model always improved performance, though improvements were modest for most variables with the exception of branch and foliage biomass. However, the relatively poor predictions of branch and foliage

Table 5.2 *Statistical comparison of the overall fit of model outputs (using all results from the 15 different combinations of thinning and fertilization) against observed data*

Statistic	Top height[a]	DBH[a]	Stem density[a]	Merchantable volume[b]	Stemwood biomass[c]	Bark biomass[c]	Branch biomass[c]	Foliage biomass[c]
	m	cm	Trees ha^{-1}	m^3 ha^{-1}	Mg ha^{-1}	Mg ha^{-1}	Mg ha^{-1}	Mg ha^{-1}
PHASE 1								
R^2	0.87	0.83	0.88	0.82	0.81	0.82	0.21	0.17
ME	0.84	0.87	0.89	0.66	0.65	0.69	0.15	0.12
PHASE 2								
R^2	0.95	0.80	0.94	0.92	0.88	0.88	0.72	0.74
ME	0.91	0.89	0.95	0.89	0.81	0.54	0.54	0.66

[a]Predicted values compared to field measurements carried out by the Canadian Forest Service (Victoria, BC) from 1971 to 2003.
[b]Predicted values compared to McWilliams and Thérien's (1997) estimations.
[c]Predicted values compared to values calculated from field measurements combined with Barclay et al's (1986) allometric equations.
Note: ME = modelling efficiency.

biomass are probably related to the fact that allometric biomass equations do not work well for these biomass components, which are strongly influenced by stand conditions (e.g. Standish et al, 1985). The benefits of site-specific calibration, however, should be weighed against the costs of obtaining such data. The intended use of the model will probably determine the level of effort expended in its calibration.

Mixed-wood management in spruce and aspen

Mixed conifer/broadleaf forests (mixed woods), which cover more than one-third of the productive forest land base in BC, are highly valuable both as sources of wood fibre and as areas rich in biodiversity (Comeau, 1996). In recognition of the multiple benefits of this forest type, management paradigms have transitioned from a focus on promoting conifer plantations (frequently pine or spruce monocultures) in mixed-wood areas to the management of intimate mixtures (e.g. Grover and Greenway, 1999). The exceptionally dynamic growth properties and species interactions in mixed-wood forests present a challenge for projecting the growth and development of different types of mixed woods (different species mixtures and different spatial scales of mixing) and their response to different silvicultural systems.

The FORECAST model was evaluated with respect to its ability to reproduce the growth trends observed in a long-term (18 years), mixed white spruce and trembling aspen silviculture trial in the sub-boreal spruce (SBS) zone in the central interior of BC (Seely et al, 2008). The evaluation included a comparison of model output against a time series of field measurements of tree growth and an assessment of the simulated relative impact of light and nutrient competition on growth dynamics.

Description of field trial

The study site was established in 1985 in an area that was clearcut in 1969, broadcast burned in 1970, and planted in 1971 with two-year-old white spruce (*Picea glauca*)

Table 5.3 *A comparison of FORECAST's predictions for selected ecological process rates against published field-based estimates from control plots (T0F0) at Shawnigan Lake*

Stand age range (yr)	24–33		33–42	
Litterfall	Foliage litterfall (kg ha^{-1} yr^{-1})	N in litterfall (kg ha^{-1} yr^{-1})	Foliage litterfall (kg ha^{-1} yr^{-1})	N in litterfall (kg ha^{-1} yr^{-1})
Trofymow et al (1991)	1890	9.44		
Mitchell et al (1996)	1612	7.70	1931	9.50
FORECAST Phase 1	1629	7.98	1864	9.13
FORECAST Phase 2	1409	6.90	2189	10.73

Foliar biomass efficiency (FBE)	FBE above-ground (kg kg^{-1} leaf)	FBE stemwood (kg kg^{-1} leaf)	FBE above-ground (kg kg^{-1} leaf)	FBE stemwood (kg kg^{-1} leaf)
Brix (1983)	1.02	0.71		
Barclay et al (1986)	0.66	0.43		
Mitchell et al (1996)	0.79	0.54	1.31	0.76
FORECAST Phase 1	0.68	0.34	0.71	0.38
FORECAST Phase 2	0.88	0.46	0.69	0.32

N uptake[a]	Total N uptake (kg ha^{-1} yr^{-1})	Net N uptake (kg ha^{-1} yr^{-1})	Foliar N retranslocated (kg ha^{-1} yr^{-1})	Total N uptake (kg ha^{-1} yr^{-1})	Net N uptake (kg ha^{-1} yr^{-1})	Foliar N retranslocated (kg ha^{-1} yr^{-1})
Mitchell et al (1996)[b]	10.0	4.5	9.7	10.0	–2.1	9.7
FORECAST Phase 1	30.1	7.8	9.9	32.3	6.1	11.9
FORECAST Phase 2	34.3	13.0	9.9	34.2	9.0	15.3

Note: Phase 1 corresponds with regional calibration and Phase 2 corresponds with site-specific calibration.
[a]Mitchell et al (1996) calculated these values with only above-ground nutrient budgets. Output from FORECAST includes nutrient budgets for both above-ground and below-ground biomass components.
[b]Average values for the age interval 24–42 years.

seedlings. Following the burn, the plantations had a vigorous regeneration of aspen (*Populus tremuloides*) such that when plots were established in 1985, the aspen canopy was well above that of the spruce. Long-term plots were installed to examine the impact of two levels of aspen removal on spruce growth. The first treatment was the mechanical removal of all of the aspen (referred to as 'brushed'), and the second treatment was a

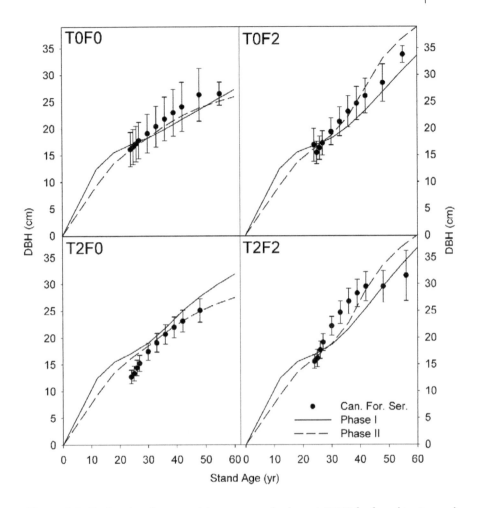

Figure 5.5 *Predicted and measured (mean ± standard error) DBH for four thinning and fertilization treatments*

Notes: Phase I: generic dataset; Phase 2: generic dataset calibrated for Shawnigan Lake stands.

Source: direct field measurements by the Canadian Forest Service (Victoria, BC) from 1971 to 2003 (after Blanco et al, 2007).

partial removal of the aspen using strip cuts ('partial brushed'). In the partial brushed plot, 5m-wide strips with no aspen removed were left adjacent to 7m-wide strips in which all the aspen had been removed mechanically. A control plot was also established in which no aspen were removed.

Model application and evaluation
Simulations were conducted to represent each of the treatments (brushed and partial brushed) and the control. In each simulation, spruce was planted in year 1 (1971) at

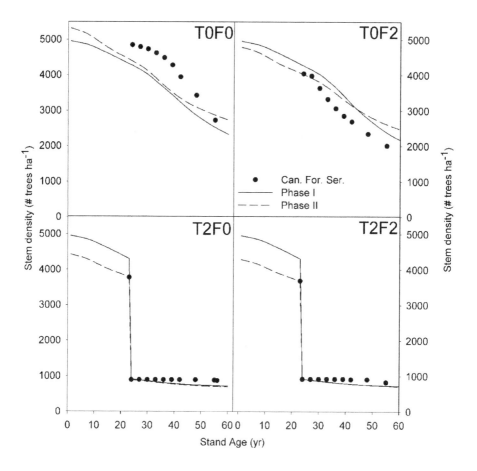

Figure 5.6 *Predicted and measured stem density for four thinning and fertilization treatments*

Notes: Phase I: generic dataset; Phase 2: generic dataset calibrated for Shawnigan Lake stands.

Source: direct field measurements by the Canadian Forest Service (Victoria, BC) from 1971 to 2003 (after Blanco et al, 2007).

1500 stems per hectare, and aspen regeneration occurred in year 1 at 2000 stems per hectare. In the case of the brushed treatment, 100 per cent of the aspen was harvested in year 15 (1985) and aspen regenerated at 5000 stems per hectare the following year (to represent suckering). In the case of the partial brushed treatment, 60 per cent of the aspen stems were removed with stems distributed evenly across all size classes. Since FORECAST has a non-spatial representation of tree distribution. The simulation was based on the assumption that the removal strips were narrow enough (7m) that the spruce would behave as if the residual aspen were distributed evenly throughout the treated area, particularly as the residual aspen matured and crown expansion occurred. It was assumed that there was some regeneration of aspen by suckering (2500 stems per

Table 5.4 *Goodness of fit of model relative to field data as measured by modelling efficiency (ME) for the spruce/aspen and Douglas-fir/birch mixed-wood simulations*

Variable	Spruce		Aspen	
	ME	n	ME	n
Average top height	0.97	12	0.93	3
Top 200 top height	0.98	12	n/a	3
Average DBH	0.76	6	0.77	3
Top 200 DBH	0.93	6	n/a	3
Stand volume	0.97	6	0.98	3
Stand stem biomass	0.88	6	0.96	3

Note: n = number of data

hectare) following the thinning treatment. The control scenario had no management interventions following the initial planting of spruce.

Model performance was evaluated using graphical comparison to examine treatment response with respect to standard growth and yield variables. In addition, overall model fit was assessed for each species and variable using the modelling efficiency (ME) statistic (see above).

Summary of results and conclusions
The overall fit of the model to observed values (as measured by ME) was good for both spruce and aspen in the simulated treatments and control. ME values were >0.9 for all variables tested, with the exception of total stand spruce stemwood biomass (ME = 0.88), average DBH for spruce (ME = 0.76) and average DBH for aspen (ME = 0.77) (Table 5.4). An example of temporal patterns in simulated growth response of spruce and aspen to the different brushing treatments relative to the control is shown for stemwood biomass in Figure 5.7. The model also allowed for an assessment of light and nutrient competition between the spruce and aspen in each of the treatments (Figures 5.8 and 5.9). This type of output is essential to help improve our understanding of the competitive interactions within mixed species stands, as well as the influence of silvicultural treatments on resource competition. While more validation work needs to be conducted in a range of different mixed-wood forest types and management interventions as datasets become available, the analysis described in this study provides a level of confidence for the use of this model as a decision-support tool in mixed-wood ecosystems.

Development of old-growth structural elements in coastal western hemlock
One of the key issues facing forest resource planners throughout BC is the conservation and promotion of OG characteristics in managed forest landscapes. The ability to develop and apply multi-objective forest management plans that include the conservation and continued recruitment of OG forest requires the ability to predict the recovery of forest ecosystems following both traditional logging activities and alternative silvicultural

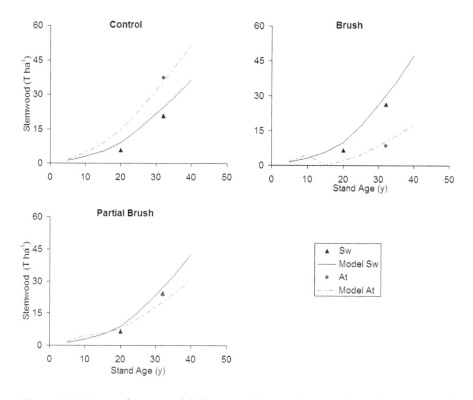

Figure 5.7 *Measured versus modelled stemwood biomass by species for each treatment and the control for the spruce/aspen mixed wood*

Note: Sw = white spruce; At = trembling aspen.

Source: after Seely et al (2008).

systems (Kimmins et al, 2008a). To be effective for evaluating the development of OG characteristics in second-growth stands, a model must include explicit representations of the structural features that can be used to distinguish OG conditions (e.g. Choi et al, 2007). Such features include measures of individual tree size, characteristics of coarse woody debris including log sizes and state of decay, and characteristics of standing dead trees including species and dimensions (Franklin and Spies, 1991; Tyrrell and Crow, 1994). It is also helpful if a model is capable of adequately representing the ecosystem processes that lead to the formation, accumulation and loss of these features. Such processes include density-dependent mortality (stand self-thinning), minor (non-stand-replacing) disturbance events such as insect-induced mortality and windthrow, harvesting events, snag fall rates, and organic matter decomposition rates (including snags and logs of different species and size classes).

The FORECAST model was evaluated for its capability to project the development of a series of OG structural attributes as measured in a chronosequence study of second-growth coastal western hemlock stands along the west coast of Vancouver Island, BC (Gerzon and Seely, in review).

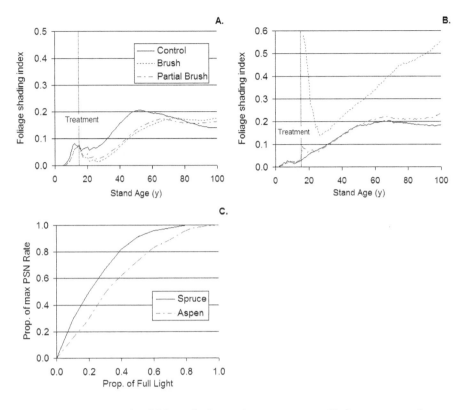

Figure 5.8 *Simulated foliage shading index as a measure of light competition for (a) spruce and (b) aspen, for the two aspen brushing treatments and the control in the spruce/aspen mixed-wood stand type*

Note: Panel C shows the relative photosynthetic rates (per unit foliar N) for spruce and aspen as a function of light levels calculated throughout the canopy profile.

Source: after Seely et al (2008).

Description of chronosequence study and OG indicators

The study was located in the economically important coastal western hemlock (CWH) biogeoclimatic zone (Pojar et al, 1987) on Vancouver Island, BC. A chronosequence including 33 sites varying in age from 60 to over 300 years was established in an area on western Vancouver Island, 380km in length (Figure 5.10). The chronosequence included 24 second-growth stands of various ages recovering from different stand-replacing disturbances (logging, windthrow and fire) with minimal residual live structure from the originating stands. The variation in disturbance history and site features raises questions about the validity of the chronosequence, but there was no other way of examining stand and ecosystem development trends over time, and it is assumed that with 24 sites, the overall trend will be valid (see Martin et al, 2002). There were nine stands in the chronosequence classified as OG. These stands could not be reliably aged, and thus were assumed to be 300 years old.

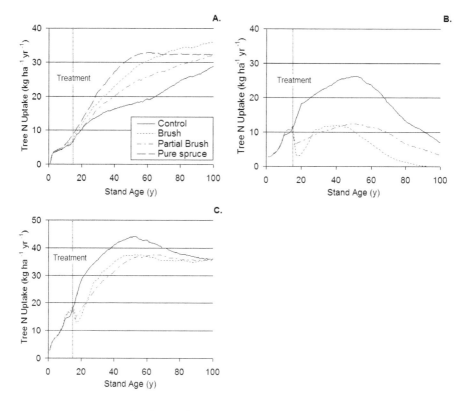

Figure 5.9 *Simulated annual nitrogen (N) uptake for (a) spruce, (b) aspen and (c) total tree, for the two aspen brushing treatments and the control in the spruce/aspen mixed-wood stand type*

Source: after Seely et al (2008).

Table 5.5 *Model output for structural variables used in model evaluation, showing criteria for inclusion of stems in specific indicator variables*

Test variables	Criteria	Units
Cumulative volume in large stems	≥50cm DBH	m^3 ha^{-1}
Frequency of large stems	≥50cm DBH	stems ha^{-1}
Cumulative volume in medium stems	25–50cm DBH	m^3 ha^{-1}
Frequency of medium stems	25–50cm DBH	stems ha^{-1}
Standard deviation of stem sizes	≥12.5cm DBH	cm
Accumulation of CWD biomass	Non-residual CWD*	t ha^{-1}
Frequency of large snags	≥50cm DBH	stems ha^{-1}

* In the case of second-growth stands, non-residual coarse woody debris (CWD) refers to logs derived from the current stand as opposed to the previous stand. Thus non-residual CWD was determined based on a comparison of log and live tree diameters.

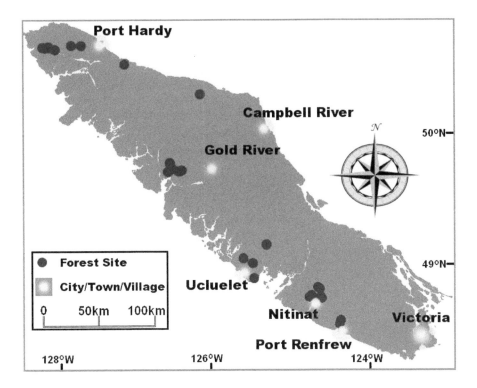

Figure 5.10 *Location of the 33 chronosequence study sites on Vancouver Island*

Field measurements of the chronosequence sites included stand structural attributes, vegetation and soil properties. In addition, tree ages were measured and disturbance history was estimated. A set of OG indices were developed (Table 5.5) that showed clear trends in the development of stand structure and other attributes with increasing stand age (Gerzon, 2009; Gerzon and Seely, in review).

Model application and evaluation

FORECAST was set up to simulate the development of a second-growth stand representative of those included in the chronosequence study. An initial condition was established based on a 250-year stand-replacing fire interval simulated over a period of 2500 years. The final set-up run consisted of a western hemlock stand regenerated at 2700 stems ha^{-1} and clearcut harvested at age 250. The details of the FORECAST simulation conducted to evaluate model output against the chronosequence data are presented in Table 5.6. The goodness of fit of model output to field data trends was quantified for the structural variables described in Table 5.5 using a series of commonly used statistics (after Vanclay and Skovsgaard, 1997). Mean model bias was calculated as the averaged difference between the model and the field data; the smaller the number, the closer the output of the model to the field data:

Table 5.6 *Model parameter values describing assumptions for natural regeneration patterns, and disturbance events simulated in the evaluation run*

Model variable	Value	Remarks
Total evaluation time	300 years	
Vegetation simulated	western hemlock, vaccinium shrub	
Shrub regeneration at first years: proportion of maximum occupation	20 per cent	
Number of cohorts	5	Three cohorts regenerate in stand initiation phase, one in understorey reinitiation and one in OG phase
Stems density regenerated	600, 400, 400, 400, 400	For cohorts 1–5, respectively
Year of cohort establishment	1, 11, 21, 171,* 211*	For cohorts 1–5, respectively
Windthrow	The disturbance is applied once in every 40 years and affects 15 per cent of the trees. The first windthrow event is applied at year 50 of the simulation	Impacts trees older than 40 years
Disease-related mortality	The disturbance occurs continuously within 20-year interval. During this interval, 3 per cent of the trees are being killed	Impacts trees older than 40 years

* The last two regeneration cohorts were included to account for regeneration that typically occurs when forest-floor light levels increase following the reduction of overstorey canopy from windthrow events in mature stands. The dates correspond with simulated windthrow events.

$$\text{Mean Bias} = \frac{1}{n} \sum_{i=1}^{n} (Y_i^{field} - Y_i^{model})$$

The root mean square error (RMSE) statistic was calculated as an estimator of model accuracy. It measures the average error associated with single prediction points of the model:

$$\text{Root mean square error (RMSE)} = \sqrt{\frac{1}{n} \sum_{i=1}^{n} (Y_i^{field} - Y_i^{model})^2}$$

The ME statistic was calculated as described above.

Summary of results and conclusions

Because the chronosequence data is from sites that varied in disturbance history and natural regeneration patterns, comparisons against model output were made against the best-fit curve to the field data (i.e. the trend line) rather than against individual data points. FORECAST was able to project the long-term development patterns of most of the measured structural variables quite well, as indicated by the predominantly high (>0.7) values for the ME statistic (Table 5.7). Moreover, the simulated temporal patterns in structural attribute development were biologically consistent.

Table 5.7 *Results from statistical comparisons of model output to trend lines fit to chronosequence field data*

Variable	Units	Trend line equation	R^2 field fitting	Mean bias	RMSE	ME
Cumulative volume in large stems (≥50cm DBH)	$m^3\ ha^{-1}$	asymptotic $Y = \dfrac{a}{X} + b$	0.48	−54 $m^3\ ha^{-1}$	70 $m^3\ ha^{-1}$	0.95
Frequency of large stems (≥50cm DBH)	stems ha^{-1}	asymptotic $Y = \dfrac{a}{X} + b$	0.50	−1.16 stems ha^{-1}	18.5 stems ha^{-1}	0.83
Cumulative volume in medium stems (25–50cm DBH)	$m^3\ ha^{-1}$	exponential decline $Y = \dfrac{a}{Exp(-bX)}$	0.59	−49 $m^3\ ha^{-1}$	143 $m^3\ ha^{-1}$	0.54
Frequency of medium stems (25–50cm DBH)	stems ha^{-1}	exponential decline $Y = \dfrac{a}{Exp(-bX)}$	0.70	−4.7 stems/ha	49 stems/ha	0.90
Standard deviation of DBH*	cm	asymptotic $Y = \dfrac{a}{X} + b$	0.28	0.063 cm	0.275 cm	0.98
CWD biomass**	Mg ha^{-1}	asymptotic $Y = \dfrac{a}{X} + b$	0.21	8.47 Mg ha^{-1}	9.9 Mg ha^{-1}	0.72

* Model fit analysis was limited to year 170, prior to the establishment of additional cohorts following windthrow events.

** Model data was represented by a trend line fit to model output to smooth the abrupt changes due to disturbance regimes.

Note: The form of the trend line fit to the chronosequence data and the goodness of fit are shown for each evaluation variable.

An example of the temporal patterns in model output for the recruitment of large live structure (stems >50cm DBH) is shown in Figure 5.11. The model was able to reproduce observed patterns in the accumulation of coarse woody debris (Figure 5.12) and large diameter snags (Figure 5.13). The evaluation presented in Gerzon and Seely (in review) and summarized here provides support for the use of the model to project the development of forest structural attributes in second-growth coastal western hemlock stands as they mature towards an OG condition. It also provides a level of confidence for the application of the model as a simulation tool with which to explore the long-term implications of alternative management scenarios, including variable retention management, for the development and maintenance of OG structural attributes.

Applications of FORECAST

Stand-level, hybrid forest growth models such as FORECAST have been increasingly employed as decision-support tools to allow forest managers to explore the long-term

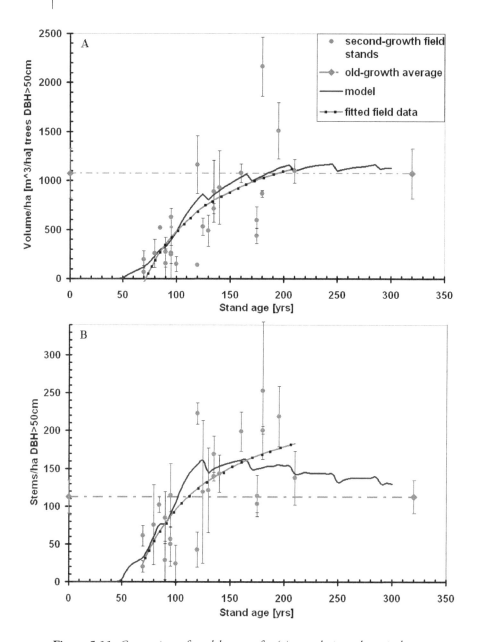

Figure 5.11 *Comparison of model output for (a) cumulative volume in large stems (≥50cm DBH) and (b) frequency of large stems against trend lines fitted to chronosequence field data*

Note: the average for OG stands measured in the field study is also shown with the 95 per cent confidence interval represented by the error bar.

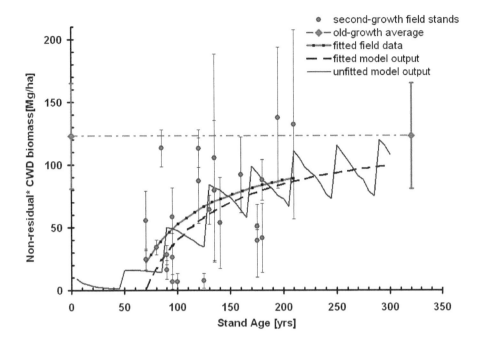

Figure 5.12 *Comparison between the CWD biomass of the field data and the model output*

Note: asymptotic trend lines are shown for both model output and field data. The error bar of OG average represents the 95 per cent confidence interval. Residual logs in chronosequence sites were excluded from CWD mass totals based on size and decay class, relative to the size of live trees.

implications of forest resource management options and alternatives without waiting for feedback from slow and expensive field trials. These models can be applied as stand-alone models or as a component of a set of linked models operating at multiple spatial and temporal scales (meta-models). In developing and applying sustainable forest management plans, models are most effective when implemented within an adaptive management cycle (Walters, 1986; Rauscher, 1999; Chapter 9), including a well-defined set of indicators, monitoring systems, and mechanisms for feedback from researchers, industry and stakeholder groups. The model plays a critical role in this process by:

- highlighting potential conflicts between competing management objectives;
- providing a common, science-based framework for stakeholders to evaluate the potential consequences of specific management options;
- conveying knowledge about the long-term dynamics of forest ecosystems; and
- providing guidance for the monitoring process by projecting expected trends in selected indicators.

In this section, we provide examples of past applications of the FORECAST model to address specific forest management issues, including forest carbon management,

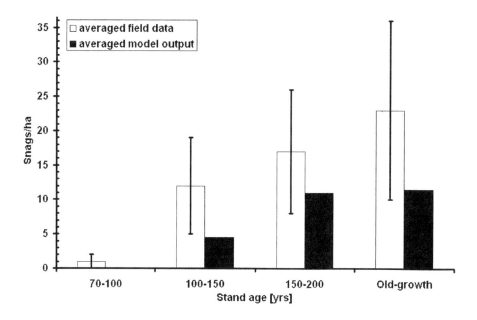

Figure 5.13 *Average frequency of large snags (≥50cm DBH) in field sites grouped into age classes compared to model output*

Note: residual snags were excluded as explained in the text. OG has 95 per cent confidence interval as an error bar, whereas other age classes have standard error of the mean bars. Simulated stand age class 200–300 were compared to OG field data.

assessing the long-term impact of alternative salvage strategies on stand dynamics following widespread insect-induced mortality, and assessing the major determinants of plantation yield decline in China. We also examine the linkage of FORECAST to landscape-scale forest management models to expand the scale of model application.

Forest carbon management

The management of forests for carbon sequestration and/or storage is attracting increasing attention as the perceived risk of climate change continues to increase and markets for forest carbon offsets continue to develop (e.g. Ecosystem Marketplace, 2010). As forests grow, trees take up CO_2 from the atmosphere through the process of photosynthesis. This carbon is subsequently stored in the forest ecosystem as live biomass (foliage, branches, stems and roots), dead organic matter (e.g. litter, logs and snags), and soil organic matter. Forest management and/or conservation that increases the amount of carbon stored in a forest ecosystem over and above standard management practices may be eligible for sale as carbon offsets. The development of forest carbon projects, however, requires the use of scientifically credible forest growth models to project the long-term sequestration and net storage of carbon within a defined forest area. To be effective, models must include an accurate representation of the relevant pools outlined above, as well as the impacts of proposed management practices and expected disturbance

Table 5.8 *Management scenarios simulated with FORECAST*

Run ID	Tree species	Rotation length (y)	Rotations (#)	Disturbance
Nat. Dist	Aspen, spruce, pine	150	2	Fire
Aspen 30	Aspen	30	10	Harvest (whole tree)
Aspen 60	Aspen	60	5	Harvest (whole tree)
Aspen 90	Aspen	90	4.33	Harvest (whole tree)
Spruce 60	Spruce	60	5	Harvest (stem only)
Spruce 100	Spruce	100	3	Harvest (stem only)
Spruce 200	Spruce	200	1.5	Harvest (stem only)
Pine 30	Pine	30	10	Harvest (stem only)
Pine 75	Pine	75	4	Harvest (stem only)
Pine 150	Pine	150	2	Harvest (stem only)
Aspen 30 Fert	Aspen	30	10	Harvest (stem only)

Note: rotations refers to the number of management cycles completed within the 300-year simulation period.

regimes on ecosystem carbon storage and long-term sequestration rates. FORECAST, being a biomass-based model including above- and below-ground components and with a comprehensive representation of dead organic matter dynamics, is well suited for such applications. The following section provides an example of the application of FORECAST to examine long-term carbon storage and sequestration rates in forest ecosystems subjected to different management regimes. An additional example of such applications is given in Chapter 10.

The effect of alternative harvesting practices on ecosystem carbon storage and sequestration was investigated for a boreal forest located in north-eastern BC using FORECAST (see Seely et al, 2002). The model was calibrated for a boreal cordilleran mixed-wood ecosystem type and used to evaluate the relative impact of planting three different tree species (white spruce, trembling aspen and lodgepole pine) with different rotation lengths on average biomass accumulation rates and long-term ecosystem carbon storage. Simulation results for managed stands were compared against a natural disturbance scenario in which a mixed-wood stand composed of all three species was subjected to catastrophic wildfire on a 150-year fire cycle. All simulations were conducted for a 300-year time period using the management scenarios described in Table 5.8.

An example of model output showing the temporal patterns of storage in the ecosystem carbon pools for different tree species is provided in Figure 5.14. This example illustrates the capability of the model to project the carbon dynamics in ecosystems subjected to periodic disturbance events. The model also allows for an evaluation of the long-term carbon storage potential of different species and rotation-length combinations, relative to the potential production of timber from this ecosystem type (Figure 5.15). This type of trade-off analysis is essential to help forest managers assess the most suitable silviculture systems to meet diverse management objectives.

Exploring salvage alternatives after widespread insect-induced mortality

The recent mountain pine beetle (MPB) epidemic in lodgepole pine forests of the interior of BC is without precedent in the recorded history of forest management in the province (Eng, 2004). Vast areas of productive forest – more than 6 million ha – that have in the

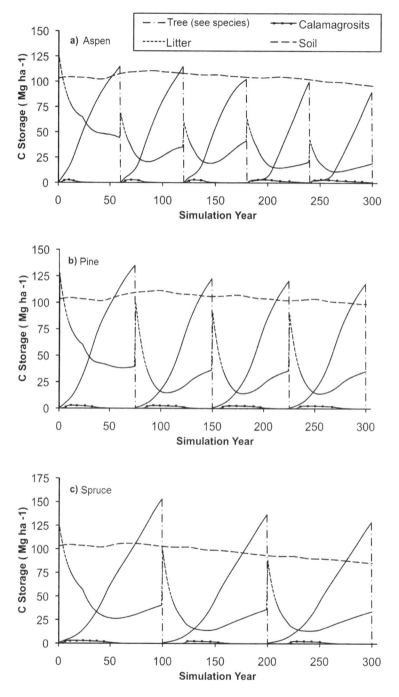

Figure 5.14 *Temporal dynamics of carbon storage in ecosystem pools: (a) aspen (60-year rotation); (b) (75-year rotation); (c) spruce (100-year rotation). The ecosystem starting condition for each simulation is representative of a natural stand.*

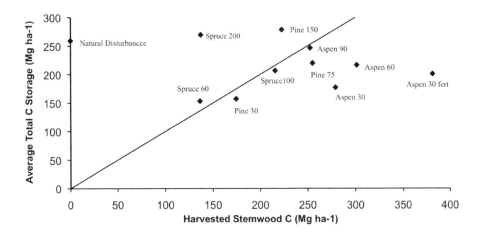

Figure 5.15 *Average total carbon storage and its relation to cumulative harvested stemwood carbon for the 300-year simulation period*

Note: the diagonal line denotes the 1:1 relationship between the variables. Scenarios falling above the line are relatively more favourable for carbon sequestration, while points falling below the line favour wood production.

Source: after Seely et al (2002).

past provided the economic foundation for the forest industry in interior BC are entering a period of transition. MPB has shown an affinity for many types of stands varying in age, structural stage, species composition, productivity and climatic zone. Consequently, the post-outbreak trajectory of forest development is uncertain, and could be steered in many different directions through salvage and other forest management activities.

To explore the consequences and developmental implications of alternative salvage and post-salvage harvesting strategies, managers require decision-support systems that allow them to examine the potential short- and long-term consequences of MPB salvage activities on both economic and ecological indicators of sustainable forest management (SFM). To be effective, stand-level models underlying such decision-support systems must be capable of projecting patterns of stand development following varying levels of attack and associated salvage/rehab systems, and following silvicultural interventions designed to reduce the susceptibility of existing stands to future attack. In addition, they must be able to track the development of stand attributes associated with both economic and ecological indicators of SFM.

The FORECAST model was applied as part of two comprehensive studies examining the potential consequences of MPB management on both economic and ecological indicators of SFM (see Seely et al, 2007a, 2007b). In both studies, FORECAST was used to explore the long-term dynamics of stand development following alternative salvage options in a variety of stand types in the central interior of BC (Figure 5.16). The different salvage options included management designed to enhance biodiversity (maintain residual live structure), increase productivity (plant fast-growing species such as pine and fertilize where feasible), or reduce the risk of future MPB outbreaks

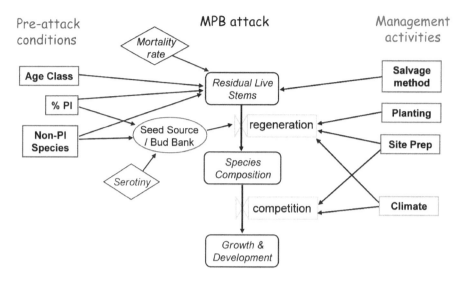

Figure 5.16 *Sources of complexity requiring consideration when projecting post-MPB-attack stand-development trajectories with and without salvage activities using the FORECAST model*

Note: Pl = pine.

Source: after Seely et al (2007a).

(planting more non-pine species). An alternative option of leaving the attacked stand in an unsalvaged condition with no planting or rehabilitation was also simulated for each stand type. Output from stand-level simulations was linked to landscape-scale models using a shared database approach to facilitate landscape-scale analyses (see Chapter 10 for a description of methods for linking FORECAST with landscape-scale models).

A series of indicators was defined to facilitate an evaluation of individual stands and larger landscapes with respect to general measures of wildlife habitat suitability and economic value (Table 5.9). Indicators of stand susceptibility to future MPB attack and ecosystem carbon storage were also included. The snag and large live structure indices were developed to function as measures of mature forest conditions. To facilitate a direct trade-off analysis, all indicators were indexed to have values ranging from 0 to 1.

Results from the stand-level analysis of salvage alternatives with FORECAST showed that there were several factors that influence the trajectory of stand development following MPB attack, including:

- the relative pine content;
- the age class of the stand prior to attack; and
- the dominant non-pine species in the stand.

An example of the simulated effect of relative pine content and non-pine species on the snag index (0–1) is shown in Figure 5.17. The relative impact of alternative salvage

Table 5.9 *A description of the indexed stand-level indicators used for trade-off analysis*

Indicator	Description	Threshold*
Snags	Index of snags (all species) greater than 25cm DBH	25 snags ha^{-1}
Large live structure	Index of large live stems greater than 25cm DBH	200 stems ha^{-1}
Merchantable volume	Index of economic value based on 10cm top diameter limit	607 m^3 ha^{-1}
MPB susceptibility	Index of MPB susceptibility based on site susceptibility index (Shore and Safranyik, 1992)	100
Ecosystem C storage	Index of ecosystem carbon storage	354 tonnes C ha^{-1}

* Threshold value used to calculate the indexed value. When the threshold is achieved the value = 1, values < threshold are scaled (0–1) by dividing the current value by the threshold value.

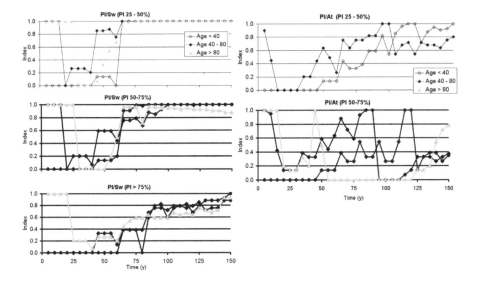

Figure 5.17 *Simulation results showing the projected post-MPB-attack trend in snag index (0–1) for unsalvaged stands*

Note: differences in index trajectories are shown for both pine/spruce (Pl/Sw) and pine/aspen (Pl/At) stand types that vary in age at the time of attack. The time axis refers to time since initial attack.

Source: after Seely et al (2007b).

methods on long-term trends in indicators is shown for snags and large live structure in Figure 5.18, and for MPB susceptibility and ecosystem carbon storage in Figure 5.19. These studies demonstrate the utility of the FORECAST model for examining the implications and trade-offs of alternative management strategies on multiple forest values.

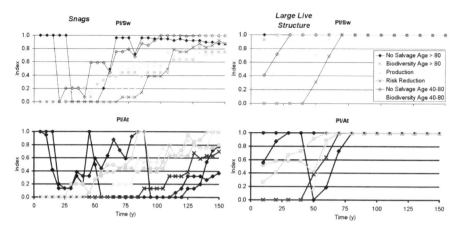

Figure 5.18 *Simulation results showing the projected post-MPB-attack trend in snag index and large live structure index (0–1) for stands salvaged with different objectives*

Note: differences in index trajectories are shown for both pine/spruce (Pl/Sw) and pine/aspen (Pl/At) stand types that vary in age at the time of attack. The time axis refers to time since initial attack for unsalvaged stands and time since salvage for salvaged stands.

Source: after Seely et al (2007b).

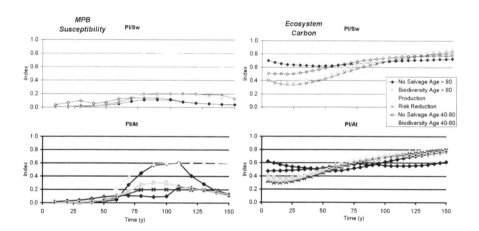

Figure 5.19 *Simulation results showing the projected post-MPB-attack trend in MPB susceptibility index and ecosystem carbon index (0–1) for stands salvaged with different objectives*

Note: differences in index trajectories are shown for both pine/spruce (Pl/Sw) and pine/aspen (Pl/At) stand types that vary in age at the time of attack. The time axis refers to time since initial attack for unsalvaged stands and time since salvage for salvaged stands.

Source: after Seely et al (2007b).

Exploring the stand-initiation/early stem exclusion stages of stand development

Many models initiate their simulations with established trees of a minimum size, assuming a free-to-grow status: the tree canopy is above the competing shrub and herb canopy. Other models may initiate simulation of growth and yield from the time of canopy closure. However, the stand establishment phase is often the most challenging, with issues such as low planted-seedling survival, competition from shrubs and herbs, ingress from either non-crop tree species, or excess regeneration of the crop species, leading to chronically overstocked stands, and browsing by wildlife or cattle. Where these are issues, foresters need a model that can explore them and investigate various management remedies.

FORECAST was used at the UBC Research Forest just east of Vancouver to examine the capability of the model to project the development of planted Douglas-fir (*Pseudotsuga menziesii* var. *menziesii*) subjected to varying levels of natural regeneration ingress, including western hemlock (*Tsuga heterophylla*) and red alder (*Alnus rubra*) (see Boldor, 2007). Field data were available for two measurement periods (20 years apart) and three slope positions: upper slope (relatively dry), mid slope (intermediate moisture), and low slope (seepage sites). The model was subsequently employed to examine the relative impact of the timing and level of western hemlock natural regeneration on

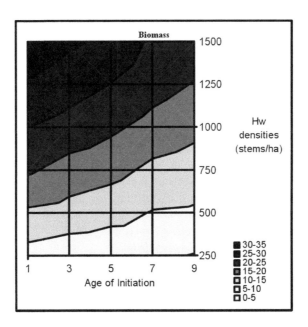

Figure 5.20 *A surface plot of impact categories showing the simulated relative impact (percentage decline) of hemlock (Hw) ingress on the growth of planted Douglas-fir stemwood biomass*

Note: the darker the shading, the greater the impact.

planted Douglas-fir growth, using a factorial analysis in which both variables were adjusted systematically.

Results from the model comparison to field data showed that FORECAST was able to project the average top height, diameter growth and biomass accumulations with good to acceptable accuracy. Model output from the factorial analysis of western hemlock competition on Douglas-fir growth (Figure 5.20) illustrated the window of opportunity for plantation establishment and growth in the face of competition from non-crop tree species.

Investigations of multi-rotation yield decline in China

One of the design criteria for the development of FORECAST was the ability to examine the sustainability of short rotation, intensively managed (SRIC) tree crops, or forest management that threatens the sustainability of ecosystem function, something that historical bioassay models are not equipped to do. This design objective came from the origins of FORECAST as a model to assess the energy efficiency and sustainability of intensively harvested, short rotation bioenergy tree crops. The model that resulted is well equipped to assess the concept of 'ecological rotations' and identify combinations of management practices that are sustainable (or not).

An example of the application of FORECAST to this issue is the exploration of the dramatic yield decline observed in Chinese fir (*Cunninghamia lanceolata*) in south-eastern China (Bi et al, 2007). A combination of ever-shortening rotations, slash burning after clearcutting, monoculture plantations of a species that grows naturally in mixed woods, and frequent off-site planting has resulted in catastrophic yield declines in some areas. In contrast, longer rotations, managing mixtures of species and limiting plantings to appropriate soils has demonstrated sustainable plantation management of this species.

The FORECAST modelling was able to reproduce the observed plantation growth decline, but only after herbs and shrubs were added, and the growth of these life-forms from both seed and rhizomes was included. This is an example in which modelling can demonstrate the importance of recognizing the complexity of such issues. Figure 5.21 presents a conceptual model of the problem of yield decline that fits with field observations. It may not include all determinants, but appears to provide an effective management tool with which to explore how such plantations can be returned to sustainability. See Chapter 8 (Figure 8.16) for additional discussion on the consequences of adding complexity to these scenarios.

Linkage to landscape-scale models

In addition to its stand-level applications, FORECAST has been linked to various landscape-level models. Scaling up from a stand to the landscape level introduces a new dimension to forest modelling, and its description is the focus of this section.

Certain issues in forest resource management can be addressed only at large spatial scales. In this respect, their resolution depends, at least in part, on landscape-level models that can serve in decision-support and scenario/values trade-off analysis. Until recently, most landscape-level timber supply and habitat suitability models were driven by conventional historical bioassay growth and yield models, or used very simple assumptions about age-dependent stand development. Unfortunately, over-simplifying stand-level processes undermined the ability of these models to address the multi-scale

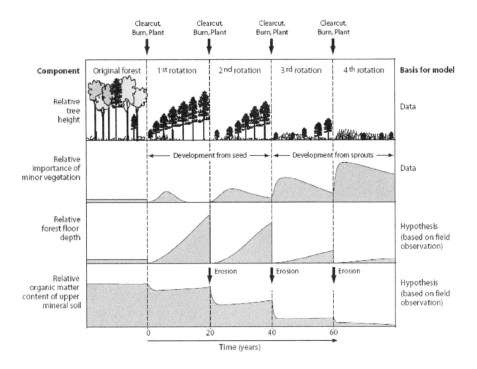

Figure 5.21 *A conceptual model illustrating the temporal dynamics of Chinese fir decline*

Note: the results of the simulations support this interpretation if regrowth of non-crop vegetation is simulated from sprouting as well as seed.

Source: after Bi et al (2007).

and multi-value issues of stewardship, certification and sustainability. To address these shortcomings and facilitate a stronger link between stand-level ecology and multi-criteria landscape analysis, researchers within the Faculty of Forestry at UBC, in collaboration with government and industry partners, have conducted a series of multidisciplinary modelling projects. These projects included a multi-criteria analysis of alternative management options in a 40,000ha landscape unit of the Arrow Lakes Timber Supply Area in south-eastern BC, and a similar project conducted in association with Canadian Forest Products for Tree Farm Licence (TFL) 48 (~500,000 in size) in north-eastern BC (see Seely et al, 2004). In both cases, FORECAST was linked to the timber supply model FPS-ATLAS (Nelson, 2003), and a habitat supply model, SIMFOR (Daust and Sutherland, 1997) (Figure 5.22). The projects also included a visualization component to provide images to stakeholders involved in the development of the SFM plans for the respective projects.

Each of the models in Figure 5.22 was linked through a shared database approach, whereby output from one model was used to provide input to another. Spatial ecological classification data (e.g. terrestrial ecosystem mapping) were used in conjunction with vegetation inventory data to develop a stand-level database of the major forest types within the target forest area. Each forest type, in conjunction with a set of silvicultural

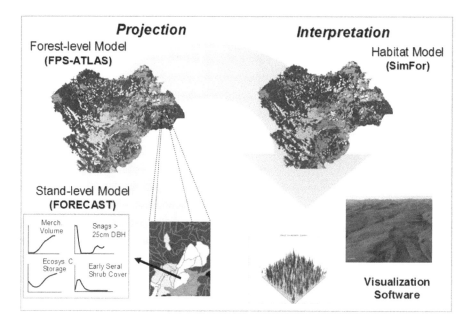

Figure 5.22 *A schematic illustration of the UBC forest management decision-support system (UBC-FM)*

Note: the principal flow of information within the modelling framework is in a clockwise direction, beginning with projection of management and disturbance events at the stand level and moving towards interpretation in the context of selected indicators at larger spatial scales.

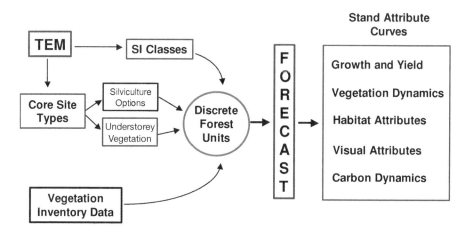

Figure 5.23 *A diagrammatic representation of the methodology used to develop a set of discrete forest units to be modelled within FORECAST*

Note: a series of stand attribute curves is produced for each forest unit for use within the larger modelling framework. SI = Site Index (site quality).

options, was considered a discrete forest unit. FORECAST was then used to create a series of stand attribute curves for each forest unit, including merchantable volume, species composition, stand structure and carbon storage (see Figure 5.23). These unique forest units were subsequently assigned to individual polygons within the landscape models using a set of criteria including TEM site types, indices of site productivity, and current forest cover. See Chapter 10 for more discussion about linking models and meta-modelling.

Development of FORECAST Climate

The impact of anthropogenic climate change on forest health and growth has been identified as a key issue with respect to the sustainability of forest management in BC (BC Ministry of Forests and Range, 2006) and other temperate forest regions throughout the world. A recent analysis of the potential effects of climate change on tree distribution in BC suggests that important timber species, including white spruce and lodgepole pine, may lose suitable habitat and suffer adversely from a combination of warming trends and reduced growing season precipitation (Hamann and Wang, 2006). In contrast, species such as Douglas-fir and ponderosa pine may actually expand their range and potentially show improved growth rates in parts of their existing range. Recent dendroclimatological studies along elevation gradients in the north cascade range of the US Pacific north-west found that lodgepole pine and Douglas-fir responded similarly to climate factors depending on elevation (Case and Peterson, 2005, 2007). At high elevation, trees responded positively to increased temperatures, while at low elevations trees showed a negative response to growing season maximum temperature and a positive correlation with growing season precipitation. Similarly, white spruce has shown variable responses to temperature variables, but generally positive responses to precipitation, particularly in drier parts of its range (Wilmking et al, 2004; Andalo et al, 2005; Johnson and Williamson, 2005).

While tree growth has been shown to be correlated to climate variables, the direct or indirect causal factors are often less clear. Climate can influence nutrient dynamics and subsequently productivity through its impact on organic matter decomposition rates. Recent litter decomposition studies have shown that temperature and soil moisture influence mass loss and mineralization rates (Trofymow et al, 2002; Prescott et al, 2004).

Modelling tools are required to help forest planners navigate the potential implications of climate change for timber supply through the use of scenario analysis and case studies. Although detailed physiological models have been useful in exploring climate impacts on tree growth and ecosystem processes, they are often data-intensive and difficult to apply for management-related applications (e.g. Grant et al, 2005, 2006). To be effective for guiding management, such tools must be able to capture the current understanding of the effect of specific climate variables on ecosystem processes governing forest growth, but still be practical for estimating impacts on tangible projections of forest growth and yield and other ecosystem values (Landsberg, 2003a; BC Ministry of Forests and Range, 2006).

The FORECAST Climate model (presently in the testing stage of model development) has been developed and designed in such a way as to give it the capability to explicitly represent the potential impacts of climate change on forest growth and

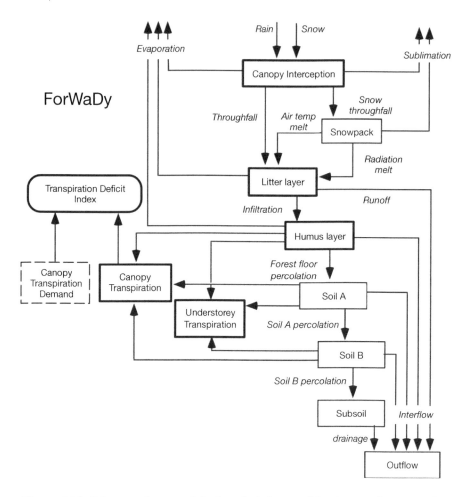

Figure 5.24 *Schematic diagram of the forest hydrology model indicating the various flow pathways and storage compartments*

Note: compartments with bold borders indicate areas of the model which facilitate feedbacks between hydrology and forest growth relationships in FORECAST.

development. In the general version of FORECAST, tree growth is limited by light and nutrient availability. The FORECAST Climate model includes an explicit representation of soil moisture and forest hydrological processes based on a linkage to the forest water dynamics (ForWaDy) model, also developed at UBC (see Seely et al, 1997).

ForWaDy is a vegetation-oriented model originally developed as a companion forest hydrology model to FORECAST. The model was designed to provide a representation of the impacts of forest management activities on water competition between different tree species, and between trees and understorey vegetation. Potential evapotranspiration (PET) in ForWaDy is calculated using an energy balance approach. Incoming radiation is partitioned among vertical canopy layers (vegetation type) and the forest floor to

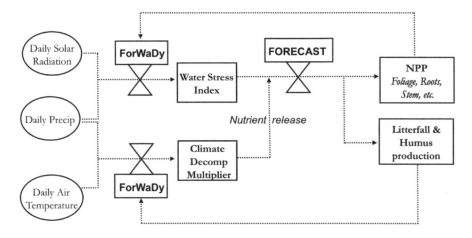

Figure 5.25 *Schematic illustration of the simulated relationship between climate inputs and the links between ForWaDy and FORECAST contributing to net primary production as simulated by FORECAST Climate*

Note: ForWaDy operates on daily time-steps, whereas FORECAST Climate simulates ecosystem processes on annual time-steps. Daily output from ForWaDy will be summarized on annual time-steps to facilitate feedback between the models.

drive actual evapotranspiration (AET) calculations. A schematic diagram of the model showing the various flow pathways represented is provided in Figure 5.24. The model is structured for portability, with minimum soil data requirements and parameter values that are relatively easy to estimate. It has a simplified representation of the soil physical properties dictating moisture availability, storage and infiltration. A more detailed description of the model is provided in Seely et al (1997).

The linkage with ForWaDy provides an additional feedback on tree growth rates based on a climate-driven quantification of tree water stress (Figure 5.25). Moreover, the simulation of soil and litter moisture content in FORECAST Climate facilitates a climate-based representation of organic matter decomposition and associated nutrient mineralization rates. These developments, in combination with a simulation of temperature effects on length of growing season and forest growth rates, will provide the foundation for the representation of climate impacts on forest growth in FORECAST. The completed model will allow users to explore the potential impacts of varying climate scenarios on indicators of multiple forest values.

Take-home message

As the needs, goals and methods of forest management shift, the types of modelling tools required to support the application of new and relatively untested management systems have also changed. Hybrid forest growth models provide a means to capture the flexibility and understanding of process-based simulation while maintaining the believability inherent in the historical bioassay type of empirical model. There is a great deal of variability in the structure and functionality of hybrid stand-level models reported in the literature. This variability is largely a reflection of the wide range of management issues and systems that are often addressed in ecosystem-based forest management. Stand-level, hybrid forest growth models can be effectively classified into three functional groups:

1 hybridized empirical models;
2 hybridized process models; and
3 reduced-form process models.

Each group has its strengths and weaknesses. In this chapter we have provided a thorough overview of the hybridized process model FORECAST, including a description of its functionality, examples of its verification and validation against field data, and applications of the model to examine different types of questions related to the development of strategies for ecosystem-based management.

Additional material

Readers can access the complimentary website (www.forestry.ubc.ca/ecomodels) to access the full versions of the following additional material:

- PowerPoint Presentation "Yield decline and sustainability in Chinese Fir plantations – a simulation investigation to analyze the possible causes.: Presented by Blanco J.A., Bi J., Kimmins J.P., Ding Y., Seely B., Welham C. at the International congress on cultivated forests. Bilbao, Spain, October 3–7 2006. Union of Forest Owners of Southern Europe, Derio, Spain.
- PowerPoint Presentation "Testing the ecosystem-level forest model FORECAST through North America, from the Caribbean Sea to the Pacific NW" Presented by Blanco J.A., Seely B., González E., Haynes P., Welham C., Kimmins J.P., Seebacher T. at the 6th North American Forest Ecology Workshop. Vancouver, Canada. June 18–22, 2007.
- PowerPoint Presentation "An integrated modeling approach for the assessment of forest growth and development in the face of climate change: a case study in the western boreal forest" Presented by Seely B., Welham C., Blanco J.A., Kimmins J.P. at the International Workshop on Simulation of Ecosystem Productivity, C-N Cycling and Multi-objective Forest Management. Lin-An, Hangzhou, China. September 22–24 2008.
- PowerPoint Presentation "Validation of FORECAST. Results from coastal and interior forests in BC, Canada". Presented by Blanco J.A., Seely B., Welham C., Kimmins J.P. at the International Workshop on Simulation of Ecosystem Productivity, C-N Cycling and Multi-objective Forest Management. Lin-An, Zhejiang, China. September 22–24.
- PowerPoint Presentation "Hybrid poplar in Saskatchewan: Projected long-term productivity and N dynamics using the FORECAST model" Presented by C. Welham, H. Kimmins, K. Van Rees, B. Seely at the Poplar Council Meeting, Saskatchewan, 2005.

- PowerPoint Presentation "FORECAST Modelling Workshop" Presented by B. Seely, C. Welham, J.P. Kimmins in Norway, 2003.
- Invited paper "Present applications of the FORECAST model" (in Spanish). Presented by Blanco J.A. at the 5th International Symposium on Sustainable Management of Forest Resource (SIMFOR). Pinar del Río, Cuba. April 22–26, 2008.

Chapter 6

Landscape-Level Models in Forest Management

Introduction

As described in Chapter 5, the present challenge in forest management is planning for a sustainable supply of timber and other forest resources while preserving the integrity of the forest ecosystem itself (discussed in Chapter 2). This ecosystem-based approach to forest management requires that we respect the hierarchical nature of forest ecosystems (O'Neill et al, 1986; Kangas et al, 2000) in developing sustainable forest management strategies. In other words, we can no longer focus strictly on the management of individual trees or stands, but rather we must also consider stands in the context of the landscapes in which they exist. This modern paradigm of multiple-value forestry implies a change in traditional management strategies, moving from a management focused on trees or stands to a management of the landscape as a whole and its multiple values and ecosystem services.

For a landscape-level, ecosystem management approach to be effective, forest managers must be able to:

- determine viable management options for relevant spatial and temporal scales;
- predict long-term effects of management actions on stand and landscape development and associated values;
- understand and evaluate management impacts on biological diversity;
- project the relationships between landscape patterns and ecological processes (e.g. connectivity and edge effects);
- evaluate habitat quality for a broad range of species;
- compare natural and anthropogenic disturbances; and
- assess the influence of global climate change on specific landscapes (Korzukhin et al, 1996).

However, all these demands are characterized by extraordinary complexity, a limited availability of mechanistic hypotheses and a scarcity of data to evaluate those hypotheses (Galindo-Leal and Bunnell, 1995).

The combination of forest management activities and natural disturbance agents acting at multiple spatial and temporal scales within forests leads to the creation of a complex mosaic of ecosystem conditions across forest landscapes. The character of this mosaic influences ecological processes, natural regeneration, tree species distributions, habitat quality and usage, and patterns of spread of future natural disturbances. For reasons discussed earlier, the only realistic option for assessing the long-term impacts of different disturbance regimes on forest landscapes is through the use of simulation tools operating at this spatial scale (Shugart, 1998). Such landscape-scale forest management and disturbance models are dependent upon data from silviculture, forest ecology, geography and remote sensing. The continuing advances in the capacity of computers, and the reduction of costs of software and equipment used in geographic information systems (GIS) and remote sensing, have facilitated the increasing use of the type of spatial simulation introduced in this chapter. However, unless spatial landscape models provide a credible representation of ecosystem structures and processes at the stand level, their predictions may not be reliable. Such models should be based on the simulation of the ecological processes that define a forest ecosystem.

In this chapter we discuss the development and application of different types of landscape-scale models used both as decision-support tools for forest managers and as research tools for landscape ecologists. Several examples of landscape-scale forest models are provided, and their structure and functionality discussed. We also present a detailed overview of the local landscape ecosystem management simulator (LLEMS), part of the FORECAST family of models. LLEMS was developed at UBC as a tool to facilitate the assessment of the short- and long-term implications of variable retention harvesting systems for indicators of sustainable forest management in coastal and interior forest types within BC and elsewhere. Examples of model applications are provided.

Development of landscape-level models for forest management

Development of computer models for forest management and ecological research started in the early 1970s (see Chapter 3), but not until the technological developments in the late 1980s (higher computer power, development of GIS systems and availability of satellite data, among others) did it become possible to address larger spatial scales and to undertake the spatially explicit simulation of ecological processes across forested landscapes (Mladenoff and Baker, 1999). In spatially explicit models, the behaviour of an individual cell or pixel cannot be predicted without knowing its location relative to other cells. The earliest model of this type may be the gradient fire model of Kessell (1979), who used spatially estimated vegetation and fire fuels data to simulate spatial fire patterns and post-fire succession. Following this work, a model of shifting cultivation and secondary forest succession was developed in the late 1980s by Wilkie and Finn (1988). These modelling initiatives were paralleled by the development of theories of landscape ecology, and the concept of cellular automatons (a cell grid where complex dynamics arise from simple neighbourhood interaction rules; Wolfram, 1984) was introduced into forest landscape models (Mladenoff and Baker, 1999; Xi et al, 2009). The first example of the incorporation of these new theories in a landscape model was Turner's (1988) model of landscape change in Georgia (south-eastern USA).

During the 1990s, the continuing development of stand-level forest ecology, disturbance ecology, landscape ecology and the mathematics of cellular automata interacted to create a wide variety of models. Earlier models emphasizing stand-level forest ecology and small-scale disturbances (the so-called gap models; see Chapter 3) were implemented in a spatially explicit way across larger areas, such as the extension of the gap model ZELIG (Urban, 1990) in FACET (Urban and Shugart, 1992). Forest landscape models developed during this period increasingly incorporated characteristics of ecosystem process simulation models, not only tracking the spatial location of individual trees, but also representing key ecological processes such as nutrient cycles and energy flow (He, 2008). They also incorporated stochasticity, and the simulation of the long-term effects of forest harvesting, wind, pests and diseases on ecosystem processes (Xi et al, 2009). Other models continued expanding the use of cellular automata and were broadened to include other ecological dynamics such as seed dispersal, fire spread, etc. Several examples of models developed using this general approach (spatially specific positions and stochasticity) are described below.

The FACET model, an extension of the gap model ZELIG, was developed and parameterized for Pacific north-western USA forests (Urban and Shugart, 1992; Urban et al, 1993). The model simulates a forest stand as a grid of tree-sized plots; each plot corresponds to a conventional gap model (Urban, 1990). The model is called FACET because it adjusts climate (temperature, precipitation and radiation) for topographic position, so that a modelled grid represents a homogeneous slope facet. This model has been used to study the potential effects of anthropogenic climatic change on conifer forests in the Pacific north-western USA (Urban et al, 1993).

FORMOSAIC is a spatially explicit, individual tree-based, stochastic model for simulating forest dynamics in landscape mosaics (Liu and Ashton, 1998). The model predicts population trajectories for individual species, species richness, stand density and timber volume (basal area) in response to management practices, as well as biotic and abiotic factors, which influence tree recruitment, growth and mortality (Liu et al, 1999). This model has been used in forest management to study the influence of cattle and fast-growing plantations on the species richness in a target tropical forest in Malaysia (Liu and Ashton, 1999).

METAFOR is a model designed using the cellular automaton approach that represents a landscape as a grid wherein each cell is assigned a cover type (one of several tree species) and age class (Urban et al, 1999). The cover and age maps are draped over a digital elevation model. The model simulates the demographic processes of establishment, aging (but not growth) and mortality, as well as fire disturbances. The physical template is expressed in this case as gradients in temperature and soil moisture over a range of elevations; biotic processes are represented by successional relationships mediated by species' tolerances of cold and drought as well as local seed dispersal; and disturbance is represented by fire spread determined by forest age (fuel load) and soil (fuel) moisture (Urban, 2005). This model has been used to explore the ecological implications of seed dispersal in the Sequoia National Park in Sierra Nevada (California, south-western USA) (Urban et al, 1999).

The DISPATCH model is a spatial landscape-model designed to simulate the effects of changing disturbance regimes on landscape structure (Baker, 1999). The model contains five major components: a climatic regime, a disturbance regime, a set of map layers that together determine the probability of disturbance initiation and spread, a user-defined equation to combine these maps, and a GIS program to quantitatively analyse the spatial structure of the landscape as the model runs. The DISPATCH model has been used to study the potential restoration of historical fire regimes following a long period of fire suppression in northern Minnesota (north-western USA) (Baker, 1994).

HARVEST is a raster-based model designed to simulate harvest methods that produce openings greater than the size of an individual cell (Gustafson and Crow, 1999). It was designed to simulate even-age silvicultural methods such as clearcutting, shelterwood and seed-tree systems. It can also simulate the uneven-age selection method and can be modified to simulate small patch harvesting. However, HARVEST is not able to simulate other uneven-age harvest systems such as single-tree selection or group-retention harvest systems. HARVEST is supported by the US Forest Service (www.nrs. fs.fed.us/tools/harvest) and it has an educational version. It has been used mainly in the Michigan area to predict the collective, landscape-wide effects of diverse management objectives in a working forest landscape, and to provide insight into the problem of practising sustainable forestry in multi-owner landscapes (Gustafson et al, 2007).

Another group of models was developed with the objective of simulating the development of forest management in time and space with a focus on disturbances, land mosaics and the ecological consequences of conservation. These include LSPA (Li et al, 1993); ECOLECON (Liu, 1993); LEEMATH (Li et al, 2000); SEPM (Dunning et al, 1995); PATCH (Schumaker, 1998); and FORMIX-3Q (Ditzer et al, 2000).

While landscape-scale models often share similar objectives, they employ a wide variety of approaches with respect to mechanisms and structures for representing spatial relationships and associated ecosystem processes. Models that attempt to incorporate multiple spatial scales with the representation of multiple ecological and landscape process often take cellular or vector models and merge them with successional models, dispersal models, movement models and disturbance models that interact in a spatially explicit format. Some examples of models using this general approach are described below.

The LANDIS model (Mladenoff and He, 1999) is a raster-based, spatially explicit and stochastic model that simulates forest landscape change over long time periods and large, heterogeneous landscapes. Within each cell, LANDIS tracks the presence or absence of species age cohorts rather than the actual number of trees. This model is supported by the US Forest Service and it has been used for several projects in the USA and Latin America (visit www.landis-ii.org for more information on these projects).

LANDSIM is a mechanistic, spatially explicit parsimonious model that focuses on capturing the most important elements of vegetation dynamics and spatial distributions while remaining simple to parameterize (Roberts and Betz, 1999). In this model there are three primary components of interest: the vital attributes of each species (Noble and Slatyer, 1980), the physical environment and the disturbance regime. Variability is generally represented by a simple classification rather than by a detailed mathematical parameterization. This model has been used to determine the consequences of changes in the historical fire regime of Bryce Canyon, Utah (western USA) (Roberts and Betz, 1999).

FARSITE is a mechanistic fire model that integrates component models for surface fire, crown fire, fire acceleration, spotting and fuel moisture (Finney, 1999). This model is being developed by the Missoula Fire Sciences Lab (http://firemodels.fire.org/) and it has been used, among other applications, to assess the effectiveness of landscape fuel treatments on fire growth and behaviour in southern Utah (western USA) (Stratton, 2004).

SAFE FORESTS is a model that accounts for human intervention at a watershed level, using forest structure, fire hazard, watershed condition and timber output as measures of ecosystem health and sustainability. In this model, two general types of human intervention can be accounted for: timber harvest and prescribed fire. It also has a stochastic module to generate wildfires, and a sub-model to calculate watershed disturbances and the need for restoring riparian areas (Sessions et al, 1999). SAFE FORESTS has been used to simulate and analyse the effects of fire dynamics and timber harvesting in the forests of Sierra Nevada, California (south-western USA), and to contribute to forest management decisions on wildfire and harvesting (Sessions et al, 1999).

The first decade of the 21st century has been marked by great advances in remote sensing (RS) technology and increasing attention to climate change in all environmental sciences. Today's landscape ecologists are able to quickly obtain time-series RS images, and, combined with related GIS data, these can be used as the initial input data for forest landscape models, allowing pixel-based simulations for the entire target area (Xi et al, 2009). The emergence of super-computing machines and the progress of computer graphics technology have enhanced the capabilities of computer hardware to perform large-scale and complex imagery and quantitative analyses. Researchers around the world are increasingly using remote sensing imagery and spatial analysis software to conduct landscape change modelling and prediction.

An example of this new generation of models is the family of models derived from LANDIS. It includes models such as FINLANDIS (Pennanen and Kuuluvainen, 2002); QLAND (Pennanen et al, 2004); LANDIS 3.0 (He et al, 1999); LANDIS 4.0 (He et al, 2005); and LANDIS-II (Scheller et al, 2007). The latter has flexible time-steps for every ecological process, allows for the optional incorporation of ecosystem processes and states (e.g. live and dead biomass accrual, organic matter decomposition), and uses an advanced architecture that allows rapid model development and easy distribution and installation of model components (Scheller et al, 2007). Another example is SIMPPLLE, which also simulates multiple spatial processes, including fire, insect, disease and harvest disturbances (Chew et al, 2004). It is a polygon-based model that uses the current state of each neighbouring polygon to adjust the probability of insect and disease processes in the next time-step.

These models are developed to simulate the repeated patterns of spatial processes in an interactive manner. They are suited to the examination of the long-term effects of spatial processes such as harvest, insect, disease and wind disturbances. For example, FRAGSTATS (McGarigal et al, 2002) and APACK (Mladenoff and DeZonia, 2004) have been developed to study the spatial pattern of ecological processes. This generation of landscape models tends to use empirical relationships to aggregate detailed dynamics in representations of succession (Xi et al, 2009). The approach is used to expand the stand EFIMOD model up to the landscape level (Chertov et al, 2005) to permit the long-term quantification of the effects of different silvicultural regimes in central European Russia. Moorcroft et al (2001) introduce a method to scale up individual-based models to simulate forest landscapes, and Peng et al (2002) linked the stand-level model TRIPLEX to a GIS model for landscape analysis.

The recent trend in landscape modelling is for physical details to be incorporated into stochastic models (Li et al, 2005; Cary et al, 2006); for example, the model TreeMig (Lischke et al, 2006). This is a raster-based model, the forest dynamics in each grid cell being simulated with a multi-species, multiple canopy-layer forest model, based on growth, competition and death of the trees in each height class. Within-cell heterogeneity is accounted for by assuming that trees are randomly distributed. This model was used to simulate changing tree species spatial distributions in the Alps.

Several reviews of landscape models are available in the literature. Keane et al (2004) presented a comprehensive classification of landscape fire succession models, provided guidelines for model selection, and interpreted differences between models for both modellers and users. He (2008) grouped forest landscape models into four main approaches to the simulation of site-level succession: no site-level succession (spatial processes as surrogates), successional pathways, vital attributes and model coupling. Xi et al (2009) reviewed the history, recent developments and main types of forest landscape models, as well as their applications in forest ecology research and forest management. They also compared the characteristics and applicability of different forest landscape models.

The LLEMS model: a multi-value, local landscape extension of FORECAST for variable retention harvesting

Introduction

Use of alternatives to clearcutting, broadly termed variable retention (VR) management, has become an increasingly popular method employed by forest managers to address non-timber management objectives. The retention of individual or groups of trees within a block (see Figure 6.1) is intended to maintain structural complexity, provide habitat for wildlife and to reduce the negative aesthetic impact of timber harvesting (Burton et al, 1999; Mitchell and Beese, 2002). It also ensures the regeneration and adequate growth of shade-intolerant tree species in forests where natural disturbance has generally involved stand-replacing disturbance, but stand-replacing harvesting is no longer socially acceptable. In essence, it is a space-for-time substitution in which forest structural elements that have a long ecological rotation are embedded in a matrix of shorter ecological rotation stand elements, thereby sustaining these different stand values. While it has shown promise in achieving these goals in the short term (e.g. Arnott and Beese, 1997; Burton et al, 1999), the long-term implications of VR are largely unknown. There is evidence suggesting that selective harvesting may alter regeneration patterns, change competitive interactions between species, and influence growth rates (e.g. Bowden-Dunham, 1998; Franklin et al, 1997; Thysell and Carey, 2000). In addition, VR has been shown to increase windthrow risk (Beese, 2000; Burton, 2001) and to increase harvesting costs (Arnott and Beese, 1997). As there are few long-term field trials for variable retention systems, managers are forced to turn to scientifically credible models to project trajectories of vegetation community development and link stand-level responses to forest-level models.

Much of the fieldwork and associated modelling with respect to the retention of structure has focused either on stand-level issues or at the landscape level (e.g. Mladenoff and Baker, 1999). Hence, from a spatial perspective, we have acquired some

Figure 6.1 *A photograph showing a variable aggregate retention-cut block in a 60-year-old mixed western hemlock and amabilis fir stand on one of the islands in Johnstone strait (between Vancouver Island and the mainland of BC)*

Note: some windthrow occurred following harvest.

understanding of structural complexity at tens of hectares, and over many hundreds of thousands of hectares. Yet little has been done to model complexity at intermediate scales (10–1500ha, for example). This is problematic given that many key forest management decisions must be made at this scale, particularly in the case of VR. Strong interest in the emulation of natural patterns of disturbance (e.g. Bergeron et al, 1999; Perera et al, 2004) has also created a need for spatially explicit models that can span the spatial scale from very small gaps to large disturbances (Kimmins, 1990c). Presently, there are few models designed to project the development of complex cutblocks and associated indicators of sustainable forest management (SFM). In the following section we describe the structure, function and application of the LLEMS model (Seely, 2005a), which was developed specifically to address the spatial aspects of forest regeneration and growth dynamics following alternative variable retention harvesting methods.

Model description

LLEMS is a key component of the FORECAST family of models developed by the Forest Ecosystem Management Simulation Lab at the Department of Forest Sciences, UBC. The relationship between this family of models is illustrated in Figure 6.2. LLEMS

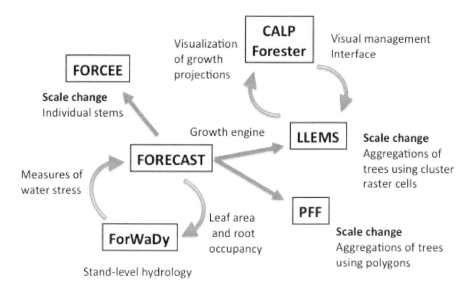

Figure 6.2 *The relationship between the FORECAST family of models described in the book, including descriptions of information transfer between models*

represents an extension of FORECAST (Chapter 5) to small landscapes (also see PFF, Chapter 7).

The LLEMS model was developed as an ecologically based, spatially explicit decision-support tool to explore the long-term consequences of the implementation of a wide variety of cutblock shapes, orientations and VR strategies for a series of economic, ecological and social (including visual) indicators of SFM. The model operates at a spatial scale of an intermediate to large cutblock or group of cutblocks (20–2000ha). Growth within LLEMS is ultimately driven at the pixel or pixel-group level, based on a modified version of the FORECAST model (see Chapter 5). The foci of the principal components of the LLEMS framework include: simulation of growth and ecosystem development of the individual stands within the landscape being represented; a visual management interface and tools for exporting model output to visualization systems; and an evaluation of windthrow risk associated with different cutblock designs. A brief description of the various sub-models included within LLEMS is provided in the following section. For a more detailed description of the model, see Seely (2005a).

The spatial framework and model initiation

LLEMS was constructed upon a raster-based GIS platform with a minimum resolution of 10 × 10m. This structure allows for varying the resolution of the simulation to capture detail where necessary (e.g. in ecotone areas near forest edges) and reduce resolution where it is not warranted (e.g. within relatively homogeneous areas) by aggregating pixels into pixel groups to increase modelling speed (Figure 6.3).

The initial step in running the model includes identifying an application area for which high-quality inventory data are available, including forest cover information and

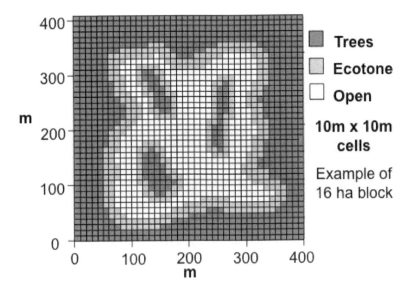

Figure 6.3 *A simplified diagram illustrating the raster-based structure of the LLEMS modelling framework*

some form of terrestrial ecosystem mapping (TEM) data. These data must be clipped to the boundary of the selected modelling area and then converted to raster format with a 10 × 10m resolution. Subsequently, the variables on which clustering is to be based must be converted to a numerical format and normalized to values between 0 and 1 to avoid biases during the clustering. Lastly, the prepared spatial data are imported into the clustering interface within LLEMS (Figure 6.4).

Using this interface, the user can conduct a clustering exercise to create aggregate groups of pixels with similar properties based on the four clustering variables, which include: TEM groups (groups of similar site types or site series; Pojar et al, 1987); age class; inventory type group; and productivity class. The interface allows the user to weight the variables differently (if so desired) and to explore the effect of increasing the number of clustering seeds on a calculated goodness of fit index (GFI), the ratio of the between-cluster sum of squares error (SSE) and the within-cluster SSE. The resulting set of clustered pixels becomes the initial set of discrete forest units to be used as the starting point for spatial modelling. Each is assigned a starting condition (including a species composition, a productivity class, an age and a soil type) for the forest growth modelling (see below) based on the mean values of the clustering variables.

Spatially explicit functionality

The spatially explicit functionality of the forest growth and development component of the LLEMS modelling framework can be divided into the following stages:

- *Definition of retention areas.* Once the model has been calibrated and initialized for a particular study area, the user can define potential retention areas and associated

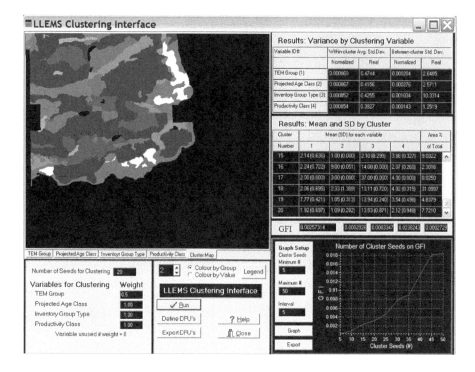

Figure 6.4 *An example of the clustering interface developed for LLEMS*

Note: the interface is designed to import spatial datasets and to prepare an initial set of discrete forest units (DFUs) for spatial modelling. A colour version of this figure is shown in the plate section as Plate 2.

silviculture using the CALP-Forester Visual Interface. The interface allows the user to explore the current inventory using real-time visualization, and to visually examine harvesting alternatives prior to growth simulation. An example of a cutblock designed using the mouse in CALP-Forester is shown in Figure 6.5. A detailed description of CALP-Forester is provided in Chapter 10.

- *Mapping light ecotones.* Following stage 1, the model will project the impacts of harvesting on spatial light profiles, and a series of light ecotones is created by grouping pixels with similar forest-floor light levels.

- *Simulation of natural regeneration patterns.* Spatial patterns of natural regeneration are projected based on the LLEMS natural regeneration sub-model, which includes an explicit representation of seed dispersal and an empirically driven sub-model to predict seedling establishment rates on different substrate types (see description of the natural regeneration sub-model used in LLEMS below).

- *Spatially constrained clustering analysis.* This analysis is conducted within LLEMS using a series of variables, including original forest-cover type, light ecotones, seedling establishment densities, silviculture applications/windthrow, etc. The clustering analysis results in a series of relatively homogeneous pixel groups, which will form the basis for the growth simulation.

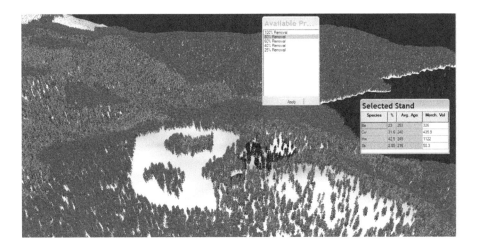

Figure 6.5 *A screen capture from the CALP-Forester visualization system illustrating its use as an interactive visual management interface for the LLEMS model*

Note: examples of both group retention (left) and dispersed retention (right) harvesting are shown. A colour version of this figure is shown in the plate section as Plate 3.

- *Simulation of future growth.* The growth of the forest is simulated for each of the pixel groups generated from the clustering analysis using FORECAST. Light conditions are evaluated on an annual time-step, and the effects of shading on the growth of vegetation in individual pixels or pixel groups by neighbours are accounted for. The user can specify how many years to simulate into the future, and the model reports results for a series of variables and associated indicators.

Natural regeneration sub-model

The objective of the natural regeneration sub-model is to provide the forest growth module of LLEMS with spatially explicit, post-harvest-disturbance tree recruitment estimates for the tree species occurring within the target ecosystem type. Projections made by the sub-model, based on a combination of process-based simulation and an empirically derived expert systems approach, reflect the relative ability of species to recruit following various levels of harvest disturbance. The species-specific regeneration life history traits used in the sub-model include seed mass, shade tolerance, and regulating factors such as density of seed trees, light levels, seedbed types, predation and competition (Figure 6.6). Seed mass and foliage biomass (of parent trees) are used to determine the intensity of seed production, while dispersal distances are determined as a function of seed mass and canopy types. Germination rates are subsequently regulated by seed abundance and seedbed types. Lastly, populations of established seedlings (recruitment to the stand model) are determined by the species-specific potentials of germinants to survive under specified light and competition levels. Calibration was accomplished by compiling relevant data from the following sources: published literature and reports, including a recent regeneration study on existing VR blocks within the Clayoquot Sound area;

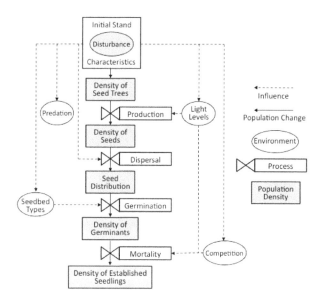

Figure 6.6 *A schematic diagram of the natural regeneration sub-model included within the LLEMS modelling framework*

Note: both process and expert system-based components (ovals) are shown.

industry regeneration surveys; and synthesized knowledge contained within ministry field guides and handbooks. Seed and seedling mortalities due to small mammals, grazers and browsers are represented deterministically by the user.

Example application of LLEMS to evaluate dispersed retention alternatives

A series of simulations conducted within the LLEMS modelling framework to demonstrate the application of the model to project the impact of varying levels of dispersed retention harvesting on the long-term productivity of retained and regenerating trees (Table 6.1). The model was set up to represent a medium (zonal) site in the very wet and humid sub-zone of the coastal western hemlock biogeoclimatic zone in coastal BC (CWH vh1 ecosystem type; Pojar et al, 1987) with a site index of approximately 24m at breast-height-age 50. In addition, to investigate the effect of stand vigour at the time of harvest, two initial stand conditions (OG western hemlock (Hw) and western red cedar (Cw), and second-growth (HwCw)) were examined. In each simulation, planting density (95 per cent Cw and 5 per cent Sitka spruce) and natural regeneration density (Hw) were assumed to vary with level of retention, with total regeneration densities ranging from 1800 to 1200 stems per hectare split evenly between planted and natural regeneration.

Results from the dispersed retention simulations (Figure 6.7) indicate the importance of initial stand vigour for the distribution of volume production between

Table 6.1 *A description of initial stand conditions and dispersed retention levels represented in the modelling scenarios*

Initial stand conditions	Retention levels (% of basal area)
Old HwCw* (225 yrs)	15, 30, 40, 50
Second growth HwCw (80 yrs)	15, 30, 40, 50

* Hw = western hemlock, Cw = western red cedar

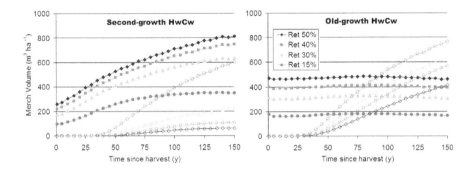

Figure 6.7 *Simulated volume accumulation for different levels of dispersed retention, starting with a second-growth or OG stand composed of western hemlock (Hw) and western red cedar (Cw)*

Note: open symbols represent total volume accumulation in new regeneration, and closed symbols represent volume in retained stems.

new regeneration and retained stems. When starting with a vigorous second-growth stand, the model suggests that growth rates of new regeneration will be marginal for levels of retention greater than 15 per cent, but that volume growth in retained stems will be substantial, accounting for most of the volume accumulation in the future stand. A different pattern emerges in the simulations beginning with an OG condition. In this case, volume accumulation in newly regenerating trees can be significant for higher levels of retention. The explanation is that there is little further crown expansion and volume accumulation in retained old trees. The fundamental difference between these scenarios is the rate and degree to which the overstorey canopy recovers and shades regeneration following harvest. Overstorey canopy recovery is relatively quick and complete in the second-growth scenarios, where tree vigour is relatively high. In contrast, recovery is relatively slow and incomplete in the OG scenarios, where retained trees have reduced vigour. Further, the properties and economic values of the logs generated and available for harvest in the next rotation will vary with initial stand conditions and retention levels.

A more complete assessment of the effects of implementing different types of VR systems on volume production in subsequent rotations requires that results for volume production be weighed against projections of the impacts of specific retention systems on aesthetic and ecological values as part of a multi-value trade-off analysis. This can be

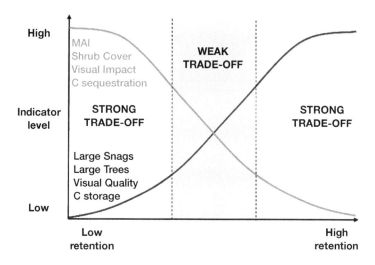

Figure 6.8 *A conceptual diagram illustrating a multi-value trade-off analysis for varying levels of dispersed retention at the cutblock scale*

done on an individual cutblock basis and/or at a larger landscape unit scale. The LLEMS modelling framework provides a tool for conducting such analyses in the context of developing sustainable forest management plans. A conceptual representation of multi-value trade-off analysis for dispersed retention systems at the cutblock scale is shown in Figure 6.8.

Ongoing development

Development of the LLEMS modelling framework is continuing on several fronts:

- The UBC development team is continuing to work with International Forest Products (Interfor) towards the implementation of LLEMS as a decision-support tool for forest planners and policy makers.
- The representation of natural regeneration within LLEMS will be evaluated and refined as part of a BC Forest Investment Agreement (FIA)-funded project to resample a series of permanent sample plots established in 2003 in VR blocks within the North Coast Forest district.
- A generic wildlife habitat sub-model has recently been completed (funded by the BC Forest Sciences Program) to allow users to construct and quantify habitat suitability indices and associated spatial metrics based on a large number of variables represented within FORECAST/LLEMS.
- New funding has recently been awarded from Natural Resources Canada to develop the LLEMS modelling framework as a tool for evaluating alternative spatial mountain pine beetle salvage strategies in interior BC.

Take-home message

The ecosystem-based approach to forest management requires that we respect the hierarchical nature of forest ecosystems in developing sustainable forest management strategies. In other words, we can no longer focus strictly on the management of individual trees or stands, but rather we must consider stands in the context of the landscapes in which they exist. The highly complex nature of forest landscapes interacting with forest management activities and natural disturbance agents requires the development and application of spatially explicit, landscape-scale forest management models. In this chapter we discuss the development and application of different types of landscape-scale models developed both as decision-support tools for forest managers and as research tools for landscape ecologists. Several examples of landscape-scale forest models are provided, and their structure and functionality discussed. A detailed overview of the LLEMS model, developed for assessing alternative strategies for variable retention harvesting, is provided as well as examples of its application.

The full value of landscape models will only be realized when their output is linked to visualization systems, and interactive versions are developed (see the discussion of management games in Chapter 7) that enable model users to explore many different possible landscape futures. This will facilitate their use by forest managers and policy makers, but more importantly will assist in the engagement of non-technical stakeholders in public participation processes that are predicted to increasingly determine how forests are actually managed in the future. This is discussed further in Chapter 10.

Additional material

Readers can access the complementary website (www.forestry.ubc.ca/ecomodels) to access the full versions of the following additional material:

- Slideshow 'LLEMS: local landscape ecosystem management simulator', presented by J. P. Kimmins, B. Seely, K. Scoullar and D. Cavens at CSC, Vancouver, BC, 2006.
- Slideshow (in Spanish) 'LLEMS: an ecologically-based, spatially-explicit tool for assessing the implications of variable retention management for selected indicators of sustainable forest management', presented by B. Seely, J. A. Blanco, J. P. Kimmins, C. Welham and K. Scoullar at the scientific workshop 'Spatial analysis in ecology: methods and applications', Alcoy, Spain, 2006.
- Slideshow 'Challenges in implementing a VR strategy: barriers to progress and tools to help navigate a way forward', presented by Gerry Fraser, Brad Seely, Duncan Cavens, Hamish Kimmins and Stephen Sheppard at the FORREX Science to Management Forum: Overcoming Obstacles to Variable Retention in Forest Management, Prince George, BC, 2007.

Chapter 7

Educational Models in Forest Management

Introduction

Models can be valuable tools for education. They provide a controlled learning environment in which to explore interactions between ecological processes and management decisions. This chapter begins with a description of the reasons for using models in education and training, and a brief review of the general characteristics of educational models. The application of ecologically based models is then discussed, with a particular emphasis on how to create integrative ecosystem-level thinking in the students, followed by a brief review of existing models within this context. The chapter concludes with a description of FORTOON (an ecosystem-level educational model, built using the framework developed for FORECAST), PFF (an educational, watershed-level expansion of FORECAST), and MRM (a tool to facilitate multiple runs of FORECAST with different management options to compare their results in an educational environment).

Schools, high schools, colleges and universities play an important role in the preparation of students for their personal and professional lives in an ever-changing job market and physical environment. This preparation involves current social values and knowledge from the biophysical, earth and social sciences. Educational institutions are expected to help students to become independent and critical thinkers: to discriminate, reflect and be able to evaluate (Posch, 1991; Jungst et al, 2003).

Rapid change has occurred over the past few decades in science, technology, social values, and the state of local and global ecosystems. This has led to recognition by the public of the need for a more balanced, sustainable use of natural resources and the protection of the environment. Global issues such as climate change, biodiversity loss and human population growth have increased the urgency for new generations to understand the nature of these issues and the possible consequences of different human responses to them. There are few more important tasks for our teaching institutions. Making both younger and older generations aware of the importance of forests, socially and environmentally, is part of this educational responsibility.

Forestry is increasingly being integrated into broader natural resource management curricula (Straka and Childers, 2006). Students who graduate from these programmes are almost certain to face controversial public issues throughout their professional careers (Jungst et al, 2003), and they will benefit from exposure to the biological, ecological and social/cultural complexities involved, and the realization of the inadequacy of fragmented approaches to complex problems.

Traditional methods of teaching in most developed societies have emphasized learning and remembering information, but have involved relatively little thinking or problem-solving. It was expected that dealing with workplace challenges would require mainly the recall of specific technical knowledge and procedures (Magee, 2006). This was an effective approach to adult learning in a static, structured and unchanging work environment. The information recalled would generally match the challenges faced. However, in the current rapidly changing world, any workplace related to environment or natural resource management involves a steadily increasing volume of information about the issues in these fields. The volume of information, knowledge and understanding involved challenges the ability of individuals to memorize and comprehend. Also, the half-life of information is decreasing due to the fast pace of acquisition of new knowledge, further challenging the ability of individuals to deal effectively with complex problems. Memorization of all important details is no longer possible or effective, and there is increasing criticism from the professionionals that graduates from traditional programmes are ill-equipped for the contemporary work environment (Straka and Childers, 2006). Many lack the critical thinking skills needed and the ability to communicate the complexity effectively (Jungst et al, 2003; Magee, 2006).

In response to these challenges, there is increasing use of simulation in professional training programmes. Training for pilots, the military, astronauts, police and fire-fighters is increasingly based on simulations of possible events, challenges and emergencies. Simulators are also used by experienced professionals to improve policies and procedures and to explore the possible outcomes for multiple values of alternative responses to issues. A similar approach to training is becoming popular in business schools, and simulators have become an important part of training and scenario analysis in a wide range of jobs.

Simulation is defined as an interactive simplification of some real event or condition in order to imitate it for education, training or research (Baudrillard, 1983; Alessi and Trollip, 2001). Educational and training simulators are one of several approaches to the development of student skills, including critical thinking (Proulx, 2004). Simulations can bridge the gap between the simplification involved in most classroom teaching and the complexity and unpredictability of the real world.

Games and simulations are terms that refer to different concepts, but they have common characteristics. Both involve a model of some kind of system, and in both the users can observe the consequences of their actions (Gredler, 1996). The distinction between games and simulation is often blurred, and increasingly 'simulation games' are used as vehicles for the application of simulation models (Jacobs and Dempsey, 1993; Akilli, 2007). We think that for simulations to be successfully applied in education and training, elements of gaming (such as winning and having fun) should be included. In this chapter we use the terms 'simulation' and 'game' as synonyms.

Reasons to use simulations in education and training

There are several justifications for such use of computer simulations:

- *Simulations provide a risk-free environment* that allows students to fail, and then provides them with the chance to learn about the possible consequences of their choices: to progressively modify their strategy until they have achieved a successful (desired) result. Failure is seen as a necessary experience for learning in a simulated environment (Aldrich, 2005). There are two advantages to removing risk during teaching. One is the ability to improve the skills of students in a way that does not affect outcomes in the real world (Walker, 1995) in situations where the student is learning to make decisions that may affect ecosystem health or human welfare. The cost of failure in real life is often high, the consequences persistent and the outcome unacceptable. Another is that students can learn to overcome their fear of failure. Failure becomes a learning opportunity by analysing what was wrong with their decisions and how they could have done better. This leads to improvement in their knowledge and understanding (Kimmins et al, 2005), rather than simply the end of their decision making due to failure (Carstens and Beck, 2005).

- *Simulations allow a learner to modify both his or her own behaviour and the model parameters* in order to observe how the simulated system changes. Most simulations are designed with a flexible architecture that allows their variables to be altered. By directly modifying a model, students can experiment with the behaviour of the models in a number of different scenarios. They can then experiment with how their own behaviour might change given the modified variables. The student-centred nature of the simulation makes the outcomes completely dependent on the player's actions (Aldrich, 2005). Also, because students learn differently, at different paces and through different mechanisms (e.g. visual learners vs kinetic learners), simulations allow instructors to deal more systematically with individuality in learning (Hertel and Millis, 2002). The objective of this level of interactivity is to provide a deeper understanding by the student of the model being used.

- *Using simulation in teaching builds problem-solving skills in students* (Magee, 2006). The ultimate test of an individual's knowledge is not simply being able to repeat that knowledge, but rather the capacity to convert the knowledge into an appropriate pattern of behaviour (Ruben, 1999). In conventional simulations, the goal is to focus the learner onto a specific set of problems that test their understanding about previously learned concepts (Greenblat, 1975). These are often scenario-based problems that reflect a situation the student would encounter in the real world. The simulation presents the context, authentic information sources and tools to let the student solve the problem and test his or her knowledge. Simulations are particularly adept at helping students acquire usable knowledge that can be applied to other situations (Hertel and Millis, 2002). Simulations encourage the purposeful use of knowledge to achieve clearly defined goals. During a simulation, the student has not merely learned some information, but has to fit it into the set of rules that define the simulation in order to achieve a particular goal (Petranek et al, 1992). The simulation then provides them with the feedback to modify their existing ideas and patterns of behaviour to achieve their objectives in that environment. This process engages the student and gives him or her insight into the subject matter, as well as the nature of his or her own skills for dealing with the problems presented in the simulation scenarios (Magee, 2006). It is not learning in a conventional classroom setting. On the contrary, the student is placed, virtually, in an authentic decision-making locale. In addition, many simulations are designed with randomness in their variables. In this way they are not predictable, and require students to examine their strategies and constantly modify them between simulation scenarios.

- *Simulations have the ability to evaluate whether the theoretical knowledge of students can actually be applied in a practical application.* Often, a passing grade in a theoretical subject is assumed to be the basis for correct decision making and problem solving in real-world situations (Jones, 1988). However, educators often find that students have become adept at passing standardized exams rather than gaining the ability to solve problems with the knowledge they possess (Magee, 2006).
- *The use of simulations as complementary tools in education and training can help to bridge the gaps between disciplines.* Appropriate simulators can integrate principles as well as knowledge and skills learned from other courses, disciplines or experiences (Hertel and Millis, 2002). This is especially important in the context of educating for sustainable forest management, where the management decisions involve biological and ecological principles, geophysical and climatic conditions, random natural processes, economic constraints, social and legal conditions, and many other diverse but important disciplines. Simulation games in conjunction with other traditional methods can be considered as a central structure around which an integrative, multidisciplinary course can evolve. Experienced educators report that students find it instructive and exciting to apply ideas and concepts from different courses in a multi-value simulation (Petranek et al, 1992). Simulations are also useful for connecting factual knowledge, principles and skills to their application within a profession. Bransford et al (2000) concluded that to develop competence in an area of enquiry, students must have a knowledge of the relevant facts, understand facts and ideas in the context of a conceptual framework, and organize knowledge in ways that facilitate its application to real work issues and challenges.
- *Games and simulations can have a significant impact on today's students*, because these young people are part of a generation that has grown up playing computer games. Learning experiences that use a familiar form of media provide this group with a recognizable tool (Ruben, 1999), one in which they have already developed problem-solving skills through years of interaction with commercial computer games. The business world has always seen the value of experiential learning, and they are watching the up and coming generation of gamers very closely (Magee, 2006). They are seeing a new generation that is much more competitive and driven to win; a generation that is more optimistic and determined to solve problems because they have learned that eventually, some combination of behaviour will result in success. This drive makes them very creative. Failure and risk do not frighten them, as they have come to learn that both are survivable. If the business world can use such professionals with success, why not apply the same ideas in sustainable forest management?
- A final benefit of using simulators is that *the social aspects of multi-player gaming equip students to be good at team activities* and coordinated approaches to problems (Carstens and Beck, 2005). Research also indicates that the work ethic of gamers in their twenties is better than non-gamers (Magee, 2006). They care more about the

organization they work for if they are presented with challenges and are properly motivated (Beck and Wade, 2004). The business world is aware that this will be a very different generation: one that will make great employees if managed properly (Silverstone, 2004). There are no reasons why we should not expect similar aptitudes in future forest managers, policy makers, researchers and environmentalists. The higher social interaction in multidisciplinary gaming also translates into better communication skills and students being more confident in public discussions of their performance. Forestry professionals have pointed out that such skills are very important in forestry careers, but are inadequately taught in many educational programmes (Straka and Childers, 2006).

General characteristics of simulations in education and training

For a simulation game be successful as a teaching tool, it has to have a set of minimum features. Thiagarajan and Stolovitch (1978) identified five critical characteristics of simulations used in the context of education and training:

1   A simulation game involves some element of *conflict*, although this conflict can be identified in different ways depending on the simulation. For example, in individual simulations, the basic conflict manifests itself in each player trying to maximize his or her personal score. In other simulations, the conflict arises when two players or teams compete against each other, trying to be the first to reach some goal.

2   The second characteristic is that of *constraints*. There is a set of rules embedded in the simulation game that constrains the behaviour of players in pre-specified ways. In the case of ecological models, these constraints are the ecological processes and their respective variables and rates of change, as defined by current scientific knowledge. All players' decisions and their consequences are affected by these rules, which try to mimic real-world natural laws.

3   A third feature is *closure*. All simulation games should come to an end sooner or later, at which time players are given feedback on their performances. Termination rules for simulation games for sustainable forest management may include a time limit, a target score, and/or the depletion of one of the simulated natural resources. In addition, simulation games used in sustainable forest management should have multiple win criteria to represent the diversity of social, economic and environmental values included in the simulation.

4   The fourth important characteristic is the *contrivance*. All simulation games are contrived situations; there are no direct real-life consequences of a player's performance in the game. This risk-free environment reduces the fear of failure and allows students to explore the possible consequences of actions that they would not take in real-life situations.

5   The final characteristic is the *correspondence* between elements of the simulation and elements of the reality it represents. This includes the roles played by the students, the rules used in the simulation, the sequences of events, and the consequences of players' actions. This is important, because the more similar the roles and rules used in the simulation are to their counterparts in reality, the more the experience gained by playing can be applied in careers in sustainable forest management.

Gredler (1992) identified a different set of characteristics of a simulation game used for education. Simulations are problem-based units of learning that are set in motion by a particular task, issue, policy, crisis or problem. The problems to be addressed by the students may be either implicit or explicit, depending on the nature of the simulation. In addition, the subject matter, context (setting) and issues addressed in the simulation are not textbook problems or theoretical questions in which answers can be formulated quickly; they can only be reached after exploring different interactions between simulation rules and students' decisions. Furthermore, participants carry out functions associated with their roles and the context of the simulation. The outcome of the simulation is not determined by chance or luck. Instead, players experience consequences that follow from their own actions. This is a key characteristic of an effective simulation game. Only if the results of the game are dependent on the players' actions can they be analysed in a post-simulation debriefing and new knowledge gained by the student. Although some elements of random events can (and probably should) be included in the simulation of sustainable forest management in order to simulate natural disturbance events such as wildfires and wind storms, the results of students' actions in the game should not be random. Finally, participants experience reality of functions to the extent that they fulfil their roles conscientiously and in a professional manner, executing the rights, privileges and responsibilities associated with their role (Hertel and Millis, 2002).

The teacher's role in using simulations as educational tools

One of the most consistent and strong claims made for using simulations as educational games has been their motivational values. It has been shown that students taught through educational games and simulations learn more content than do students taught in a conventional manner. Every teacher naturally wishes to create a classroom atmosphere in which students complete a course with a better feeling toward the subject than when the course began, and both students and instructors respond positively to multiple teaching strategies (Lauer, 2003). However, games should be used in moderation. Usually, repeating the number of plays beyond two does not produce a significant increase in student learning. If the same simulation is played more than two or three times, it is very likely that the students will lose interest and their motivation can drop quickly.

Using simulations generally requires that a teacher shed his or her traditional expert role, and become instead facilitators of learning (Lederman, 1984). Teachers must encourage learning by creating simulation experiences through which students will explore substantive content, develop discipline-specific skills and apply what they know to real-world issues.

At the beginning of the game, the teacher should explain the rules of the simulation (usually linking them to the content explained in the course through traditional lectures), the roles students will assume, the goals that they have to achieve in the simulation, and how to get the action started. During the simulation, teachers should give students the autonomy to determine how the simulation will progress, and students should have the opportunity to help each other by sharing their knowledge or experience gained from running the simulation (McKeachie, 1994). However, there are no reasons for teachers not to participate in the simulation; teachers can play some role in order to maintain some control over the events (Hertel and Mellis, 2002). In simulations of sustainable management, a teacher can play the role of a consultant in order to provide advice and guidance for students, while allowing the players to make decisions by themselves and then deal with the consequences. At the end of the simulation, the teacher's role is to debrief. A well-orchestrated debriefing generates and reinforces much of the learning acquired during the simulation, encouraging discussions and analysis of the events simulated and letting the teacher link those events with the theoretical content explained earlier in the traditional lessons. The teacher should carefully design and facilitate this last opportunity for learning.

The pre- and post-activities are as important as the game itself. Before playing, it is important that the teacher is intimately familiar with the game. A preliminary discussion on the related topics to be explored helps students to obtain a more fulfilling experience when playing. In addition, in the follow-up discussion it is important to share with the class the best and worst strategies, the simulation outcomes as perceived by the students, and to test whether students have explored all the possibilities of the game. This is especially important in games that simulate complex situations, such as sustainable forest management, when sometimes events and consequences that happen during the game are counter-intuitive or unexpected by the students. Assessments of the behaviours of the students – how they responded to opportunities and challenges in the game – are often external to the actual simulation, and they can be done by the teacher and the students as a post-game debriefing session or as an evaluation of student performance by the teacher.

The use of ecological models as educational tools

Students in forestry programmes today will be the forest stewards of tomorrow. Sustainable forest management requires respect for society's multiple values, but must be soundly based in ecosystem science that helps to determine what is possible and what is not (Kimmins, 2008a). We protect what we care about, and we care about what we know well and understand. If students are encouraged to explore the natural world, to learn about local plants, animals, microbes, soils and climate, to observe and anticipate their seasonal and longer-term patterns of change, and to get their feet wet in local rivers, they are more likely to develop a lifelong love of nature and commitment to environmental stewardship (Grant and Littlejohn, 2004). This caring for the natural world, combined with knowledge of how forest management can utilize the latest technology and scientific knowledge, will provide future professionals with an improved ability to manage forest ecosystems for the variety of values desired by society (Kimmins et al, 2007). However, the big question in teaching sustainable forest management is: how can students gain experience of the consequences of forest management decisions over the spatial and temporal scales involved in stand and landscape ecosystem processes and dynamics? How can they learn by trial and error, if these consequences can only be seen at spatial scales up to millions of hectares, and temporal scales of centuries? One possible tool that teachers can use to solve this problem is games based on ecosystem-level ecological models.

Games based on ecological principles can be as simple as board games. For example, the Floristic Relay Game (Ortiz-Barney et al, 2005) has been successfully used in the USA to teach high school students plant community succession and disturbance dynamics. In this teaching activity, students play a board game in which each student represents an imaginary plant species. Each time the game is played, the students are conducting a type of theoretical experiment or simulation. Students explore plant community dynamics by playing the game and interacting with each other (as different plant species) and responding to chance events. At the end of the game, students report on the results and discuss with the class what they have learned. To apply their new knowledge, students predict changes in the plant community and attempt to make the community change in specific ways. Students enjoy the game, successfully learn the basic concepts of disturbance and succession, and assimilate the rules of the game as representations of the current ecological understanding of the topic (Ortiz-Barney et al, 2005). However, this game lacks the role-playing simulation ingredient (discussed above). In this game, students only observe how imaginary plants follow different successional paths, but they do not take decisions on managing the forests plant communities that are represented. They do not learn the consequences of such actions, unless some additional component is included in the game. To rectify this shortcoming of board games, there is increasing use of ecology-based computer games that can deal with more complex rules and provide more interaction.

The use of computer models in teaching the sustainable management of natural resources is not a new phenomenon. In the early 1970s, Walters and Bunnell (1971), helped by a group of graduate students and faculty at UBC, developed and used the computer management game FARMS to simulate patterns of land use in BC. This early computer game allowed resource users to make year-to-year changes in basic grazing, forestry and wildlife management tactics, and to see the probable short-term and cumulative consequences of their actions. The model included simulations of plant productivity, wildlife habitat selection, food intake, population age structure, reproduction and death, and how management options affected all these variables. Even with the limited capabilities that computers had 40 years ago (including limited memory and no visual screen output), Walter and Bunnell (1971) reported FARMS as being valuable as a teaching tool, because its simulations could be used as an ecological laboratory, introducing the students to such pervasive ecological phenomena as time lags, non-linearities and thresholds. A few years later, Pelz (1978) reported that most of the forestry schools in Canada and the USA used some type of computer simulation tool to teach topics in forest management, silviculture, harvesting and wildlife management.

In the 1990s, interactive programs such as Tree Tales (BC Ministry of Forests, 1996) became common. This program was designed to provide an exciting learning environment in which students could play the role of an explorer in the forest, meeting eccentric characters to learn facts about forests by interactive conversations (rather than by listening to lectures), solving mysteries and choosing their own adventures. Computer programs such as Tree Tales usually target late primary or early secondary school students, guiding their first steps through concepts of forest ecology and management. During the 2000s, the quality of visual and audio output from computers has increased exponentially, increasing the appeal of working with computer-based simulations. During this decade, computer games have been developed to teach many ecological concepts, such as evolutionary biology with Critters! (Lathman and Scully, 2008) or water quality in streams with EnviroLand (Dunnivant et al, 2002). Educators have advocated the use of commercial computer games as a way to create discussions of the biology behind such games and their biological credibility (MacKenzie, 2005).

Possibly the most successful computer software specifically designed for teaching ecological concepts is EcoBeaker (SimBiotic Software for Teaching and Research, 2009). EcoBeaker is an ecological simulation program designed primarily for teaching ecological interactions (Meir, 1999). Focused on late undergraduate students majoring in ecology, or on graduate students already familiar with basic ecological concepts, this state-of-the-art computer game takes advantage of the multimedia resources that present-day computers offer. The exercises help students understand basic concepts like population growth, succession, predation, competition, island biogeography, and even effects of fragmentation and disturbance on populations. The game addresses critical aspects of sampling populations, such as how estimates are influenced by size, number and spatial arrangements of sub-samples, and EcoBeaker also lets students investigate different types of spatial models (Williams, 2000). One of the most interesting aspects of this software is its flexibility, which allows the teacher to create different exercises oriented to groups of students majoring in different topics, such as forestry, agriculture, natural resources and biology. Exercises carried out with this software are more successful when they are played as a game and the students face a challenge, such as achieving a given score or beating the score of other players, which greatly increases their motivation (J. B. Imbert, Universidad Pública de Navarra, personal communication).

The creation of an ecological model can be a very educational activity by itself. Blackwood et al (2006) used model building successfully to increase the knowledge of students about ecological interactions. Students constructed qualitative models of an ecosystem and used these to gain a better understanding of direct and indirect ecological interactions. Qualitative modelling was used in two applications, which differed in educational goals and student backgrounds. The first involved non-major or beginner-undergraduate ecology students, and was intended to improve understanding of the ecosystem of interest and provide a framework for the instructor to assess student learning. The second was designed for more advanced students of ecology, and involved the use of modelling software such as POWERPLAY to design hypotheses concerning the analysis of ecosystem responses to simulated disturbances (Blackwood et al, 2006). Use of this software resulted in improved skills of qualitative reasoning and understanding of the dynamics of complex systems. Students gained insights into their own understanding of the ecosystem they were studying, and used the models they created to predict possible experimental outcomes or patterns in the ecosystems they were studying.

Creating ecological models benefits teachers as well as students (Dresner and Elser, 2009). Using a series of conceptual ecological models demonstrated an improvement in ecological understanding on the part of the teachers during a short course. The sequence of models developed by the teachers portrayed their initial intuitive explanations, often containing misconceptions, through to the post-test elaboration of a more complex and accurate understanding of ecological phenomena. This illustrated shifts in the teachers' thinking and understanding, while the ecological models provided them with a means to visualize their conceptions of ecosystem processes. Their understanding was further enhanced through collegial discussions.

Ecological models for education in forestry

Ecological interactions in forests are multiple and complex. While some management decisions have direct, visible and almost instantaneous consequences (e.g. clearcutting, removing the understorey), there is usually a less obvious sequence of events that can span from stands to regional landscapes, and from days to centuries. In addition, society is increasingly asking for sustainable, ecosystem-based forest management for multiple values (Kimmins, 2008a). Management options can no longer be judged solely on their effects on any one value, such as volume of timber produced. Effects on other values such as wildlife habitat, various measures of biodiversity and soil fertility must also be considered. This increases the complexity of decision making in forest management, and this complexity should be included in the simulations (Kimmins et al, 2008b). However, more complex decision-making tools also mean that decision makers need to know more about the ecological links between different parts of the ecosystem and the long-term consequences of forest management. Ideally, decision makers should gain this knowledge as part of their professional training, but the use of ecosystem models of appropriate complexity can be an important complementary tool.

Models designed originally for management have been used in education for sustainable forest management. For example, the forest vegetation simulator (FVS; Wykoff et al, 1982; Dixon, 2002) is a distance-independent, individual-tree forest growth model widely used in the USA to support management decision making. For the 30 years since the model was initially introduced, the development team has anticipated and provided needed enhancements and maintained a commitment to working with and training users (Crookston and Dixon, 2005). This model is routinely applied to support the silvicultural training and certification of professionals, and has been used as a teaching tool in several schools of forestry in the USA (Crookston and Dixon, 2005). Stands are the basic projection unit, but the spatial scope can be many thousands of stands. The temporal scope is several hundred years at a resolution of 5–10 years. Projections start with a summary of current conditions as defined in the input inventory data. Similar to FORECAST, FVS contains a self-calibration feature that uses measured growth rates to modify predictions for local conditions. Component modules predict the growth and mortality of individual trees, and extensions of the base model represent disturbance agents including insects, pathogens and fire. The component sub-models differ depending on the geographic region as represented by region-specific model variants. Applications range from the development of silvicultural prescriptions for single stands to landscape and large regional assessments. Key issues addressed with FVS include forest development, wildlife habitat, pest outbreaks and fuel management. The predictions are used to gain insights into how forested environments will respond to alternative management actions. The developers of this model (Dixon, USDA) indicate that future work will focus on adopting recent biometric techniques and including new information linking geomorphology to mortality and growth, extending the model to more closely represent biophysical processes, adapting this model so that it is more relevant to management questions related to predicted climate change, and providing ways to dynamically link FVS to other models

From 1980 until 1996, the FORCYTE (Kimmins, 1993b; Kimmins et al, 1981) ecosystem management model (replaced by FORECAST; Kimmins et al, 1999) was used annually in the Pacific North-West's US Forest Service Silviculture Institute advanced training programme at Oregon State University, and from 1986–1996 in a similar application at the BC Silviculture Institute. These training programmes were part of the certification of silviculturists in these two regions, until they were discontinued.

FORECAST has also been used in Silviculture Institute workshops in Ontario, in university courses at UBC, and in training courses in Scotland (UK), Norway, Spain, Cuba, Australia, Japan and China. The reason for using these models was because they are at the ecosystem rather than population or plant community level, they were able to represent all the major silvicultural practices of the time and the major natural disturbance regimes, and they provided abundant graphical and tabular output to help users to understand how the end-of-simulation predictions were derived in terms of key ecosystem, community and population processes. They permitted the user to examine the 'temporal fingerprints' (Kimmins, 1990a, 1990b) of key ecosystem processes and the associated time trends in a variety of social values: wood supply, employment, stand-level economics, and energy-use efficiency.

The main challenge to the use of these types of management model directly in high schools and undergraduate courses is that they are not designed as tools for novices in forest ecology, stand dynamics and silviculture. They can be a complementary tool for advanced courses, but they are not really attractive to beginner students because previous knowledge of both the scientific basis and the practice of silviculture is needed. They are seen as complicated, difficult to use and not fun for novices. Management models are obviously not games, and although they can be considered simulations of real forest ecosystems, they are difficult to use as learning tools. For this reason, the Forest Ecosystem Management Simulation Group at UBC created a user-friendly interface for FORECAST (multiple run manager, or MRM) with which to examine the consequences for ecosystem functioning and social values of a wide array of different management strategies. MRM presented a choice of 81 scenarios representing the interacting effects of three different rotation lengths, three different levels of harvest utilization, three levels of fertilization, and three intensities of post-harvest site preparation by slash burning. Working with prepared libraries from the 81 scenarios, MRM proved very effective in rendering FORECAST accessible and useful as an educational tool for college/university and professional forester education and training (examples of these outputs can be found in Kimmins, 2004d, p578, Figure 21.5).

While MRM facilitated the use of FORECAST in educational applications, this did not address the needs of high school students new to the concepts of forest ecology and forestry. To address this need, the UBC forest ecosystem modelling group developed a simulation game called FORTOON (Kimmins et al, 1997), and then a more advanced version, Possible Forest Futures (PFF), a spatially explicit and interactive educational tool for the analysis of sustainable forest management at the landscape level.

FORTOON: a high-school-level, introductory, multiple-value forest management game

FORTOON as a simulation game

FORTOON (Kimmins et al, 1997) is an educational forest management game driven by FORECAST (Kimmins et al, 1999, described in Chapter 5). It was designed for use in the education of students interested in the sustainability of both the environment and the economy, and also for beginner forestry students. FORTOON was field-tested in a high school (Grades 10–12, ages 16–18) with good results, and has been used successfully as an educational tool in graduate and undergraduate courses in forestry schools. Olarieta et al (2000) found it to be a very useful teaching tool in a Spanish forestry school because it integrates the principles of sustainable forest management, and focuses in particular on two important but frequently neglected principles: the need to define overall, rotation-length management plans as precisely as possible, and the need to analyse for multiple values into the long-term future. While the game was developed for pre-forestry and forestry courses, students with an agricultural, natural resource management or environmental sciences focus have found it useful (Olarieta et al, 2000). Using FORTOON, students can explore some of the consequences of different management choices without waiting a lifetime to see the possible outcomes, and what combinations of management practices result in desired outcomes for six different values. The teacher can use the students' success or failure in meeting the established objectives as a teaching opportunity, and there is an extensive library of pictorial, graphical and tabular output designed to assist the educational process.

To play FORTOON, a student assumes the role of Chief Forester in an imaginary forest and 'signs' a contract agreeing to manage the forest for six of the multiple forest values that contemporary society is demanding (Figure 7.1). For each of eight harvest blocks, the player selects from a set of 48 pre-generated management scenarios, and as the model runs he or she learns about their effects on three management and three social/environmental values. During the game, the student must maintain all six values at acceptable levels by varying the management between the eight harvest areas, and changing management after each timber harvest as needed: adaptive management. If any one value falls too low, the game ends and the student is scored on their performance. In addition to the game, there is an extensive library of definitions of ecological and forestry terms, and descriptions of the habitat of various plants and wildlife species that may live in the area represented by the game. The performance of the different management combinations can be explored and compared pictorially, graphically and in tables and a score card. The objective of the game is to keep a balance of the six values across the eight cutblocks for 240 years, and to achieve as high a value as possible for the overall score and also for each individual value. The challenge of the game is to avoid being fired for not keeping one or more of the values above the minimum level (Figure 7.2).

The FORTOON software, a detailed user's manual (in English and Spanish) and a teacher's and a student's manual can be downloaded from the companion website (see below).

The challenge for the players is to balance the need for economic profit and raw materials with the social need for jobs and environmental quality. In the simulation, the players are required to make decisions about how to manage eight different harvest areas (referred to in the game as cutblocks) (Figure 7.2). Each area is managed independently, and FORTOON tracks the values for each of the six variables in each of the eight areas, and for the entire forest. For each area, the player has to decide if and when to log or keep it as a reserve. If the stand is logged, the player has to decide the rotation length (40, 80 or 120 years); the level of utilization (whole tree or stems only); whether to plant a monoculture of Douglas-fir or depend on natural regeneration that results in a mixed stand of Douglas-fir and alder; and whether or not to use thinning (to remove the competing alder in the mixed stands, and/or to reduced the density of Douglas-fir to increase tree size and log value at the time of harvest), and fertilization (adding nitrogen fertilizer, nitrogen generally being the limiting nutrient in these forests). All these

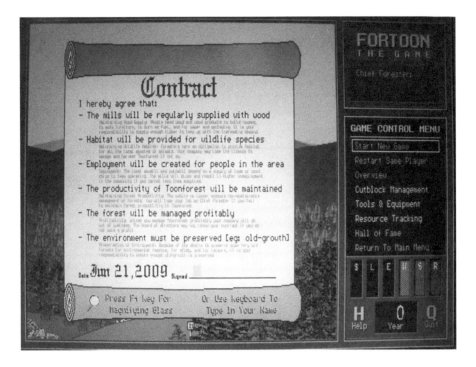

Figure 7.1 *In FORTOON, the player has to sign a contract as Chief Forester of Toonforest agreeing to maintain acceptable levels of six different values: economic (timber supply and profits for the company), environmental (wildlife habitat, levels of productivity and old-growth areas) and social (employment)*

management choices will affect the values of the six target variables: three social (net revenue, supply of logs and employment), and three environmental (wildlife habitat, soil fertility and area of mature forest reserve). The overall score and performance of the players is evaluated based on pooling the values from the eight management areas. This management scenario is an over-simplification of what happens in the real world, but is sufficiently realistic to illustrate key management choices and consequences.

FORTOON keeps track of each stand, follows the management plan decided by the player and simulates its ecological consequences, such as creation or destruction of wildlife habitat, levels of nutrients in the soil, the workforce needed to implement the actions and the timber produced by them. The simulation engine used in FORTOON is FORECAST, an ecosystem-level forest model (see Chapter 5 for a complete description).

Analysis of FORTOON management scenarios
The second module of FORTOON is a set of analysis tools with which to explore 64 management scenarios, each of which is simulated over 240 years. These are defined by six broad management questions: rotation length (harvesting 40- or 80-year-old trees), weeding (early competition control), monoculture conifer or a hardwood/conifer

Figure 7.2 *FORTOON simulates forest growth and development (using the same basic simulation rules as FORECAST) for each of the eight harvest areas that comprise Toonforest*

Note: students have made management decisions in the eight harvest blocks in order to keep high levels in the six values over the entire forest (lower right corner of the screen) so as to avoid the game ending.

mixture, thinning to control stand density, stand fertilization, and level of harvest intensity (whole-tree or stem-only harvesting). The results can be examined pictorially, graphically, as a tabular presentation or in histograms. The player can select up to 9 of the 64 management scenarios at a time, and compare their graphical outcomes for the simulated period.

In the visualization presentation, values are presented for deer habitat, carbon storage, total organic matter in the soil, available nitrogen, available phosphorus, available potassium, wood harvested, money earned, bio-energy produced and employment. In the graphical presentation, values are given for conifer (Douglas-fir) stem and foliage mass, hardwood (red alder) stem mass, herb and shrub foliage mass, forest floor mass, humus mass and an index of nutrient site quality. In the tabular presentation, wood harvested, money earned, bioenergy produced, employment provided, wildlife habitat (for deer), carbon storage on site, the mass of forest floor and soil humus, and an index of soil fertility are scored for performance or as histograms. In the final section ('Analysis of results'), total site biomass, total site nitrogen, available soil phosphorus, available soil potassium, economic performance and energy benefit–cost ratio are presented. As in the game module, all these criteria are realistic representations of the variables used by forest planners to asses the sustainability of a given forest management plan, and they provide the students with a plausible estimate of the possible future levels of the values they will have to work with in their future careers. This claim of plausibility is made on the basis that the scenarios used are generated by FORECAST, which has been validated for these values for the type of forest represented in the game.

FORTOON is currently calibrated for a particular forest type in coastal BC. However, the degree of detail presented in FORTOON is much higher than in similar educational forest management games (Olarieta et al, 2000), and because the issue of long-term sustainability is generic to nearly all forests, the game is a widely applicable educational tool. It gives students the chance to explore the differences between evaluating management outcomes as a 'temporal fingerprint' of management consequences, versus the final outcome at the end of the simulation period, the approach used in most games and simulation models. Working with averages gives a static picture of the forest management system, whereas it is often the minimum and maximum values that are important, such as whether the amount of money earned falls below a level that causes the saw mill to close, resulting in unemployment; or whether the volume of wood harvested goes over the maximum that the mill can handle. These are the types of problem that students can experience when playing FORTOON. A teacher can use the analytical module in the briefing stage to emphasize the extent of the powers, duties and responsibilities of the players, and consider the relationship between the fictional Toon Forest and the real world (Jones, 1980).

The analytical module can also be used at the debriefing stage, when the students explain how well they did in the game and the reasons for success or failure. Contrary to the experience with simpler games in which user interest declines after two or three runs of the model, field testing has shown that students enjoyed and continued to learn after many attempts to win the game. This is attributed to the complexity of the game and the richness of the supporting material. Finally, the teacher can ask the students to compare several of the pre-generated management scenarios that varied from those that they used in the game, and carefully analyse how the six values change over time. In this way, the report from each student contributes to the overall learning experience of the group (Jones, 1980).

PFF: Possible Forest Futures

Possible Forest Futures (PFF) is a game with which to explore what we mean by sustainability and stewardship in forestry in a spatial framework and at a landscape scale. Essentially, it is a more advanced and realistic version of FORTOON, but with a different type of output. It provides an ecologically based vision of future forest landscape mosaics and stand conditions, and of the values that could occur under alternative landscape management scenarios. Different site and stand types that exist because of soil and topographic differences (and different climates, if the game is set up for mountainous topography) respond differently to natural disturbance and forest management, and these differences should be reflected in site-specific management methods. This is not addressed in FORTOON but is in PFF.

PFF is currently calibrated for a coastal forest landscape in BC, but as noted above, a lot of forestry issues are common to many forests, and the questions model users are faced with in PFF are relevant in many forested landscapes. PFF uses FORECAST as the simulation engine for different scenarios selected by the user, and consequently the set of ecological rules underlying the simulation of individual stands in PFF is the same as for FORTOON. There are ecological differences between different site types within the landscape represented in PFF, and these influence the way different cutblocks within the simulated forest in PFF respond to the management choices the players make.

PFF is designed to simulate small- to medium-watershed (up to 10,000ha) issues that involve the stand level, but with spatially explicit representation of the stands (cutblocks or harvest areas) across the landscape. This spatial dispersion of cutblock areas involves issues such as road construction, harvesting schedules, landscape pattern and riparian forest management, among others. Similar to LLEMS (Chapter 6) in terms of polygon structure and interactions (minus the detailed light profiling and seed dispersal between stands see Figure 7.3), PFF can include a hydrology model if FORECAST Climate is used as the driver rather than the basic FORECAST, and can track road development. The model also includes extensive output, including economic costs and benefits, timber harvests, carbon budgets and various other social and environmental variables, the variation of which over time can be examined graphically for individual stands or for the entire watershed.

PFF can prepare rotation-length movies of different landscape scenarios for subsequent analysis, a feature that can be used by the teacher to prepare in advance several scenarios and then ask students to try to recreate them and/or analyse the differences in outcomes. Because it is an ecosystem management model that can simulate many aspects of landscape-level issues in forest management, PFF has the potential to be used to examine land-use patterns that include mixtures of forest management, forest reserves and other land uses such as agroforestry and agriculture. As in FORTOON, there is a series of management choices the user can select, but in PFF these are site-specific.

PFF will provide a small library of prepared set-up files: a variety of histories of past disturbance and hence of starting ecosystem conditions. Because a 120-year simulation can take 10–20 minutes (depending on computer speed), the teacher can choose to view a pre-prepared set of movies in which previously prepared runs are replayed in a couple of minutes. The students will also be able to make new movies and view them. The student can then choose between a large number of alternative management strategies (scenarios) and watch the change in the maps of a variety of values, and graphs of these values. As in FORTOON, at the end of the run, the student can review any of these graphs for the whole watershed, or for each individual cutblock.

Multiple run manager for FORECAST

The complexities of ecosystem structure, seral stage, soil condition and site type, combined with that of different types of natural disturbance and alternative silvicultural strategies, make the use of FORECAST somewhat overwhelming for a beginner. As noted above, to lessen the difficulties of the relatively steep learning curve of such a model, we developed a user-friendly interface that permits a selection from 81 different management scenarios (a 3 × 3 × 3 × 3 matrix). The details of this matrix can be created by the administrator of a workshop using MRM/FORECAST to render the forest simulation and management/natural disturbance options appropriate for the region,

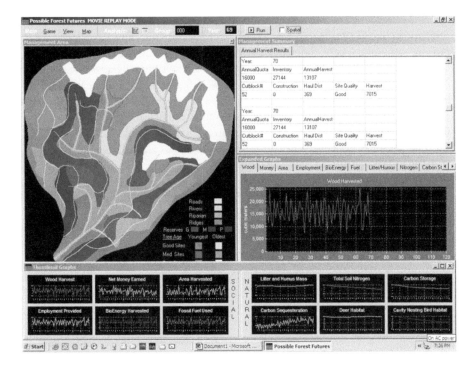

Figure 7.3 *User interface of PFF*

Note: A colour version of this figure is shown in the plate section as Plate 4.

forest type, site type, natural disturbance regime and stand-level management strategies familiar to the workshop participants. Alternatively, an educator wishing to have a specialized matrix for her or his area could have a matrix prepared under contract by Life Science Programming of Naramata, BC.

Take-home message

What we want to accomplish by using games in education and training for sustainable forest management is create a learning environment that engages the learner. The best learning experiences generally available are either simulators such as FORTOON or PFF, or interactive sessions with skilled facilitators on field trips or in participatory laboratory sessions. However, most learning experiences currently involve the heavy use of text, too many knowledge-test assessments, and facilitators who may care but are not as skilled as is needed (Quinn, 2005).

As suggested by Portney and Cohen (2006), unless you can put yourself in the shoes of the decision maker and understand the context in which the decision is made, you have not really understood the decision. We should be creating a learning environment in which we transform our learning objectives into important behaviours learnt or adopted by the students, where contexts are meaningful to the learner, and where the decisions are consequential. Such learning environments can be created with simulations like FORTOON where the student has the opportunity to experience the possible consequences of different management decisions in a virtual world defined by the current scientific knowledge of forest ecology. We strongly advocate the use of some type of simulation, because it is the only way that complex non-linear interactions between ecosystem variables can be experienced by the students over very long-term timescales, and they can get a glimpse of the complex ecological and social consequences that a simple management decision can have in a changing world.

Additional material

Readers can access the complementary website (www.forestry.ubc.ca/ecomodels) to access the full versions of the following additional material:

- FORTOON full program (for DOS);
- FORTOON user manual (versions in English and in Spanish);
- FORTOON student handbook;
- FORTOON teacher's package; and
- slideshow presenting PFF.

Chapter 8

How to Develop a Model for Forest Management[1]

Introduction

Why develop new models for forest management when we already have so many?

Models are simplifications of reality. They are designed to explore and analyse some aspects of the real system that they try to emulate. The creation of a model for forest management is linked to the problem that the modeller is interested in. This chapter begins with a description of the reasons for developing a new model for forest management, and provides a brief review of the current trends in good modelling practice. The chapter later provides a step-by-step guide to model development, covering issues such as parameter calibration, sensitivity analysis and scenario analysis, with each step accompanied with examples from real models. The chapter ends with a discussion of how to deal with complexity and uncertainty, key issues in forest management models dealing with the simulation of broad time and spatial scales.

Forests are complex ecological systems, and forestry is a complex social, political, economic, management and biophysical activity. Multi-value sustainable forest management involves complex planning, management, public involvement and the application of knowledge from biological, ecological, earth, atmospheric, management, engineering and social sciences. This makes forestry one of the most complex of human activities.

Decisions regarding forest conservation and management across large spatial scales and long timescales are fraught with great uncertainty unless this complexity is addressed. *The Limits to Growth* (Meadows et al, 1972, 2004) reports current concerns over climate change, and numerous environmental and social challenges underscore the dangers faced by humanity when it fails to comprehend and incorporate adequate complexity in decision making. Increasingly it is recognized that faced with a complex, changing and uncertain future, which by definition we have not yet experienced, knowledge-based predictive models, imperfect though they always are, constitute what is often the best way of supporting decision making.

Traditionally, models have focused on a specific, often narrow objective. They have generally made forecasts for a single value (e.g. tree growth, hydrological regimes, wildlife habitat, recreational opportunities) or a limited group of values. This is typical of the administrative and timber-focused phase of forestry evolution, but, surprisingly, this has continued in the implementation of the ecosystem-based management (EBM) phase. That is because although EBM is conceptually the management of forest ecosystems, implementation on the ground generally reverts to management of individual ecosystem values, albeit on the basis of their ecology. Single-value decisions are often made in the absence of adequate consideration of their effects on other ecosystem values. Only in true ecosystem management does one find ecosystem-level consideration of whole-system consequences of individual value decisions. Impediments to true ecosystem management include political and institutional structures, and either the lack of suitable decision-support tools that can address the complexity of policy and management decisions in a multi-criteria environment, or a failure to use those that are available.

A model is a substitute for a real system. Models are used when it is easier to work with a substitute than with the actual system (Ford, 1999). An architect's blueprint, a geologist's map or a child's paper aeroplane are all models. They are useful because they help the model users to learn something about the systems they represent, and to make decisions related to them, without having to address the full complexity of the real system represented by the model. However, the simpler the model relative to the full complexity of the system it represents, the less information about the real system is provided by the model and the less reliable it is as a basis for policy with respect to that system. For example, a simple paper aeroplane provides little guidance to engineers designing improvements to a modern passenger jet.

At its most basic level, modelling is simply a process for thinking systematically about a problem (Jakeman et al, 2008). Modelling in forest management involves the organization of data, assumptions and knowledge for a specific purpose. It is undertaken for knowledge generation and communication in order to improve policy and management actions. As an approach to enhancing understanding of complex systems, modelling can take many forms. It can be purely qualitative (word models, for example), heavily quantitative (mathematical models that use equations to represent the interconnections between the components of a system) or some mixture of the two (Kimmins, 2004d). Traditionally, modelling has involved simplification of complex systems, but when this simplification omits key information about the response of forest ecosystems to different management approaches or environmental change, model output will often be misleading and can result in non-sustainable management. Modellers working in forestry should always be aware of the dangers of over-simplification.

The major challenge in forestry modelling is development of models that are as simple as possible but as complex as necessary to address ecosystem and social complexity. Output from such models should be complex enough to provide a basis for model evaluation and understanding of model forecasts, yet simple enough to be usable by busy foresters or other target audiences. Without adequate complexity, the predictions of possible outcomes of changes in policy and practice or the environment become an illusion.

Forest managers and planners are relying increasingly on access to models of forest dynamics that can be used to generate credible options for improved resource management, but unless the output of these complex models is relatively simple, they may be unusable. This challenge, which we assert is beyond the capability of many forest models today, creates the need for new models that can provide reliable scenario and value trade-off analyses of both environmental and social values while providing user-friendly output. It is widely accepted that although relatively simple growth and yield models remain fundamental to forest management, the global community now demands more sustainable use of multiple values and a more comprehensive understanding of forests and forest products (Vanclay, 1994). A broader, ecosystem-level approach is needed in the development of forestry decision-support models (Kimmins, 2004d; Kimmins et al, 2005), and these should be linked to advanced, interactive visualization systems (see Chapter 10) as a complement to traditional graphical and tabular output formats to increase the utility of the output.

Forest modelling has been slower than forest management and policy in its response to rapidly changing public perceptions of, and demands for, the many values provided by forests. Scientists have acquired detailed knowledge of individual components and processes of forest systems but have in the most part failed to integrate and synthesize this into a form that can provide appropriate ecosystem-level decision-support tools. The need is to develop ecosystem-level forest management models based on the latest scientific knowledge that can be integrated with traditional knowledge and other management tools and techniques.

There is a well-developed literature on how to create models for ecology and natural resource management. In this chapter we will cover some basic concepts that should be taken into account when creating a new approach to such modelling. We illustrate the process with examples from the models described in Chapters 4–6 and others that our team has developed. Readers interested in detailed modelling techniques and philosophical and practical considerations in modelling natural resources such as forest ecosystems can consult Box (1979), Starfield and Bleloch (1991), Vanclay (1994), Hilborn and Mangel (1997), Ford (1999) or Shenk and Franklin (2001). In addition, the journals *Ecological Modelling* and *Environmental Modelling and Software* provide a summary of issues and progress in ecological and natural resource modelling.

Good modelling practice

With the availability of larger datasets covering longer periods, and with the greater power of computers and software, the number of ecological models has increased dramatically (Monserud, 2003). The result has been a plethora of models and modelling approaches, ways of describing and testing them, and methods for evaluating their predictions. From this has arisen the need to standardize approaches to environmental modelling (Van Waveren et al, 1999; Council for Regulatory Environmental Modeling, 2009). Many forest managers and researchers are using models developed by others without fully understanding their strengths and limitations. This can lead to the misapplication of otherwise credible models, and suggests the need to clearly identify the capabilities of individual models and what they are not capable of.

Failure of a model to perform as desired can be a function of an incomplete conceptual model, incorrect model formulation and/or errors in the software coding, while errors in model use can result from lack of (or incomplete) users' manuals, careless treatment of input data, insufficient model calibration and validation, and/or using the model beyond its intended scope, any of which can lead to unreliable model output (Van Waveren et al, 1999). This can have far-reaching consequences considering the important and increasing role being played by models in today's forest management. In order to avoid such problems, a consensus about good modelling practices is starting to emerge. The most comprehensive guidelines are those provided by Van Waveren et al (1999) and the US Council for Regulatory Environmental Modeling (2009). These recommend that model developers and users should:

- subject their model(s) to credible, objective peer review;
- assess the quality of the data they use;
- corroborate their model(s) by evaluating the degree to which they correspond to the system(s) being modelled; and
- perform sensitivity and uncertainty analyses.

These guidelines help to determine when a model, despite its uncertainties, can be used appropriately to inform a particular decision.

Crout et al (2008) provide an additional definition of good modelling practice. They assert that the primary requirement is that a model has a clearly specified purpose. What is the model for? What is the reason for creating a new model, and what is the model expected to do? In the absence of a definition of a model's purpose, its degree of success cannot be judged and the necessary level of structural complexity cannot be defined. The more explicit the statement of model purpose, the better.

Models can be used to:

- represent variables and rates of change;
- describe ecosystem structure and both temporal and spatial patterns of individual ecological processes;
- reconstruct the past or predict future behaviour of the ecosystem;
- generate and test theories and hypotheses about ecosystem organization and function;
- display, encode, transfer, evaluate and interpret knowledge;
- guide development and assessment of policies; and
- facilitate collective learning and settlement of disputes (Morton, 1990; Beven, 2002).

While future applications of a particular model may be hard to predict, this does not change the importance for model development of defining the intended purpose (Crout et al, 2008).

The second key aspect of good modelling practice is to clearly report on the data used for model development, how the data are used in the model, all the model's assumptions, and the model formulation. Explicit listing of assumptions reveals the thinking process of the modeller(s) and facilitates testing of these assumptions at a later stage (Crout et al, 2008). The modeller should provide brief but explicit statements to justify the assumptions, and document references that describe them in detail, including listings of any subjective preferences and opinions (Wagener et al, 2003).

A third key component of good modelling practice is model evaluation (Council for Regulatory Environmental Modeling, 2009), which should be a continuous process, recognizing the difference between evaluating model assumptions, model implementation and model performance (Crout et al, 2008). Ultimately, models are evaluated against the original objective of the modelling activity, but there are several other approaches that are discussed later in the chapter. Model evaluation, which traditionally, has involved some measure of predictive performance and uncertainty, should be a central part of the model development process, not something considered only after the model is finished. The evaluation phase should also include assessment of the data used (Crout et al, 2008) and whether the patterns simulated by the model make sense from an ecological view (Blanco et al, 2007).

A common understanding and consistent use of model terminology is required for the communication of model development and evaluation to others. New, confusing or ambiguous terminology should be clearly defined, and any redefinition of particular words made clear (Crout et al, 2008). Much of modelling theory has its origins in the statistical literature, and where possible modellers should use the original terminology (see the glossary in Ripley, 1996). Additional modelling-related terminology can be found in Council for Regulatory Environmental Modeling (2009).

Documentation of models should include the conceptual model that the model developers are representing, the mathematical formulations and the assumptions on which it is based, the model's parameterization and parameter values, its implementation including operating instructions, and the analysis undertaken to evaluate the model. The importance of this issue is exemplified by the editorial policy of *Ecological Modelling* that reporting these model features is a prerequisite for acceptance of papers describing new ecological models (Jørgensen et al, 2006). Documentation of models is also important for the development of databases on environmental models (e.g. Table 8.1).

Table 8.1 *Electronic databases on environmental and forest models*

Organization	Type of models	Web address
Council for Regulatory Environmental Modeling (EPA, USA)	Models developed by EPA	www.epa.gov/crem/knowbase/index.htm
ECOBAS server (Kassel University, Germany)	All types of ecological models	http://ecobas.org/www-server/index.html
Forest Model Archive (University of Greenwich, UK)	Forest models	www.forestmodelarchive.info
National Commission on Science for Sustainable Forestry (NCSE, USA)	Biodiversity models	http://ncseonline.org/ncssf/dss/Documents
Forest growth models database (Wageningen University, The Netherlands)	Forest growth models	http://models.etiennethomassen.com/index.php

Working to the standards of best modelling practice is the responsibility of all model developers. Adequate documentation of models makes their characteristics and assumptions explicit and facilitates integration with other models (Villa et al, 2006). However, even a model developed under best practice may not be suitable for a given purpose. It is the responsibility of the users of a model to be aware of its limitations and to use it appropriately.

Finally, the gap between model design and model application should be minimal. To achieve this, model developers should work closely with their ultimate clients: model users (McIntosh et al, 2008). The insights gained by model developers, forest managers and stakeholders through a participatory development process is one of the greatest, and most frequently overlooked, benefits of modelling projects. Such processes can clarify and resolve the sometimes contradictory objectives of model design and use, expectations and perceptions between science and practice; it can play a fundamental role in mediating compromises on both sides. McIntosh et al (2008) also consider that establishing and maintaining credibility and trust is a key issue for model developers. Model results are much more likely to be used to inform forest policy or management when forest managers and other relevant stakeholders trust the scientists developing or running the models. The key to developing this trust is openness and transparency about underlying model assumptions and limitations, and developing and maintaining professional relationships with forest managers.

Different objectives lead to different modelling approaches

There is a wide variety of modelling objectives that represent variations in the backgrounds, knowledge and experiences of modellers. There are many specific objectives addressed by forest models, but most could be assigned to one of two main classes: explaining how the ecosystem works, or predicting numerical values of variables defining ecological processes. The objective of a model will determine how it is built and how it can be evaluated.

Models that are designed for prediction without any need for explanation about ecosystem function can be as simple as the mean of a set of values, which could be useful to represent the set but has little conceptual value. These models are typically without a mechanism, although they hypothesize relationships between variables, answering the question: 'How much does Y change as X changes?' Even with simple equations, this question can be answered in a number of different ways. In other words, there are many possible numerical solutions to provide a prediction for a given phenomenon, depending on what the modeller decides the best way to represent the numerical relationships between variables. For this reason they are specific to the location and the system studied. Their details are not intellectually transportable, although the general approach may be (Mangel et al, 2001).

On the other hand, when the objective of the modeller is to explain how the forest ecosystem works, a specific description of the key ecological processes and how they are linked would be needed. Although such explanatory models could be general and applicable to a wide range of forest ecosystems (e.g. a textbook on forest ecology explaining the general interactions between ecosystem structure and function), if the modellers intend to use explanatory models in forest management they will need to include a numerical representation of the ecological processes in order to test the quantitative relationships between ecological variables. This is an important part of the modelling process: to clarify what is known and what needs further elaboration. Since explanatory management models should be formulated in biologically meaningful terms, they indicate what needs to be measured in order to establish parameter values (Mangel et al, 2001).

As in other branches of environmental sciences and resource management, models are not usually used for just one objective. Therefore, the distinction between predictive models and explanatory models is often blurred. In forest management, pure predictive models are usually created by summarizing large databases of field observations in permanent plots and are usually called 'empirical models'. These models are intended for use as predictors of patterns of change in values of selected variables in the future, under conditions that are essentially similar to those of the past. Earlier in the book we have called these 'historical bioassay' (HB) or 'rear-view mirror' models. They are based on empirical relationships between variables of interest and time, or some other independent variable. Commonly called 'statistical models', they involve the fitting of mathematical equations to datasets that describe what has happened in the past, such as tree growth, water yields or production of non-timber forest products. However, there is no explanation of why the temporal pattern recorded in the data occurred in terms of plant, animal, microbial and/or soil/site processes; the data simply provide a temporal pattern of what happened.

On the other hand, models that are intended for the prediction of possible futures for the ecosystem or some subset thereof, based on a representation of key processes, are referred to as 'mechanistic models' when based on a representation of the key biological components and processes that determine their temporal patterns of change. In general, these mechanistic models provide a deeper level of understanding of forest ecosystems (or some subset thereof, such as individual species populations) than empirical models, but this is not necessarily reflected in the degree to which output from the two types of model conforms to reality. They consist of a series of equations that define rates of key processes, the consequence of which is change in the size of ecosystem or sub-ecosystem compartments. Such explanatory models can be used to predict how an ecosystem or some sub-component thereof will function and change under a given set of future conditions that may be different from those of the past.

Many recent models have a mechanistic component that is based on an empirical model: so-called 'hybrid models' (Kimmins, 2004d), which are a mixture of causal and empirical elements at the same hierarchical level (Johnsen et al, 2001). The popularity of this modelling approach is increasing, and in the past few years some modellers who have developed models based on theoretical relationships between components of the forest ecosystem have seen the need to make their models more useful for forest management (Mäkelä et al, 2000; Monserud, 2003). Development of hybrid models will be accelerated if it can be demonstrated clearly that empirical models can be improved through the incorporation of causal or theoretical functions (Monserud, 2003); and that mechanistic models can be improved by incorporating system-level empirical elements and constraints (Mäkelä et al, 2000).

The selection of any of the previous modelling approaches depends on the specific objectives selected by the modeller. One can conclude from this that science will generally favour more mechanistic models because of their explanatory power and ability to explore the possible consequences of changed future conditions (Giske et al, 1992; Giske, 1998). If modellers are more interested in relatively short-term predictions, foresters may prefer the empirical believability of HB models. However, because changing environmental conditions and management methods have not been factored into HB models (other than through operational adjustment factors), thereby posing a significant risk to the accuracy of the predictions, there is increasing interest by foresters in models that can account for the effects of future changes in ecosystem processes. Therefore, if the objective is to forecast a possible range of future system behaviours under a variety of possible future circumstances that are significantly different from the past, one should use a mechanistic model that accounts for those processes that are expected to be different in the future than in the past.

There are many alternative classifications of forest models (Kimmins, 2004d), some of which focus on the objective (Brugnach et al, 2008); some on the structure (Chertov et al, 1999a; Mäkelä et al, 2000); and others on the mechanics of the model (Monserud, 2003; Landsberg, 2003b). As noted in earlier chapters, both the empirical and the mechanistic approaches have their merits and weaknesses, and we advocate a synthesis of the two approaches in hybrid simulation models.

Basic steps to develop a model of forest management

Building a model is an iterative, trial-and-error process. A model is usually built in steps of increasing complexity until it is capable of replicating the observed behaviour of the system (Ford, 1999). Then it is used to learn whether the simulated behaviour can be improved by changes in the system variables. There are many books, papers and reports on how to create forest models. Readers can consult Kitching (1983), Hanks and Ritchie (1991), Vanclay (1994), Shenk and Franklin (2001), Canham et al (2003), Hasenauer (2006) and Jakeman et al (2008), among others. Here we provide a simple step-by-step guide, inspired by Ford's (1999) detailed description of how to develop models for environmental sciences and resource management.

The modelling process starts with the statement of the forest management problem (Figure 8.1). As noted before, every model is developed to fulfil one or more objectives, and only by comparing the model with the problem that it has to solve can we decide if the model is adequate. The steps are illustrated here with examples from two forest models developed with the STELLA software (ISEE Systems, 2009). The first, PINEL, is a model of nutrient cycling and tree growth developed to analyse the long-term effects of thinning on stem productivity in a Scots pine forest (*Pinus sylvestris* L.) in the western Pyrenees, on the border between Spain and France. A full description of the model can be found in Blanco et al (2005). The second was developed to explore the theoretical connections between tree growth rates and climate change in the southern interior of British Columbia (BC) and is fully described in Lo (2009). The STELLA files for both models and other related files can be downloaded from the book's companion website (www.forestry.ubc.ca/ecomodels).

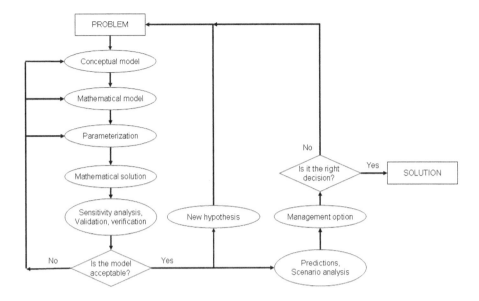

Figure 8.1 *The overall process involved in the development of a model for forest research and management*

Step 1: define the objective, the reference mode and the conceptual model

The first task of any forest modeller is to become familiar with the system to be modelled, and with the people who have identified the problem; it is vital to understand the needs of the eventual model user (McIntosh et al, 2008; Ford, 1999). Greater progress in forest modelling will come when modellers cooperate with and are responsive to the needs of forest managers, and model developers should move from a technology-push to a demand-pull orientation (Reeve and Petch, 1999). This continues to be an issue for forest management modelling, although progress is being made (Monserud, 2003). As a corollary to understanding user needs, model developers need to be clear about why they are creating a model and ensure that potential users understand that the model may suggest changes to existing management policies and/or practices. The costs involved in such change can be significant, and therefore close collaboration is required to ensure that the model is capable of suggesting how management could be changed while minimizing additional costs or even providing a benefit (White, 2001). This is a vital step, but one which is often missed.

Once the reasons for modelling are clear, model development can start. Firstly, the modeller should ask if the problem is 'dynamic': one that needs to be analysed with a dynamic model. If yes, a graph of anticipated time trends in important, interesting or target variables should be prepared. This qualitative graph, known as the 'reference mode' (Ford, 1999), serves as a target pattern of the known behaviour of the real system that the final model should be able to emulate. Preparation of the reference mode at the outset forces the modeller to be clear about the dynamic patterns that are to be predicted and/or explained (Ford, 1999). The reference mode will always have time on the horizontal axis (the time horizon) and one or more important system variables on the vertical axis. The time horizon should be sufficiently long to allow the expected full range of dynamic behaviour to be depicted. If the reference mode infers a highly volatile, oscillatory system, the time horizon should allow sufficient time to see two or three cycles of change so that any trends, such as a tendency for the oscillations to dampen out or increase over time. This is also the time when the modeller has to decide the time-step of the model (hourly, daily, weekly, monthly or yearly).

Once the type of forest ecosystem and the characteristics that the model should emulate have been established, the modeller should determine what is already known. This can be in the form of graphs, a flow chart or a picture: a 'conceptual model', which should be the focal point of the development effort and should address the model as a 'dynamic hypothesis' (Richardson and Pugh, 1981). The conceptual model should be accompanied by a clear statement and description (in words, functional expressions, diagrams and graphs, as necessary) of each of its elements and the science behind it (Crout et al, 2008). By creating a detailed word model of the diagrammatic conceptual model, identifying all the system components (compartments or state variables) and all the process components (system variables) that link the system components, the modeller can create a framework for computer code development. Separating the word model into individual components creates a structure for all the lines of the code that represent the components and processes, and for establishing variable names and equations defining transfer or control processes. Crout et al (2008) also recommend a consideration and evaluation, whenever possible, of competing conceptual models or hypotheses. This may identify existing acceptable conceptual models that have already been represented in computer code, obviating the need for a new model.

Conceptualization defines the structure of the model, and is the stage at which state variables and processes are defined (Richardson and Berish, 2003). Interactions between them may be identified from field studies and/or existing knowledge of ecological systems in general, and of the specific system being modelled in particular. Data requirements for this step can often be fairly general. Mangel et al (2001) offer practical advice concerning pitfalls that can be encountered while developing conceptual models.

When creating models for forest management, mechanistic models that depict causal relationships between variables are preferred over descriptive or phenomenological models (Nichols, 2001). The latter two types define relationships between variables without explicitly representing the underlying mechanisms. Mechanistic models which do so are likely to provide better predictions when simulated state and system variables assume values outside their observed historical ranges, or outside ranges used in estimating model parameters. Simple numerical models of observed past relationships between variables are useful if the conditions that prevailed when those relationships were described recur in the future for which the model is making predictions. However, given the anticipated global climate, global and economic changes, this is unlike to be true for forest management models (Kimmins, 2008a). The modeller should always remember that models which are mechanistic in ways that are not essential to the purpose of a particular modelling project have no intrinsic value for that project (Nichols, 2001).

An example of reference mode for the model PINEL, which was developed to assess the long-term effects of thinning on site fertility (Blanco et al, 2005), is shown in Figure 8.2. Site fertility changes with time (Kimmins, 2004d), and after reviewing the literature

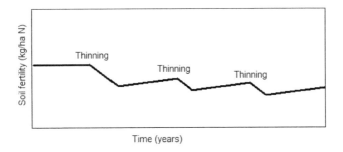

Figure 8.2 *Reference mode for the model PINEL*

on the topic, Blanco et al (2005) expected that thinning would reduce it by exporting part of the site nutrient capital. However, natural nutrient inputs such as atmospheric deposition or N-fixation will augment the site nutrients capital over time. Given the long-term nature of the issue (changes in nutritional site quality may take decades) and the complex of processes involved, it was decided that a dynamic model was needed to explore the balance between inputs and outputs and the consequences of this balance for tree growth.

The next step was to create a conceptual model of the system (Figure 8.3). Because site fertility is related to soil nutrient content and its availability to plants, it was decided that the model should focus on nutrient cycles in both trees and soil. Understorey vegetation was not included, and only one tree species (Scots pine) was simulated. Scots pine was the most important species in the ecosystem (95 per cent of trees in the stand), and the minor vegetation was not very developed, so this simplification was felt to be acceptable. Although Figure 8.3 provides an overall description of how the ecosystem works, it lacks sufficient detail to permit analysis of the issue being modelled. Consequently, an expanded conceptual model was developed (Figure 8.4) incorporating sub-models and variables needed to address the complexity of the Scots pine plantations. This included a sub-model that simulates management practices such as thinning. An existing model (SILVES, Del Río and Montero, 2001) was used to simulate tree growth, with PINEL focusing on nutrient cycling and nutrient regulation of growth (Blanco et al, 2005).

A second example of model development is that of Lo (2009), who explored the possible consequences of climate change on tree productivity. Physiological theory and empirical evidence suggested that tree species adapted to warm environments would experience reduced growth during cold years and increased growth in warm years, whereas trees adapted to cold environments have unchanged growth in cold years and are negatively affected during warm years (Lo, 2009; Lo et al, 2010a, 2010b). A second climatic variable, precipitation, was considered to be as important as temperature, with trees producing less biomass during dry years. The reference mode of this dynamic problem is illustrated in Figure 8.5.

Once the reference mode was prepared, a theoretical conceptual model of the response of tree productivity to climate change was developed (Figure 8.6). This assumed that:

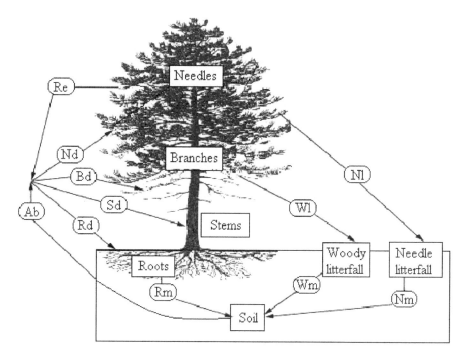

Figure 8.3 *Conceptual model of the nutrient flows (arrows) and stocks (boxes) in a Scots pine stand in the western Pyrenees, as described by Blanco et al (2005)*

Note: Nd = needle demand. Bd = branch demand. Sd = stem demand. Rd = root demand. Ab = root uptake. Re = retranslocation. Wl = branch litterfall. Nl = needle litterfall. Wm = branch mineralization. Nm = needle mineralization. Rm = root mineralization.

- tree productivity has one of three relationships with temperature: adaptations to a warm, a moderate or a cold climate (displayed in Figure 8.6 as 'species optimum temperate scenarios');
- tree species have one of three adaptations to annual precipitation levels: humid, moderate and dry climates (the 'species optimum precipitation scenarios' in Figure 8.6); and
- tree physiological processes are slowed or cease during periods of sub-zero temperatures. This required a 'frost index', defined by a variable that regulates the delay in recovery of process rates following a freezing period (Figure 8.6).

All three of these theoretical relationships are defined by the user rather than by field data, because the objective of the modelling was to investigate qualitatively the interaction between climate and tree growth rather than to make quantitative predictions for a given ecosystem.

Some other examples of conceptual models include Kimmins et al (1999 – the conceptual model of FORECAST, Figure 5.1); the ForWaDy model, Figure 5.24, Seely et al, 1997); the regeneration submodel in LLEMS, Figure 6.6; Blanco et al (2009a – a

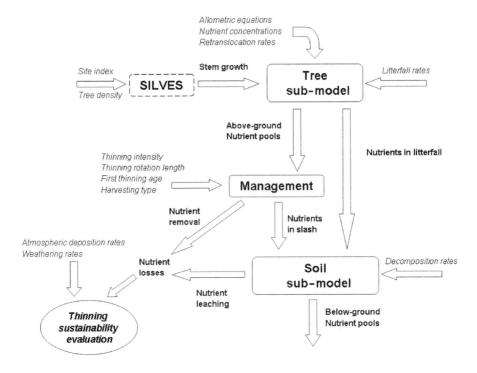

Figure 8.4 *Expanded conceptual model of PINEL*

Note: input variables are in italics, and the internal variables that connect different sub-models are in bold case. These are shown as boxes with solid lines, whereas sub-models external to PINEL are boxes in dashed lines.

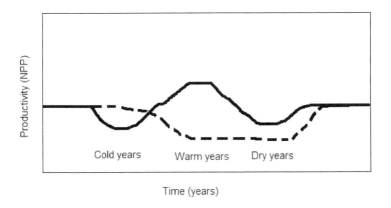

Figure 8.5 *Reference mode for tree productivity and climate model (TPCM)*

Note: Hypothetical productivity of tree species adapted to warm weather follows the solid line, whereas the dashed line represents the productivity of tree species adapted to cold environments. NPP = net primary production.

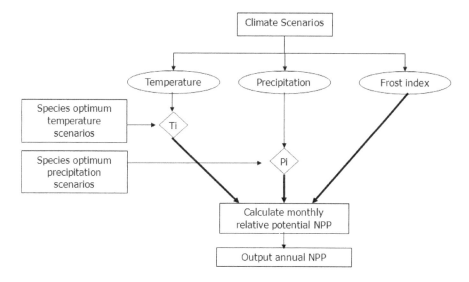

Figure 8.6 *Conceptual model of the TPCM model*

Note: squares represent the input variables and output data. The three oval-shaped boxes are climatic data inputs. Ti and Pi are the temperature and precipitation multipliers that modify monthly NPP. Three thick black lines are the major controlling factors of monthly NPP.

model of forest regeneration in boreal forests); and Blanco (2007 – a conceptual model of how allelopathy affects forest ecosystems). A poster on the last-mentioned is available online from the companion website to this book (see below).

Step 2: construct the flow diagrams

Flow diagrams represent in detail the general relationships described in the conceptual model, identifying sub-components of the major system variables and how major and minor sub-compartments are connected. Programs such as STELLA (ISEE Systems, 2009) or VENSIM (Ventana Systems, 2009) are very helpful during their preparation, but they can equally be created with pen and paper.

Development of the flow diagram starts with the compartments (also called reservoirs, pools, stocks or state variables), adds the flows (also called transfers, fluxes or system variables) between them, and identifies the regulating variables (also called controlling variables or converters) that influence the flows. Flows can be actual transfers of something material (such as water or nutrients in a forest ecosystem, or timber harvests in a managed forest). The flows connect different compartments, and are controlled by various regulators. The flow diagram should include all the variables shown in the conceptual model, expanded to the detail required to deal adequately with the complexity of the system being modelled. It is important to check the units of each variable. If the model is being designed to analyse the effects of different forest management practices, the flow diagrams should include variables to represent the management scenarios the modeller wants to explore (Ford, 1999). If the modeller has trouble identifying the stocks and flows in the system, insufficient time was spent understanding the system and preparing the conceptual model: it is time to go back to Step 1.

The advantage of having a clear flow diagram is that it facilitates implementation in modelling software such as STELLA or VENSIM, or direct translation into computer code. A STELLA flow diagram for the tree sub-model of PINEL is illustrated in Figure 8.7, which shows the major nutrient stocks identified for this part of the conceptual model (Figure 8.3): leaves, branches, stems and roots, woody litterfall and leaf litterfall. All the flows described in the conceptual model have been included, together with a set of regulators that determine the actual rates of nutrient transfers. Finally, several regulators have been included in the right side of the flow diagram to define the management options (type of thinning, year of first thinning, thinning intensity), and a new compartment has been included (harvested trees) to permit simulation of different management options. If these new variables, flows and compartments had not been included in the model, it would have been a useful representation of nutrient cycling in a forest ecosystem, but would have had no value as a management model to investigate the impacts of thinning on site fertility.

Figure 8.8 presents the STELLA flow diagram for TPCM, with accumulated NPP modified by temperature, precipitation and frost. In contrast to PINEL, TPCM is an example of a model with no mass flows; there are only relationships and modifiers between state variables. The conceptual model has been expanded to include different simulation options. In this case, the model is not focused on management; it is more of a research tool. The simulation options included in the flow diagram are a combination of the selection between different climates and adaption strategies of trees to temperature and precipitation. Climate can be set for the ESSF (Engelmann spruce, sub-alpine fir), a high elevation zone with low temperatures and high precipitation, or the IDF (interior Douglas-fir), an ecological zone at lower altitude with high temperatures and low precipitation (Pojar et al, 1987). The model user can also select trees adapted to warm, medium or cold temperatures (T effect on NPP rate 1, 2 or 3) and to humid, medium or dry conditions (P effect on NPP rate 1, 2 or 3). TPCM is a hybrid model, because it uses historical climate records for the ecological zones (ESSF or IDF), but hypothetical functions for the temperature, precipitation and frost effect on tree productivity.

Another example of flow diagram is that of FORECAST (Figure 5.1 and 5.4). FORECAST was not built in STELLA, but directly programmed into computer code from the diagrammatic conceptual model and a word model representation thereof.

Step 3: prepare causal loop diagrams

Causal loop diagrams identify key feedbacks. They can be constructed from the completed flow diagram (Ford, 1999) or during its development, or even after the initial model has been completed. While some (e.g. Richardson and Pugh, 1981) recommend that they be prepared early in the conceptualization process, this is not essential given the iterative nature of modelling. The flow diagram and accompanying causal loop diagrams are often revisited several times during model development. Because causal loop diagrams are a communication tool rather than an analytical tool, it is not necessary to show every relationship included in the conceptual model or flow chart.

In some systems, the loop diagrams may become rather complicated. When this happens, it may be helpful to draw several diagrams, each one highlighting a different

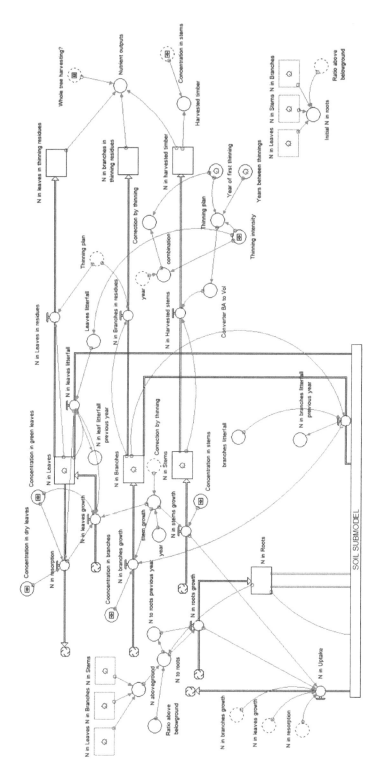

Figure 8.7 *STELLA flow diagram for the tree sub-model of PINEL*

Note: Stocks are drawn as boxes, nutrient flows are drawn as thick straight arrows, and regulators of nutrient flows are drawn as circles connected by thin curved arrows to the flow that is affected.

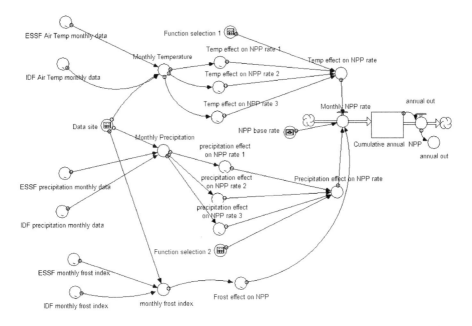

Figure 8.8 *Flow diagram for TPCM*

Note: there is only one flow (NPP), which is modified by a set of regulators (the effects of temperature, precipitation and frost), which in turn collectively determine the single state variable (cumulative annual NPP).

loop in the system. If no feedback loops can be identified, the modeller should return to the flow diagram in Step 2, and ask whether some of the model inputs could be changed from exogenous to endogenous variables (Ford, 1999). However, because of the complexity of forest ecosystems, ecosystem and management models will generally have several or many feedback loops. If the modeller fails to incorporate any, a reconsideration of the conceptual model and its representation in the flow diagram is probably needed.

Causal loop diagrams can be developed automatically from the flow diagram if modelling software such as STELLA or VENSIM is being used. A more detailed description of how to prepare casual loop diagrams, with practical tips and examples, can be found in Ford (1999). An example of a partial causal loop diagram for PINEL is shown in Figure 8.9. It illustrates the concept of positive or negative interactions between the compartments, flows and regulators that are defined in the flow chart (Figure 8.8), and shows several small feedback loops, such as positive feedback between nutrients in uptake and nutrients needed for leaf growth. The more leaves grow, the more nutrients they need, and the more nutrients taken up by trees, the more leaves they can grow. On the other hand, there are some negative feedbacks loops. The more nutrients available in the soil in excess of plant uptake and soil exchange capacities, the more nutrients may be leached away. The more nutrients lost by leaching, the fewer nutrients available in the soil. There is also a second-order positive loop: the greater the uptake of nutrients, the more leaves can be grown, leading to greater leaf mass and therefore more leaf

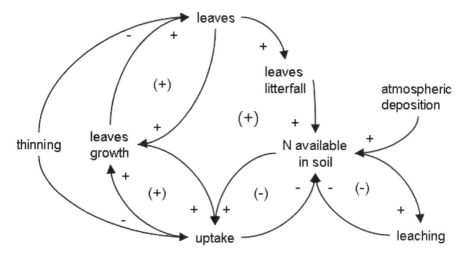

Figure 8.9 *Partial casual loop diagram of the PINEL model*

Note: arrows accompanied with symbols + and - denote positive or negative influences, whereas the symbol between brackets in the middle of a full loop represent positive or negative feedback loops.

litterfall. After decomposition, this results in increased nutrient availability in the soil and potentially more uptake by the trees. However, this second-order loop is modulated by the negative loops related to the nutrients available in soil.

Step 4: estimate the parameter values and starting conditions

Parameters are typically constant values that describe relationships between variables, maximum potential values, etc. Values of *variables* change over time as a model runs. However, when running a model these variables have to have an initial value. These values are the *starting conditions* of the model and they describe the initialization of the system. The process of estimating model parameter values and initial values of variables is usually called *calibration*. This defines the coefficients that determine the magnitude of compartments, flows and regulators. Model performance and outputs are highly dependent on the selection of input data parameters and the starting conditions of variables, and erroneous parameter values can result in significant over- or under-prediction (Larocque et al, 2008).

Wherever possible, model parameters and starting conditions should be derived from field measurements based on good experimental and sampling design. All calibration data should meet acceptable data quality criteria (Crout et al, 2008), and must be relevant and accurate. The range in values for rates of ecological processes are frequently not well known, complicating their calibration, and events at time or spatial scales larger than those at which rates were measured may modify the recorded rates (Richardson and Berish, 2003).

Modellers have to decide the degree of integration of model parameters. Should the several determinants of a process be aggregated into a single parameter, or represented by separate parameters (Nichols, 2001)? The modeller should seek *sufficient parameters*: those that represent implicitly all the factors that result in the functional capabilities required of the model (Levins, 1966), as long as this simplification does not constrain model performance when addressing complex, dynamic issues. Aggregation of parameters in models for forest management should include a structure that accommodates the simulation of forest management practices, and results in predictions that are useful in discriminating among alternative management plans.

Modellers should be prepared for a wide range of uncertainty in estimating parameter values and starting values for variables. Some values may be known with high accuracy (e.g. conversion factors, or well-known physical constants). Others may be known within a narrow range of instrumental or measurement error, and some may be highly uncertain. The answer should be based on how sensitive the behaviour of the model is to that parameter or starting value of a variable, and this should dictate how much effort should be put into estimating a given parameter value (Richardson and Pugh, 1981). However, to know this the modeller should run the model, which in turn requires parameter values and a definition of the starting conditions of the model variables.

To resolve this dilemma, the modeller can pick a range of preliminary values and run the model to establish its sensitivity to this range. The initial values should certainly be selected with concern for accuracy, but they can be adjusted on the basis of the sensitivity analysis, and choosing them should not be allowed to delay the modelling process, even for highly uncertain parameters. Generally, modellers should try to avoid building too many uncertain parameters into their model (Mangel et al, 2001). When values for physical or biological parameters are not known or are measured with great uncertainty, it is important to limit the number of such parameters. With well-defined and independently measured parameters this is less critical, but there will always be a trade-off between model simplicity, the desired level of mechanistic description and number of parameters used.

Field-work in the area where the model is going to be applied is probably the best way to obtain values to define parameters and starting conditions. However, it is usually expensive and time-consuming. Useful information for parameter estimation may be obtained from a variety of sources: scientific journal articles, science-based textbooks or reports of governmental agencies, experts in the field and people with reliable local knowledge. Frequently, the experts are part of the model-building team and they can provide best estimates for uncertain parameters. In some cases, it may be useful to call on a panel of outside experts to advise on the most likely values for parameters. Defining the most suitable initial values for the variables in the model usually implies a very detailed assessment of present and past ecosystem conditions. This is usually very time-consuming and even impossible, especially in complex models that account for multiple ecosystem compartments and ecological processes. A technique that can be used to solve this problem is to run the model in a set-up mode. In this way, the model is forced to run matching the observed patterns of some variables for which there are historical records (usually tree growth), and at the same time the model is producing the values for the other variables that should have existed to generate the observed record. The end of this set-up run provides the starting conditions for the real simulation. This is the approach used in FORECAST (Seely et al, 2002).

Personal intuition is the final resort for parameter values, and may be the only option where measurement of actual values may be difficult or impossible (Kitching, 1983). Such values must be treated tentatively. Modellers may be tempted to omit a parameter to reduce uncertainty in the model, but they should remember that excluding a parameter can be as large or even greater an error than incorporating a parameter with a best estimate of its value and subjecting it to sensitivity analysis.

The importance of parameter estimation in forest management models can be seen in Figure 8.10, which shows the performance of the model FORECAST when simulating the growth of a Douglas-fir plantation in Shawigan Lake (south Vancouver Island, BC, Canada) under different fertilization and thinning scenarios (Blanco et al, 2007). The model was calibrated with two different datasets of values for the same parameters. The first was estimated based on published data for the region, but none of the parameters were specific for the Shawnigan Lake plots. Using this dataset, the model produced acceptable simulations for the control and fertilized sites, but it clearly under-predicted the performance of the thinning treatments. The second dataset, which was a combination of the regional dataset and parameter values estimated by measurements from the control plots at the Shawnigan installation, improved model performance, especially for the thinning treatments.

The sources for parameter estimation for PINEL are described in Table 8.2. Most of the values were obtained from field measurements. However, some crucial data to simulate nutrient cycling (such as woody debris decomposition rates or weathering rates) were obtained from the published literature. The key parameter of atmospheric deposition was obtained from unpublished but public databases maintained by the Spanish Ministry of Environment.

Parameter calibration for TCPM was based on Figure 8.11, which shows hypothetical relationships between climate variables and NPP used to simulate the effects of climate on tree productivity (Lo, 2009). The shapes of the curves and values of the multipliers were a best guess developed from sources such as Daubenmire (1974), Salisbury and Ross (1992) and Cai and Dang (2002).

Modelling objectives influence the methods used in model calibration. In the FORECAST application at Shawnigan Lake, the objective was to predict how thinning and fertilization affects Douglas-fir growth. For PINEL, the objective was to predict the effect of thinning on future soil nutrient pools. Therefore, accurate field measurements were needed to calibrate the models to obtain valid predictions. In contrast, TPCM is a theoretical model with which to explore possible consequences of climate change on tree productivity, and therefore prediction and accurate parameter estimation are not as important as in PINEL or FORECAST.

Step 5: run the model and conduct sensitivity analysis

After preparing all the diagrams described above, estimating the parameters and initial values of the variables and translating them all into computer code, the next step is to test the model: run it to see what happens. Does the model output match the target pattern established in the reference mode (Step 1)? If it does, the modelling project has reached a major milestone and the dynamic hypothesis established early in the process – the conceptual model – can be confirmed (Randers, 1980; Ford, 1999). If the simulation

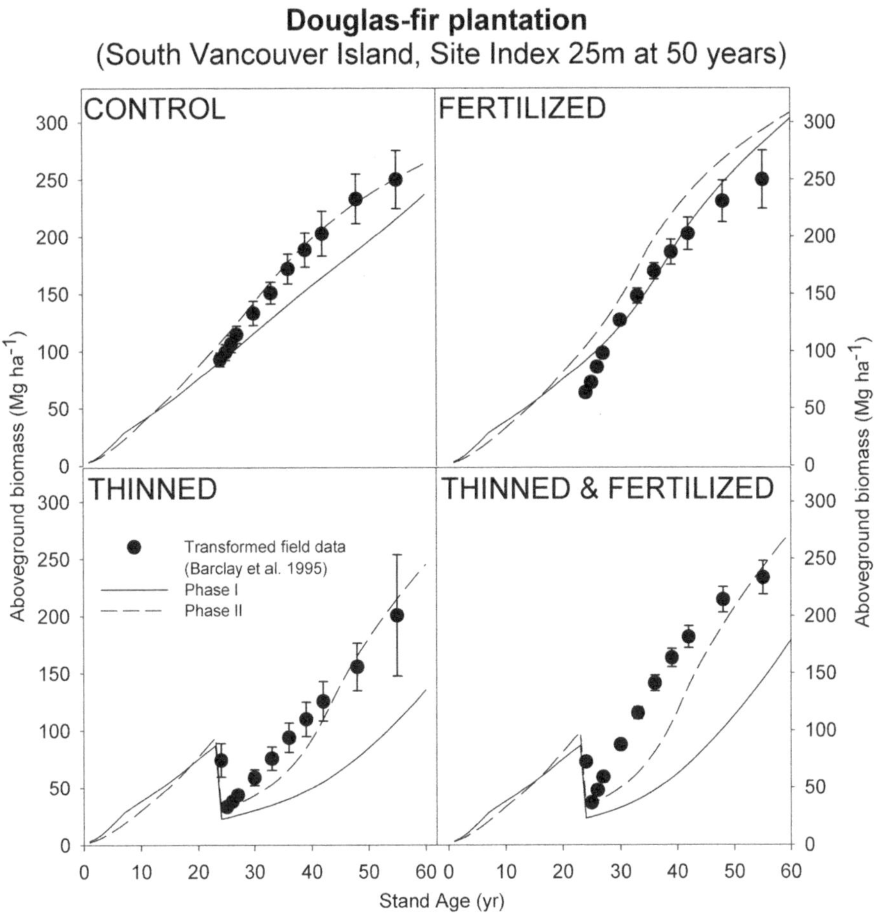

Figure 8.10 *Impact of the specificity of input data (source of parameter estimation) on model performance*

Notes: Circles are field measurements transformed with allometric equations. Solid lines (Phase I) correspond to FORECAST output calibrated with regional data. Broken lines (Phase II) are FORECAST predictions based on site-specific calibration data obtained from the Shawnigan Lake plots.

results do not match the target pattern, it is time to return to Step 2 (modify the flow diagram) or possibly to Step 1 (draw a new reference mode or get more acquainted with the system being modelled).

One run of the model only provides information about how the model behaves with a specific set of parameter values and starting values for variables. To better understand model performance it is necessary to run the model several times with variations in the parameter values and starting states. This is called *sensitivity analysis* and its goal is

Table 8.2 *Sources of parameter estimations values used in the PINEL model*

Parameter	Units	Reference
Nutrient concentration		
green needles	%	Blanco et al (2006)
dry needles	%	Blanco et al (2006)
branches	%	Blanco et al (2006)
stems	%	Blanco et al (2006)
soil	ppm	Blanco et al (2006)
Decomposition rate		
needles	years^{-1}	Blanco et al (2006)
branches	years^{-1}	Ägren and Bosatta (1996)
roots	% year^{-1}	Malkönen (1974)
Litterfall fraction		
needles	% year^{-1}	Blanco et al (2008)
branches	% year^{-1}	Blanco et al (2008)
Atmospheric deposition	kg ha^{-1} year^{-1}	Ministerio de Medio Ambiente (2003)
Weathering rate	kg ha^{-1} year^{-1}	Kimmins (2004d), Fisher and Binkley (2000)
Nutrient resorption rates	%	Blanco et al (2009b)
Ratio above:below-ground biomass	%	Malkönen (1974)
Initial above-ground nutrient stocks	kg ha^{-1}	experimental
Initial soil nutrient stocks	kg ha^{-1}	experimental

to learn if the basic pattern of the results is sensitive to changes in the uncertain parameters and initial conditions. Sensitivity analysis should be done early and often in the modelling process, and not only when the model is considered a finished product (Crout et al, 2008).

In sensitivity analysis, it is common to produce a ranking of the factors influencing the target variable (Håkanson, 2003). The basic idea is to identify the most sensitive part of the model: the part that is most decisive for the model behaviour. It is useful to check if the reference mode is supported after each test. If it is, the modelling process has reached another important milestone; the modeller can assume that the model is 'robust', which means that it generates the same general pattern despite the uncertainty in parameter values (Ford, 1999). If it is found that the model changes its basic pattern of behaviour with changes in parameter estimates for uncertain parameters, it is time to return to Step 4 (parameter estimation).

Initial sensitivity analysis should be conducted in a one-at-a-time fashion, varying each parameter that has a significant uncertainty factor by, for example, 5 per cent, 10 per cent, 15 per cent or more, up and down. This is not necessary for parameters that are considered to have high certainty; there are major differences among variables, and model objectives will also help determine which parameters and variables are subjected to such analysis. Variables related to tree morphology (height, diameter, etc.) can often be determined very accurately, while some other variables (such as woody debris decomposition rates, foliage mass, animal migration rates) have to be estimated from laboratory tests, fieldwork or theoretical considerations, which means that the values for such variables or parameters are often very uncertain. Modellers should pay particular

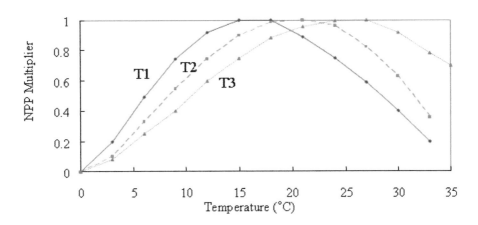

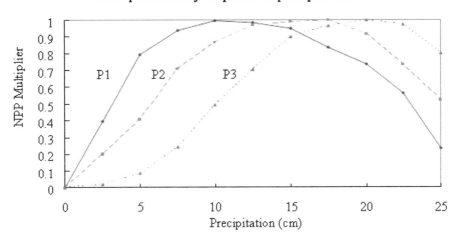

Figure 8.11 *Hypothetical relationships between climate variables and tree productivity (expressed as a multiplier of a base rate of NPP)*

attention to the best guess parameters and be sure to allow these parameters to vary across a wide range of uncertainty (Ford, 1999).

Comprehensive sensitivity analysis can be conducted by creating factorial experiments, in which several parameters are changed at the same time, especially if there may be interactive effects (Starfield and Bleloch, 1991). For stochastic models, which can yield different results in every run even with the same set of input values, sensitivity analyses are more accurate if the stochasticity is switched off (Grant et al, 1997). After the model sensitivity is established for individual parameters, the modeller can change several parameters at the same time to look for possible interactions and synergic effects.

Step 6: model testing – validation and evaluation

Before using a new forest management model, it should be tested to see if the output is useful for the intended objective (*evaluation*) and to test if the conclusions obtained with the model can be confirmed by independent data (*validation*). Model output must be evaluated by comparing it with one or more datasets that define how the real system being modelled has changed over time under various natural and/or management-induced disturbance regimes. Because models are simplified, incomplete representations of reality, it is not necessarily to prove that a model is exactly 'correct': i.e. that it precisely duplicates reality (Oreskes et al, 1994; Sterman, 2002; Oreskes, 2003). Model evaluation is an exercise to show that predictions are close enough to independent empirical data to render them useful for specific applications, and that decisions based on model output are reliable and defensible (Popper, 1963; Soares et al, 1995). In a strict sense, the evaluation is acceptable only for the condition under which it was conducted. Nevertheless, the greater the number of cases of good agreement between observed and predicted values, the greater the confidence in the model (see Oreskes et al, 1994; and Rykiel, 1996, for further discussion).

Model evaluation is a continuing process (Mayer and Butler, 1993). A model should be evaluated as soon as it is operational and in each of the successive iterative stages of development towards the ultimate modelling goal (Grant et al, 1997). Evaluation tests can be applied at the design stage (Step 3), before scenario analysis (Step 7) and also during model operation (Rykiel, 1996). Forest management models should be revaluated and revised or recalibrated as new information becomes available, or if the model is applied to a new situation or issue that is beyond its original intended application.

Model validation criteria that apply to models intended to be used for explanatory purposes include: qualitative agreement between the real system and the model (Power, 1993), confidence that an inference about a simulated process is equally valid for the actual process (Van Horn, 1969), reasonableness of the model's structure, mechanisms and overall behaviour (Grant et al, 1997), ability to explain the system's behaviour (Marcot et al, 1983), and the absence of known flaws (Oreskes et al, 1994). The emphasis in these criteria is on the model itself and its veracity, not on output from the model. For some modellers the issue is not the truth of the model, but whether or not it generates relevant and testable hypotheses (Levins, 1966): a scientific rather than a practical objective. The validation criteria for models intended for predictive and decision-making purposes are: the usefulness and accuracy of predictions from the model (Marcot et al, 1983) and the quantitative correspondence between the model's behaviour and the behaviour of the real system (Grant et al, 1997). Criteria that apply equally well to explanatory, predictive and decision-making models are the model's usefulness and whether the accuracy of its predictions is sufficient for its intended use (Rykiel, 1996).

Appropriate methods for validating models have engendered considerable discussion (Gardner and Urban, 2003), but there is general agreement that predictions should be tested against data derived independently from those used to develop the model (Aber, 1997). Simply matching the data used in the model's development provides only replicative validity (Johnson, 2001). It is also bad modelling practice to validate the model with the data that were used for calibration and verification (Council for Regulatory Environmental Modeling, 2009). Statistical tests can be performed on both predictions and empirical data, and indices of model performance can be computed based on them.

Ensuring that the temporal and spatial characteristics of the model output match the independent validation dataset is important to prevent errors that result from using the model under conditions or in ecosystems that differ in important ways from those for which the model was developed. Success in predicting for these new situations assures what has been termed *predictive validity* (Power, 1993). In some cases, two datasets are used, with one being used for calibration and the other one for validation. Examples of this validation technique are assessments of the predictive performance of FORECAST in plantations in BC (Blanco et al, 2007) and Cuba (Blanco and González, 2010).Validation datasets can also come from other studies of similar systems that have provided time-series data (Richardson and Berish, 2003). This method was used by Bi et al (2007) to evaluate FORECAST performance when simulating long-term changes in soil organic matter in Chinese fir plantations in south-eastern China. Statistical tests can be performed on both predictions and empirical data, and indices of model performance can be computed based thereon (Larocque et al, 2008).

Sometimes model results are compared not to actual data, but to output from another model (Rykiel, 1996). Getting comparable results from different models may be reassuring, especially if the credibility of one of the models has already been established. And if two models built on different sets of assumptions lead to similar conclusions, we have more confidence in the robustness of these conclusions. However, different results from two models indicate that at least one of the models is inadequate (Johnson, 2001), but it can be difficult to decide which one in the absence of field data. One example of this technique is provided in Kimmins et al (2008a). In that work, the authors compared the outputs from FORECAST with the model TASS (DiLucca et al, 1999) at different levels of FORECAST complexity, calibrated for boreal forests of *Picea glauca* in north-eastern British Columbia. Both models give similar results when FORECAST is simplified to the level of TASS (running as a light-competition model only), but deviates as complexity is added (Figure 8.12). Over longer periods (more than 100 years of simulation), there is little difference between models and complexity does not add much. However, for conventional management-length timespans that are usually shorter than 80 years, adding complexity greatly changes FORECAST behaviour. Thus, validation is relative to the interacting effects of complexity and time. The fact that the two models agree over the long term and at equally low complexity is not a useful validation for shorter timescales and more realistic representations of the ecosystem (Kimmins et al, 2008a).

For evaluation, the modeller must specify the purpose of the model, the criteria to be met for the model to be deemed acceptable for a particular use, and the context in which the model is intended to be used (Rykiel, 1996). Statistical validation is not always required, depending on the model objectives. The quality of a model depends not on how realistic it is but on how well it performs in relation to its purpose (Starfield and Bleloch, 1991). Therefore, the results of model evaluation are always relative, depending of the model user's needs for accuracy.

Johnson (2001) recommends following several steps to achieve acceptable model evaluation. First, it is necessary to examine the reasonableness of the model's structure and its representation of mechanisms. The predictions should be checked to see if the

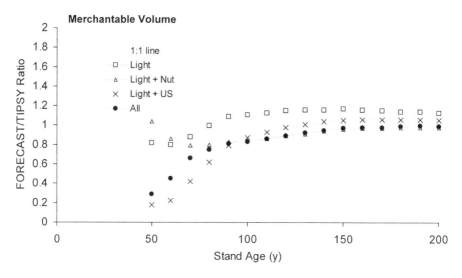

Figure 8.12 *FORECAST/TIPSY output value ratios for merchantable volume of white spruce when light competition, nutrient limitation and/or understorey herb competition, and nitrogen and grasses, are added to the FORECAST simulation*

Note: Light = light competition; Nut = nutrient limitation; US = understorey herb competition (*Calamagrostis*). TYPSI is a version of the TASS model (DiLucca et al, 1999)

model predicts impossible values and if it can produce plausible results under extreme conditions. For forest management models, it is also necessary to test if the simulated patterns make ecological sense. The modeller should then examine the quantitative correspondence between the model's behaviour and the real system (Grant et al, 1997). To do this a model should produce output that describes the temporal patterns of variation in key ecosystem variables, something that is emphasized in the FORECAST output.

While several evaluation techniques have been proposed, they can be grouped into four types (Mayer and Butler, 1993):

- *Subjective assessment.* Using expert opinion, asking experts their opinions about the model behaviour, or having a Turing test with experts to distinguish modelled from real values can provide a subjective evaluation of model performance (Rykiel, 1996). However, all these techniques are subject to personal bias and should be avoided if possible and when other options are available (Johnson, 2001).
- *Visualization techniques.* This category includes graphs and plots of real versus modelled variables, differences between actual and predicted values, comparing quantile plots or time series of observations and results (Johnson, 2001). Such methods are still subjective and may be misleading (Mayer and Butler, 1993; Rykiel, 1996), but they can be a good start and provide a general sense of how the model is performing. They can also be directly compared with the reference mode (Step

1). These techniques have been used to evaluate forest management models such as FORSKA (Linder et al, 1997) and FOREST-BGC (Lucas et al, 2000).

- *Measures of deviation.* When observed and predicted data can be paired by time, location, treatment or other variable, diverse measures of deviation can be used, such as average and absolute bias. Root mean square of the bias reflects the absolute difference between actual and modelled values using the same units and accounting for both bias and imprecision (Johnson, 2001). Examples of the use of these measures of deviation can be found for FORECAST (Blanco et al, 2007; Blanco and González, 2010) and MTCLIM (Lo, 2009).

- *Statistical tests.* A goodness of fit test between observed and predicted values provides a quantitative validation test (Power, 1993; Vanclay and Skovsgaard, 1997; Yang et al, 2004). Evaluation of linear regressions using the coefficient of determination (r^2) has been widely utilized for assessing model performance. Examples include PBRAVO (Soares et al, 1995); PnET-II (Ollinger et al, 1998); CenW (Kirschbaum, 1999); SORTIE (Beaudet et al, 2002); 3-PG (Landsberg et al, 2003); FOREST-BGC (Lucas et al, 2000; Merganicová et al, 2005); PINEL (Blanco et al, 2005); and FORECAST (Blanco et al, 2007; Seely et al, 2008). Theil's inequality coefficient (Theil, 1966; Power, 1993) can also be used to asses the adequacy of the model's fit. This coefficient has been used to evaluate FORECAST (Blanco et al, 2007) and MTCLIM (Lo, 2009). Use of t-tests, f-tests, Kolmogorov-Smirnoff tests, contingency tables and/or correlation coefficients are other possibilities for model evaluation (Sterba and Monserud, 1997; Lasch et al, 2005). Measures of model accuracy can be assessed with tests described in Freese (1960) or Reynolds (1984), and examples for FOREST and SHAF can be found in Ek and Monserud (1979). The aforementioned are tests based on the null hypothesis of similarity between the system and the model, an assumption that is arguably too strict in many cases (Robinson and Froese, 2004; Yang et al, 2004). Tests of equivalence have also been developed that use a null hypothesis of inequality of means (Welleck, 2003). This approach was used to test FVS (Robinson and Froese, 2004) and FORECAST (Blanco et al, 2007).

An example of graphical and statistical evaluation of a model is illustrated in Figure 8.13. In these graphs, six ecosystem variables are predicted by the model PINEL and compared with independent data. The original experimental sites contained nine plots in each site: three were reference plots, three were under light thinning and three under heavy thinning. The model was calibrated and developed with data from the control plots, and its usefulness was evaluated by simulating the same two types of thinning in the control plots and comparing the results with the data from the actual thinned plots (for details, see Blanco et al, 2005).

As can be seen in Figure 8.13, additional tests would be needed to check model performance for nutrients in soil and nutrients in leaf litterfall, because the graphs indicate obvious non-linear relationships between observed and predicted data, contrary to what would be expected. However, the r^2 value, which provides a measure of the variance in observations explained by the model, was high (>70 per cent) for all variables.

Similarly, a detailed evaluation of the model FORECAST for a Douglas-fir plantation in southern Vancouver Island can be found in Blanco et al (2007), and in the

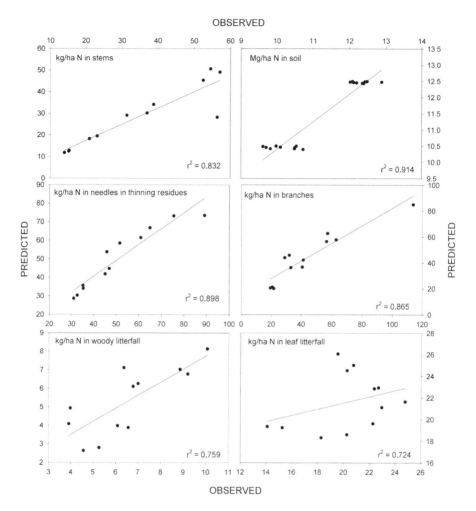

Figure 8.13 *Example of graphical and statistical evaluation of PINEL output; observed values for six variables compared to model predictions*

additional material for this chapter available online (see below). Additional evaluations of FORECAST for boreal conditions can be found in Seely et al (2008), and for tropical conditions in Bi et al (2007) and Blanco and González (2010).

Step 7: scenario analysis

At this point in model development, the new forest management model has been proven to be ecologically consistent, logically well founded, and robust to uncertain parameter calibration; and the difference between real and simulated systems have been evaluated. Therefore, the modellers can assume that they have a tool that is ready for decision-support and scenario analysis. Scenario analysis develops data and information that can be used to support management decisions with respect to choices between alternative management policies and practices. In contrast to validation, which requires that predictions be limited to the same temporal and spatial scales as the input data, scenario analyses are often extended to larger areas and timescales. However, modellers should always remember that predictions outside the time and spatial domain of the input data may not accurately reflect conditions in the real world.

Scenario analysis involves running a model several times with variations in management variables. This will reveal which management option is likely to result in the most desirable changes in the simulated system (Ford, 1999). Once a promising management option is identified, the modeller returns to Step 5 (sensitivity analysis) to learn if the policy performs well under a wide range of values assigned to the uncertain parameters. Also, if significant random variations in parameters or natural disturbance events were ignored up to this point, the modeller should now introduce such randomness to produce a more realistic setting for the policy analysis. If the simulated management outcomes are encouraging, the modeller will probably wish to define the policy variables in a more detailed manner. At this stage, it is possible to return to Step 2 (flow diagram) to describe policy in more detail.

Model-based decision making is strengthened when the underlying science is presented via comprehensive documentation of all aspects of a modelling project and effective communication occurs between modellers and decision makers. This re-emphasizes the basics of good modelling practice, discussed above. Proper documentation enables decision makers and other users of models to understand the process by which a model was developed, its intended area of application, and the limitations of its applicability (Crout et al, 2008).

An example of scenario analysis with PINEL calibrated for two different Scots pine stands in the western Pyrenees (northern Spain) is shown in Figure 8.14. Accumulated outputs on N (mesh line in Figure 8.14) are compared with accumulated N inputs from atmospheric deposition (the contour thick line in Figure 8.14). If the accumulated losses are above the thick line, the ecosystem will lose nutrients in the long term. The scenarios were different combinations of thinning intensity, type of thinning and time between thinning events, with the target variable the amount of N lost from the system after 100 years. The main conclusions were that thinning that only removes the stems will not deplete N soil reserves in the long term (the accumulated losses are always below the accumulated inputs), whereas more intensive biomass removal during thinning poses a real threat to nutrient reserves (some scenarios are above the plane of accumulated inputs), and therefore to long-term site fertility. This threat was predicted to be greater in site 1 than site 2, suggesting that thinning practices should be site-specific. The model revealed non-linear relationships between thinning intensity, thinning rotation and nutrient removal. Had thinning regimes been based on simple linear relationships, these non-linearities would have not been detected (Blanco et al, 2005).

Uncertainty and complexity in model development

Developers of models for forest management face two major challenges. First, the problem of many uncertain parameters can paralyse a modelling project if a compromise between accuracy of parameter estimation and model objectives is not reached. The trust that model users will have in a model will depend on how uncertain they perceive its

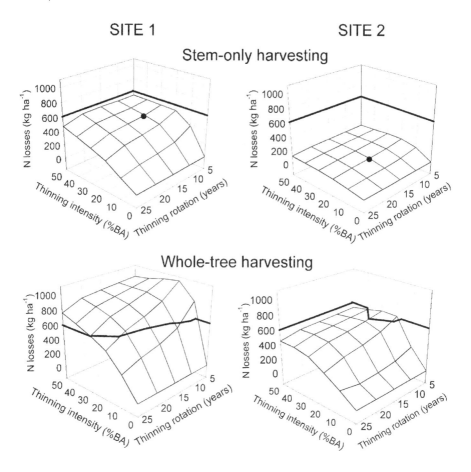

Figure 8.14 *Results of scenario analysis with PINEL show the amount of nitrogen lost from the forest ecosystem for different combinations of thinning intensity (basal area removed), time between consecutive thinning events (thinning rotation) and type of thinning (stem-only or whole-tree) for two different sites*

Note: The contour thick line marks the accumulated N deposition (in kg/ha) in 100 years in each site.

predictions to be, and much of the criticism of models originates from disagreement over the importance the model developer gave to uncertain parameters and their calibration. Second, when developing a new model for forest management there is always the temptation to create a model that accounts for many or even all of the known process and compartments in the ecosystem. Excessively complex models have little practical value for managers, but, equally, overly simple models will provide an unreliable basis for planning sustainable forest management. If key processes are excluded, the model may prove to be unsuitable for scenario analysis, as was discussed in Chapter 2.

Dealing with model uncertainty

The term 'model uncertainty' is generally used to describe uncertainty about whether a particular model is capable of emulating the system it represents. Using a model that lacks the ability to address key management issues and questions because of an incomplete conceptual model or a structure that is not appropriate for the user's objectives, or calibrating a model with unreliable calibration data are contributors to this uncertainty. Model uncertainty analysis investigates the effects of an incomplete conceptual model relative to the modelling objectives, and the uncertainty associated with model parameter values (Council for Regulatory Environmental Modeling, 2009). When conducted in combination, sensitivity and uncertainty analyses can inform model users about the confidence they can have in a model's predictions (Mangel et al, 2001).

When developing predictive models in forest management, it is important that the user is aware of uncertainties associated with the model being used. A predictive model is often assumed to be valid within the spatial and temporal domain for which the model has been calibrated and validated. For the model to be valid outside this domain, it must be assumed that its conceptual foundation is valid. Using models for prediction with limited knowledge of model structure and complexity relative to the system and issues being modelled poses substantial uncertainty, which must be reduced or at least be known for the model to be useful (Refsgaard et al, 2005).

Predictive models are generally subject to input, model and parameter uncertainty. Uncertainties in model inputs are due to measurement errors and natural variability. Sources of error include: poor sampling design (e.g. too few or too small numbers, size and dispersion of plots), measurement error, errors in data summary and errors in data extrapolation.

Clearly, when models are used for prediction, uncertainties should be explicitly recognized and their consequences for the predictions evaluated. This is especially important when evaluating complex forest management models with multiple output variables. For example, Blanco et al (2007) found that FORECAST performed well in predicting trends in variables directly measured in the Shawnigan Lake plantation study, but not as well for some variables (branch and foliage biomass) whose values were obtained from regression models. This raised the question of whether differences between FORECAST output and field-based estimates were the result of poor ecosystem model performance, poor regression model performance, error in field measurements, or some combination thereof.

In the case of models for explanatory purposes, uncertainty analysis can inform users about potential risks as well as potential opportunities, and the conditions under which certain outcomes are most likely to occur (Brugnach et al, 2008). Thus, in explanatory models uncertainty may be regarded as a source of innovation.

The sources of uncertainty are multiple (Table 8.3). Traditionally, the focus has been on uncertainty in data and model structure. One type of uncertainty that has received limited attention in the literature is the uncertainty associated with human input, although this type of uncertainty can have a significant impact at all stages of the environmental decision-making process. However, there is an increasing recognition that the uncertainties associated with human factors, such as modeller experience and expertise, also need to be taken into consideration (Maier et al, 2008).

Table 8.3 *Sources of uncertainty in forest management models (modified from Maier et al, 2008)*

Category	Example
Data	Measurement error
	Type of instrument
	Quality and frequency of calibration of instrument
	Data reading and logging
	Data transmission and storage
	Type of data recorded
	Length of data recorded
	Type of data analysis and processing
	Data presentation
Models	Conceptual model and modelling method
	Type, quality and length of data record available
	Calibration method and data used
	Evaluation method and data used
	Input variability
Human	Knowledge, experience and expertise of modeller
	Political influence and perceived importance of stakeholders
	Knowledge, values and attitudes of stakeholders
	Values and attitudes of managers/decision makers
	Current political climate

Uncertain parameters imply uncertain predictions, and uncertainty about the real world being modelled implies uncertainty about model structure and parameterization. Because of these linkages, model parameterization, uncertainty analysis, sensitivity analysis, prediction, testing and comparison with other models should include some quantification of uncertainty.

There is structural uncertainty associated with forest management models (Larocque et al, 2008). Simple models often suffer from being too simplistic for management, but can nevertheless be illustrative and educational in terms of thinking about ecosystems. Complex models can reproduce the complex dynamics of forest ecosystem in detail. Complexity makes their uncertainty evaluation difficult, as it is difficult to isolate and analyse a multitude of potential uncertainty sources. However, complexity should not be an excuse to avoid carrying out uncertainty analysis. The need to quantify output uncertainty and identify key parameters and variables remains important.

Dealing with complexity

Like other branches of science, forest science has long been dominated by the principle of parsimony: when faced with two alternative hypotheses, choose the simpler (Kimmins et al, 2008a). The need for traditional experimental testing of hypotheses has required that the complexity of real ecosystems be reduced to a level at which the requirements of statistical techniques are satisfied, and the spatial, temporal and fiscal demands of experimental design can be accommodated. Advances in ecological understanding, computer power and modelling have provided an alternative to this traditional approach to science and provided the ability to test complex hypotheses in which differences between individual organisms, species, and ecosystems can be represented. This fact contrasts with the opinion of some modellers who think that when building a model, there should be a focus on the similarities and generalities (Nichols, 2001). While this may be appropriate for theoretical modelling, we believe that the right approach to modelling for sustainable forestry must include multi-value resource management in which there is never a single 'right' answer, but only difficult choices to be made between alternatives that can be explored through scenario analysis.

Modern justifications of parsimoniousness have presupposed the old principle: 'Do not posit more entities than necessary' (Sober, 1981). However, Occam's razor has two edges, and the key phrase for ecology and forest ecosystem modelling in Occam's admonition is 'more than necessary', a thought echoed by Albert Einstein when he said 'make everything as simple as possible but not simpler'. Both these quotes can be summed up as: 'as simple as possible but as complex as necessary'.

Peters (1991), in his *Critique for Ecology*, noted that the ultimate measure of success for ecology is prediction. Rowe (1961) asserted that prediction in ecology must be done at the ecosystem level, the important ecological sub-disciplines of autecology, ecophysiology, population ecology and community ecology being more to do with understanding of individual ecosystem components and processes at these levels of biological organization than the prediction of entire ecosystems (Kimmins, 2008a) (see Chapter 2).

Prediction in forestry requires prediction at the ecosystem level over multiple temporal and spatial scales (Kimmins et al, 2008a). This in turn requires predictive tools that contain sufficient complexity to address the planning needs of forestry across these scales. Most issues in forestry are complex because they are related to ecosystems and not merely to their sub-components, which do not exist in the real world outside of their ecosystem context. They are also complex because forestry is as much about people as it is about ecosystems (Westoby, 1987). The principle of parsimony has served many branches of science well, but can be a double-edged sword for both forestry and ecology; it can act as a barrier to reaching the ultimate goal of prediction because it promotes fragmentation of what is a system, and it encourages the tendency to over-simplify. In this context, modelling can be seen as a filter which retains the variables and processes that are essential to the modelling objective.

Kimmins et al (2008a) explored the effect of the level of complexity at which the FORECAST ecosystem model was run on predictions of stand productivity. FORECAST was run as a light-driven, a light + nutrient-driven, a light + minor vegetation competition-driven, and a light + nutrient + minor vegetation competition-driven model for a coastal Douglas-fir plantation in BC by activating or deactivating different sub-modules. The model output was compared with long-term field measurements, and for four different stand management treatments (Figure 8.15). The modular structure of FORECAST that facilitates such explorations is described in Kimmins et al (1999) and in Chapter 5.

The predictive ability of FORECAST increased when more complexity was added to the model, with the full model and the light + understorey version generally showing the closest approximation to the field data (Figure 8.15). However, the ranking of the different versions of the model varied with stand age and treatment, and in some cases the gains from adding complexity were relatively small. Clearly, the level of complexity required for the simulation of this Douglas-fir plantation is a trade-off between the

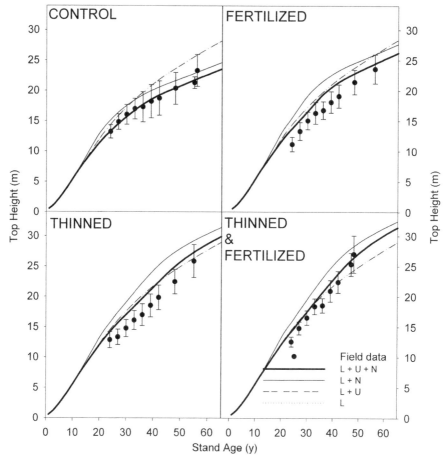

Figure 8.15 *Impact of model complexity level on model output*

Note: circles are field measurements. Lines are model predictions when different modules of FORECAST are activated: L = light; N = nutrients; U = understorey.

objectives of and the accuracy required by the model user, and the resources available for model calibration.

This exercise was repeated for a Chinese fir plantation in south-eastern China, looking at sustainability of stem mass and height growth in successive rotations, with six different levels of FORECAST complexity (Figure 8.16). The objective of this analysis was to explore the different factors that may contribute to the the rapid yield decline observed in successive short rotations in some plantations (Bi et al, 2007). There was a significant variation in model predictions at different levels of complexity, the single most

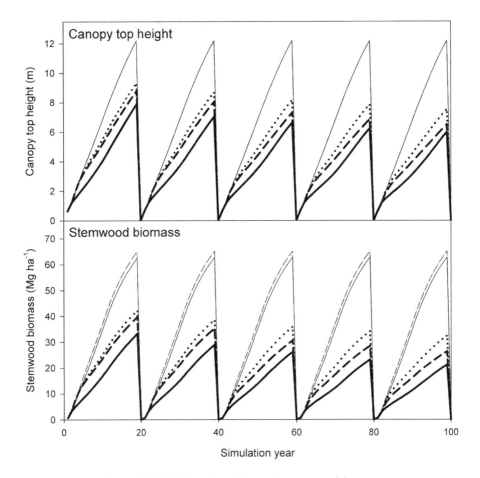

CHINESE FIR PLANTATION

Poor site (SI 17m at 50 years)
Rotation length 20 years
Slash burning after harvesting

Level of complexity simulated

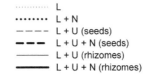

Figure 8.16 *Effects of model complexity on model output*

Note: Lines are model predictions when different modules of FORECAST are activated: L = light; N = nutrients; U = understorey growing from seeds or rhizomes.

important complexity factor being understorey competition for light and nutrients. The understorey regeneration type (growth from seed germination vs vegetative growth from rhizomes) had clear implications for model predictions.

The conceptual model of this phenomenon is shown in Figure 5.21. It suggests that Chinese fir plantations quickly become shrublands through a process of site fertility depletion caused by slash burning, litter raking, conversion of mixed woods into a conifer monoculture, and soil erosion. For tree height and stem biomass, the only projections that matched the field observations and the conceptual model were provided at the maximum level of complexity (light and nutrient competition between trees and understorey, and understorey regenerating from rhizomes). Note that the model run as a light model with understorey regenerating from seeds showed no yield decline, in strong disagreement with field measurements.

The results from these two applications of FORECAST supported Occam's razor: forest management models should be as simple as possible but as complex as necessary for the particular modelling assignment, ecosystem type and values being predicted. Kimmins at al (2008a) noted that one of the benefits of FORECAST is its implicit simplicity, which is provided by its explicit complexity; the model provides a rich library of output describing a large number of tree, plant, soil and ecosystem variables. This greatly facilitated the assessment of the comparative runs and the interpretation of the predictions. The weakest component of the datasets used to calibrate FORECAST for these investigations was the minor vegetation, which traditionally has received less documentation than the trees. This suggests the need for improved datasets for understorey vegetation in ecosystems subject to frequent natural or management-related disturbance, because the competitive and other negative effects of these life forms on trees are often much more important than has been assumed. The need for careful assessment of the importance of understorey in forest management models is also important in systems where allelopathy could play a role in interspecies competition, such as conifer forest with an ericaceous shrub layer (Blanco, 2007).

Models that lack detailed output render comparisons between ecosystem simulations and field data difficult. Verification of ecosystem models requires a rich output of soil and plant variables, because rarely there are long-term records for more than a few tree variables. In the absence of such records, richness of output becomes a key component of model verification for ecosystem models, and provides a means by which the behaviour of the simulated output can be compared with our current understanding of various ecosystem components and processes.

The take home message from Kimmins et al (2008a) was that evaluation of the effects of increasing model complexity is not independent of the simulation of management-induced or natural disturbance regimes, and the effects vary according to stand age and timing of disturbance events. The addition of any individual additional factor may not have a significant effect without the inclusion of other additional factors. These interactive effects suggest that while adding complexity to simulations of forest succession in the long-term absence of disturbance may not add much resolution to a model, for forest ecosystems subjected to relatively frequent and severe ecosystem disturbance (either natural or management-related), the level of complexity of a forest stand model may be of great significance.

The simulation of complex management scenarios (e.g. short rotations, slash burning, and resilient and highly competitive understorey) requires a relatively high level of model complexity. For the simulation of climate change or management treatments that alter site moisture, this additional complexity should be added (Lo, 2009). The determination of the optimal level of model complexity is made by making appropriate trade-offs among competing objectives (Crout et al, 2008).

Take-home message

Modelling is a personal exercise that involves the knowledge and expertise of the modeller in the system to be modelled, the definition of the system, the availability of the data to represent the ecosystem processes and the intended use of the model. Guidelines for good modelling practice can be followed to ensure the maximum possible credibility of the modelling process and the results of scenario analysis. In addition, although there are different approaches to creating new models for forest management, any new model should be subjected to sensitivity, uncertainty and evaluation tests prior to its use for scenario analysis in real forest management situations.

Additional material

Readers can go to the companion website to this book (www.forestry.ubc.ca/ecomodels) to access the full versions of the following additional material:

- PINEL model: STELLA files, user manual and documentation.
- TPCM model: STELLA files and documentation (Lo, 2009).
- Poster on the conceptual model and scenario analysis of the effects of allelopathy on forest ecosystems (Blanco, 2007).
- Slideshow on the evaluation of FORECAST for Douglas-fir plantations in Canada and Caribbean pine in Cuba.
- Slideshow on the need for complexity in forest management models (Kimmins et al, 2008a).

Note

Yueh-Hsin Lo (School of Forestry and Resource Conservation, National Taiwan University, No. 1, Sec. 4, Roosevelt Road, Taipei, 10617 Taiwan, ROC) collaborated in writing this chapter.

Chapter 9

The Role of Ecosystem Management Models in Adaptive Management, Certification and Land Reclamation

Introduction

The forestry practices that characterized much of the second half of the last century are currently under broad revision, fundamentally because of changes in social values, economic conditions and technological capabilities. Based on the principles of ecosystem-based management (EBM; Kimmins 2007b; http://canadaforests.nrcan.gc.ca/article/eco systembasedmanagement), this 'new forestry' paradigm is slowly gaining acceptance and broader application (Kimmins, 2008a). To implement EBM, ecological knowledge from a wide range of disciplines must be integrated within a decision-making framework to evaluate the synergistic and antagonistic effects and trade-offs among components of an ecosystem under different management strategies. This integration can be difficult, either because individual human experience is generally restricted to one or a few disciplines (Yamasaki et al, 2002), or because the prevailing institutional arrangements either do not allow or encourage it. Failure to achieve such integration can result in the different components of the ecosystem being managed in relative isolation, albeit on the basis of their individual ecologies. Widely accepted as a concept, EBM often fails to achieve its full potential as a result. Hybrid, multi-value, ecosystem-level computer simulation models that combine empirical information with the key processes that drive ecosystem behaviour represent a tool for integrating a diverse array of information and data into a common tool for analysing forest ecosystems and their management at various scales.

Effective use of simulation modelling in the analysis of living systems requires knowledge of their key biological and ecological processes, and a means of expressing the relationships between these processes. A well-constructed model is thus more than simply a descriptive tool; it also possesses explanatory power through its integration of empirical work on ecological processes, and provides insight into the dynamics of complex ecological systems.

Three issues to which hybrid ecosystem-level forest management models have been applied, and which are discussed in this chapter, are adaptive management, forest certification and land reclamation.

The role of forest ecosystem management models in adaptive management

The origins of adaptive management

Until recently, many or even most forests were managed simply as a source of solid wood and wood fibre. Public expectations and needs have changed, however, and most forests are now managed for a variety of ecosystem values, goods and services. We lack a long period of such new and as-yet untested and unproven management systems that could provide experience-based guidance to foresters. Traditionally, forestry has relied on such experience to suggest which management practices will sustain a diversity of desired values across landscape mosaics of diverse types of ecosystem. As a consequence, the new management approaches that are being advocated must be considered essentially as experiments. This is the core concept of adaptive management.

The concept of adaptive management (AM), originally developed for fisheries and wildlife management by Walters (1986), is based on the idea of learning by doing, and trying different approaches in different areas and different types of ecosystem. As management experiments, Walters suggested that each specific management recommendation requires a hypothesis, based on the best available science, as to what the outcome of the recommended management is likely to be. The results of the management are monitored periodically and compared with both the objectives and the hypothesis. This leads to modifications of the practices for those objectives in the type of ecosystem in question where the outcomes are not what was desired, and a revisiting of our current understanding of the ecosystem response to the management where it is clear that the hypothesis is not predicting what in fact occurred.

Walters's conception of adaptive management was thus careful design of management experiments, creation of models that constitute the hypothesis, monitoring the results, and modifying – adapting – the management and/or the models following a comparison between outcomes and both objectives and model predictions. Modelling has thus been a central component of AM from the outset.

The earliest stages of AM in fisheries and wildlife management involved bringing resource managers, modellers and scientific experts together for a series of brief, intense workshops to produce a simulation model of – a hypothesis about – the system being managed (Holling, 1978). Simulation models require the explicit description of ecosystem interactions in a quantitative, measurable way (Duinker, 1989) thereby improving the managers' understanding of the system.

Adaptive management in forestry

Because the timescales considered in AM analyses of human endeavours such as fisheries are generally fairly short – one or two decades for example – the concept of 'mid-course corrections' is both appropriate and effective. Changes in number of fishing boats, net sizes and the length of fishery openings can be expected to result in measurable changes in fish populations and catch success within two to three natural cycles of abundance (3–4 year cycles for most species of salmon) covering an adaptive management period of perhaps 6–12 years, in the case of salmon. Management can be changed, models improved and new predictions made and tested within a subsequent period of equivalent length. In contrast, in forestry the natural dynamics of ecosystem processes and stand development involve periods of many decades or even a century or more before the outcome of new management practices for a wide range of ecosystem structures, processes and values can be measured directly. The period involved in two or three disturbance–recovery cycles so far exceeds the normal planning horizon, public understanding and tolerance and human longevity that the empirical component of AM (mid-course correction) is much less effective in forestry than in shorter timespan endeavours (such as fisheries, wildlife management or agriculture). As a consequence, the modelling and predicting component in AM in forestry is much more important.

Simple models are generally regarded by many people as more appropriate than more complex but potentially more realistic models. Complex models may be more vulnerable to mis-specification than simpler ones, whilst simple models require fewer data, are quicker to develop, and may be easier to compare with alternative models. Complex models are not amenable to being developed in the short-term workshops originally proposed for AM planning processes for fisheries and wildlife managers. The purpose of modelling in such processes was not to build realistic representations of reality, but to develop simplifications of reality that were useful for the specific purposes identified (Sainsbury et al, 2000). However, as we have noted elsewhere, a model should incorporate explicitly the complexity that drives the behaviour of the target variables in the system in question. Management based on over-simplified models will almost certainly fail (discussed in Chapters 2 and 8), and this is particularly true in forestry. The difficulty with the long timescale of forestry is that the failure of overly simple models and of the resulting unsustainable management may not become measurable in sufficient time to prevent a significant period of non-sustainable practices and associated losses of desired conditions and values.

Because sufficiently complex models are unlikely to result from short workshops with non-modellers, such AM workshops in forestry, where they occur, will generally utilize existing models of adequate complexity, or meta-models (see Chapter 10). The focus of such forestry AM workshops is 'gaming' (playing) with one or more simulation models of the system of concern as a means of understanding the ecology of such systems and their values, and examining the possible consequences of different management options. Essentially, the planners use the model as a learning tool to evaluate alternative decision strategies.

Using projections derived from simulation models allows the management team to predict the possible impacts of management alternatives without causing potential harm to the ecosystem or particular values. The modelling process is relatively inexpensive and usually does not require equipment more complex than a desktop computer. By examining a range of possible futures, such simulation gaming can help forest managers and policy makers to understand uncertainty, recognize that we seldom know for sure the best thing to do, and that evaluating possible strategies in a virtual forest reduces the risk of unacceptable mistakes. Duinker and Trevisan (2003) point out that mistakes will be made when testing new approaches to managing complex forest landscapes, but if these can be minimized by gaming in advance of action using credible ecosystem-level models, and if we learn from the mistakes and do not repeat them, the social licence to manage public forests may not be revoked.

Modelling forces the recognition of errors in the management team's understanding of a system (Baskerville, 1985). Thus, if the model produces simulations that are out of line with the observed behaviour of the system, it is usually an indication that the model and some of its underlying assumptions require adjustment. Predictions based on an explicit model can be tested by comparing them against performance of the real system that is determined through monitoring of selected variables (Blanco et al, 2007).

While modelling is a powerful tool with which to make predictions, these will only be as good as the model(s) used (Kimmins, 2004d). Many foresters tend to regard simulation models as esoteric and abstract, driven by parameters of which they have little knowledge and often no way of measuring. The future of simulation models in AM in forestry lies in producing models which are relevant to the practitioner, with output information that can be understood and applied in everyday practices. In short, models must be reliable and produce answers in a form that is understandable and applicable for foresters. They should be as simple as possible, but as complex as necessary to fulfil this requirement. As the paradigm shift from tree production to ecosystem management takes hold, foresters will need to deal with many values, such as species composition, stand structure, landscape fragmentation, soil nutrient reserves, water quality and aesthetics, as well as sustaining the economically and socially important supply of wood products and rural employment.

Forestry is first and foremost about people's needs, desires and beliefs, and therefore decision-support systems in AM should also address social values (e.g. economics, employment, wealth creation and environmental services) that are directly related to the biophysical characteristics of forests. The case has been made that forest growth models used for sustainability assessment and AM should include a simulation of key processes that are expected to change in the future. These include climate change impacts on forest growth processes, the consequences of alternative silviculture systems for soil fertility, competition for resources between crop and non-crop plants, and pests and diseases (Landsberg, 2003b). However, this need must be balanced against issues of model complexity, accuracy of predictions, and the relative feasibility of applying such tools.

One of the strengths of hybrid models is that they can be used to explore the consequences of combining several sources of uncertainty and investigate the propagation of uncertainty through time, specifically in relation to achieving defined management goals (Schreiber et al, 2004). Traditional forestry has managed relatively simple stand-level systems and assumed that the effects of management on the forests are known when the management plans are prepared. In contrast, AM calls for managing a more complex system: the forest ecosystem, not just populations of trees. Foresters need different and more complex tools at both small and larger spatial and temporal scales to deal with this complexity.

Ecological computer models for adaptive and sustainable forest management need to address multiple values and span multiple scales. It is very complex to develop a model that covers all ecological processes from leaves to landscapes, and if such a model were created it would probably be a 'dinosaur': very hard to calibrate, understand and use (Kimmins, 2004d). However, these modelling objectives can be accomplished by linking stand-level models to landscape-level models (meta-models; see Chapter 10). Stand-level models should address issues of soil resources, climate and water as well as sustainability of biological production, and should contribute to analyses of wildlife habitat and measures of biodiversity. They should be capable of addressing the effects of both natural and management-induced disturbance on ecosystem form and function (Messier et al, 2003; Seely et al, 2004).

Although mechanistic and deterministic models are useful at the stand level, at the landscape scale empirical and stochastic models are needed, but these need to be informed by stand-level processes or hybrid models. At the stand level, one can assume that most of the important processes and environmental conditions can be treated as part of a relatively closed system. This is not the case at the landscape scale, where natural disturbances can propagate into a region from its surroundings (i.e., landscapes and landscape processes are more appropriately portrayed as an open system; Messier et al, 2003). Landscape models should include an empirical framework with stochastic/deterministic disturbance representations, but the stand-level components be driven by process-based or hybrid empirical-process models if they are to be credible for AM and other applications.

Many forest management practices occur on scales that make replication impossible and in landscapes where ecosystem diversity prevents spatial replication. Given this limitation, the management team must decide on the best means of providing a control for their management experiment. Developing a model to test the effects of alternative strategies and using a 'no intervention' option as a control reference provides the most defensible means of preparing predictions (Duinker, 1989). However, this still does not address the issue of uncertainty and stochastic events, and consequently models for AM are becoming increasingly stochastic, and are often designed to portray an array of possible outcomes, even when run on spatial databases. Stochasticity may be incorporated directly into a model, or a deterministic model may be run under a range of plausible starting and environmental conditions to represent stochasticity and the resulting uncertainty.

If credible models can be linked to visualization systems, models can also help to communicate the possible outcomes of alternative management scenarios to non-technical stakeholder groups (Messier et al, 2003; see also Chapter 10). This is very important when developing AM plans in which communication between different stakeholders with different levels of technical skills may be challenged by difficulties in transferring information. Since the main users of the models discussed here are forest managers, it is important that the flow of information between scientists building the models and forest managers using them be improved in order to build more goal-orientated models. We need to provide managers not only with credible quantitative predictions that are easy to understand and to transfer to other stakeholders, but also with information about the certainty of our prediction (Blanco et al, 2007). A precise prediction that is very uncertain might not be better than an imprecise prediction that is more certain.

It is important when using models in the development of an AM strategy to determine and control the error associated with them. The models must therefore be evaluated in relation to specified objectives and their capacity to realistically simulate the main processes that are of interest (Messier et al, 2003; Blanco et al, 2005, 2007; see also Chapter 7).

Too frequently, ecological impact predictions in environmental assessments consist of vague generalizations that are so imprecise that they can never be shown to be wrong (Beanlands and Duinker, 1984). In contrast, forecasts from appropriately complex ecosystem management models can be quantitative states of the system and therefore constitute testable hypotheses (Baskerville, 1985). For adaptive managers, it is preferable to be quantitative and wrong than qualitative and untestable. The former can be corrected, the latter cannot.

Examples of the application of forest models in adaptive management

FORECAST (Kimmins et al, 1999) and LLEMS (Seely et al, 2008) are especially well suited for use as decision-support tools in adaptive forest management because of their ability to produce a wide range of quantitative output variables in tabular and graphical formats. Predictions of future temporal variations in these variables can be compared with real data obtained from monitoring programmes to track how the ecosystem is responding to the management activities. If the models have been validated for the forest type being assessed, deviations between predicted and monitored values for variables of interest are found, and corrections to the management plans can be generated (the adaptation part of AM). Alternatively, if the models have not yet been validated for the system in question, such deviations may suggest the need for modifications to the model. This emphasizes the importance of model validation over appropriate temporal and spatial scales for the ecosystem in question.

FORECAST and LLEMS can also be linked with other models that simulate the effects of management for different variables and at different scales. An example of the use of FORECAST to project temporal evolution of a target variable (temporal fingerprints; Kimmins et al, 2007) is illustrated in Figure 9.1.

Trends in large-diameter snags (an indicator of habitat quality for cavity-nesting birds and other secondary cavity-using groups) are projected for use as a guide for future monitoring plans. Figure 9.1 shows the expected long-term trends in snag density for two stand types in BC. Trends in snag density were compared for a series of partial harvesting strategies with varying levels of retention. This establishes a context for monitoring, and indicator targets and thresholds, by comparing alternative silviculture systems against natural disturbance baselines. Such stand-level ecosystem management modelling can be scaled up to timber supply areas or larger units by linking the models with large landscape models of various types (meta-modelling).

Seely et al (2004) shows how FORECAST can generate a library of predictions for different variables that can be used as output for landscape-level models such as FPS-ATLAS (a harvesting schedule model) or SIMFOR (a habitat analysis model). Alternative management regimes can be considered to assess their relative sustainability in a series of scenario and value trade-off analyses (Morris et al, 1997; Wei et al, 2003; Seely, 2005a, 2005b; Seely and Welham, 2006). These studies have demonstrated that not only can such an approach be practical and operational for the assessment of the relative sustainability of different management strategies; it is also useful for testing the efficacy of existing indicators of sustainability and suggesting new indicators that can provide practical and affordable monitoring. This multi-tool approach to AM provides the scientific background to predictions of long-term sustainability, but it also allows managers to play and explore different outcomes to alternative forest practices (Chapter 10 continues this discussion).

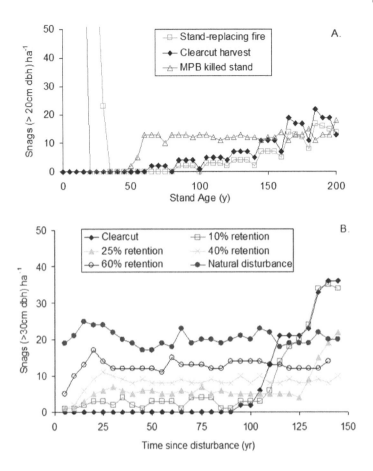

Figure 9.1 *An example of the projected temporal patterns of the population of large-diameter snags in two ecosystem types within British Columbia*

Note: A) an even-aged, lodgepole-pine-dominated stand subjected to a stand-replacing fire, mountain pine beetle attack (unsalvaged), and clearcut harvesting; B) an uneven-aged coastal western hemlock/western red cedar stand subjected to endemic 'gap-dynamic' disturbance agents and harvested with varying levels of dispersed retention (0–60 per cent).

Hybrid models are also being used at the landscape scale for the simulation of natural disturbance. For example, new timber management guidelines are being developed in Ontario to deal with landscape-level planning, and as part of the development process alternative policy options are simulated using the PATCHWORKS harvest-scheduling program (Lockwood and Moore, 1993; Baskent and Keles, 2005). The projected future forest conditions are then evaluated in terms of biodiversity conservation using multiple-scale spatial habitat models for a focal species group. Harvest projection models such as PATCHWORKS are generally empirical in nature and not process-based. Therefore, to place the evaluation of policy options in the context of ecosystem processes, a process-based natural disturbance model, BFOLDS (Perera et al, 2004), was used to generate a simulated range of natural variation for future landscape patterns. The effects of policy options on future composition, pattern and habitat suitability were thus examined using a forward- rather than backward-looking approach. At present, weather parameters in BFOLDS are being adjusted to evaluate the expected consequences of climate change on future forest conditions. Although stand development attributes (e.g. snag density and crown closure) were initially modelled using empirical methods, future iterations of the work in Ontario (Canada) will utilize FORECAST. A similar approach was used for development of forest management plans by Louisiana Pacific in Manitoba (Rempel et al, 2006).

A key feature of any AM programme is the feedback between monitoring results and management. Monitoring support systems are an extension of the system of stand and landscape management tools that companies and government agencies are using in resource management planning and public review processes. To assist the latter, the output from such a support system should be linked with advanced visualization systems, a trend that is being applied increasingly in forestry (e.g. Kimmins, 2000a; Sheppard and Harshaw, 2001; see Chapter 10).

The UBC Forest Ecosystem Simulation research group has linked the non-spatial stand-level ecosystem simulator FORECAST to several models to produce spatially explicit predictions of future growth and other variables at the stand level, and to scale up to the landscape scale (see Seely et al, 2004). These predictions can be linked to FPS-ATLAS (Nelson, 2003), a landscape-level model which uses polygons of different types to predict future timber harvests in a specific area, and also to simulate each of the pixels used by SIMFOR (Daust and Sutherland, 1997), a landscape-level biodiversity model, to simulate future measures of biodiversity in a given landscape. In addition, FORECAST output can be used to feed the stand visualization system (McGaughey, 1997) to create visual representations at the stand level. Finally, FORECAST is used as the core engine of LLEMS (Chapter 6), simulating forest growth for each pixel of a local landscape in a more integrated fashion than when externally linking FORECAST to other models in a meta-model. Similarly, LLEMS output has been linked to CALP-Forester (Cavens, 2002), to create landscape-level visualizations to improve the capacity to communicate model output for presentations to public stakeholder groups. Using such visualizations, groups other than scientists and forest managers can gain a vision of possible future forest conditions under different management alternatives, and their participation in the process of adaptive management is greatly improved.

The need for credible forecasts of possible forest futures as an essential component of forest certification

Forestry has always been concerned with sustaining a defined set of desired values, except during the early exploitative stages that may occur when new land is colonized. The traditional focus on maintaining a constant flow of timber products has been widely replaced by concerns over the sustainability of the entire ecological system, and its many components, processes, environmental services and values for human society (Thomas, 1995; Messier and Kneeshaw, 1999; Rauscher, 1999; Kimmins, 2002). Sustainability is usually defined as a feature of management systems that encompasses three components: economical sustainability, social sustainability and ecological sustainability. However, debate continues over both the definition of sustainability and how it might be achieved in practice, especially when dealing with the ecological or biophysical component (Oliver et al, 2000; Nemetz, 2004; Kimmins, 2007b).

Sustainability has been the subject of many international processes and initiatives, such as the UN's Brundtland report (World Commission on Environment and Development, 1987) or the United Nations Conference on Environment and Development, held in Rio de Janeiro in 1992 (UNCED, 1993). The Rio conference expanded the set of values requested from forests, required greater public participation in management decision-making, and opened the doors for formal forest certification procedures (Bunnell and Boyland, 2003). As a consequence, forest certification emerged in the early 1990s as a private governance mechanism to set and police sustainable management of forests (Overdevest and Rickenbach, 2006).

The ultimate objective of any forest certification scheme is to ensure the sustainability of forest management (Cashore et al, 2005). Certification systems use independent auditors to verify and monitor compliance with defined standards of forest management. Representatives from civil society set standards that are intended to define sustainability and identify the process for monitoring. After agreeing to implement the standards and pay for periodic audits by accredited independent agents, certificate holders gain the right to use green labels (eco-labels) to differentiate their product in the market as being 'well managed'. Supply chain actors (e.g. retailers, processors) who want to sell wood and other forest products as certified must also have their operations audited by the certification agency. In return, each certificate holder along the supply chain may sell a product that is labelled as socially and environmentally well managed, presumably increasing the value and market price of the products in the process (Overdevest and Rickenbach, 2006).

Certification programmes use two fundamental approaches to evaluation. The first is the product or outcome approach, where the primary focus of assessment is the forest resource itself. It is evaluated by measuring or ranking specific characteristics (criteria) attributed to the origins of the products according to ecological, social and economic performance indicators. The second approach focuses on processes, or whether the company or landowner has adopted quality management procedures that are consistent, repeatable and conducive to continuous improvement. The company or landowner is evaluated based on objectives, goals, planning, quality control measures, record keeping, staff responsibilities, regulatory compliance, training and education of employees and safety programmes.

In reality, most programmes are a combination of these approaches, although a particular programme may clearly emphasize one or the other (Haener and Luckert, 1998). Almost all certification programmes are alike in that they advocate the use of some set of standards and indicators to measure and verify sustainable forest management. Indicators may consist of general information about the ecological and socio-economic conditions of forested areas, while standards may require management plans, including regulation and enforcement measures (Cabarle et al, 1995).

Certification systems require extensive planning to maximize long-term productivity and sustainability. In these plans, the forest manager is required to assess not only the immediate impacts of the proposed timber operation on various resources, but also the cumulative effects in a reasonably foreseeable possible future (Dicus and Delfino,

2003). Such resources could include watershed, soil productivity, biological resources, recreation, visual impacts, traffic and others.

Envisioning the consequences of forest management 20, 50 or 100 years in the future is difficult, and helping to solve this issue is the main role of forest management models. The increasing importance of forest certification and the realization that certification of sustainability and stewardship entails more than simple evaluation of snapshots of present conditions extrapolated into specific future years should promote the development of a more dynamic view of forest ecology (Kimmins, 2008a; see Chapter 2). Indeed, new modelling approaches are needed to effectively identify, collect and relate the social context and components of forest ecosystem management in order to enhance and guide management decisions (Burch and Grove, 1999; Villa and Costanza, 2000; Twery, 2004).

As a consequence, the simulation tools used in certification should be rooted in the best ecological science available. Using only statistical models, like the traditional growth and yield tables, denies the advances carried out in forest ecology in recent decades, but the use of purely abstract (theoretical) and conceptual process-based models can also produce predictions with very little similarity to reality. Hence, hybrid models are particularly suitable for certification applications. The certifying organization has to field-check the submitted inventories by the forest company to ensure accurate predictions, but it also has to evaluate the ability of the models employed to predict futures based on plausible representations of the ecosystem processes responsible for sustainability.

The Sustainable Forestry Initiative (SFI) certification scheme requires that participants demonstrate that they have an appropriate method to calculate growth and yield. Similarly, the Forest Stewardship Council (FSC) requires proof that post-harvest activities maintain soil fertility, structure and functions. These certification agencies also recommend paying attention to future habitat modelling (Gordon, 2006). In some cases, the certification agencies have denied approval of the sustainable yield prediction because they did not contain credible model projections (Dicus and Delfino, 2003).

The current certification of sustainable forest management depends heavily on evaluations by experts (Mendoza and Prabhu, 2000; Lindenmayer et al, 2000). Growth and yield forecasts using hybrid forest models and multi-criteria decision analysis can help to reduce subjective errors and misjudgements in expert evaluation, and by doing so, improve the quality of their decisions (Huth et al, 2005). Furthermore, formalizing the decision problem in multi-criteria decision analysis allows a large number of criteria and timber-harvesting options to be analysed, which is important in the current situation, where the relevance and relative importance of criteria and indicators for sustainable forest management are still unclear. We also think that while the complexity involved in assessing soil fertility, biodiversity and wildlife impacts requires complex tools, this complexity cannot be presented directly to a variety of stakeholders. While complex output is a requirement for assessing the performance and forecasts of these complex models, this output must then be reformulated for documentation that is easy to understand and can be used to communicate with the public, government and certifiers (see Chapters 2 and 7). Our hybrid ecosystem-level forest management models were specifically designed to meet this need.

The need for a method of assessing timber-harvesting impacts on other forest values will probably increase in forest management, as forest certification that considers social values (such as outdoor activities or spiritual values) becomes necessary for doing business (Sheppard and Salter, 2004; Chapter 10). In addition, the involvement of the public in how forests should be managed has been increasing, and as a consequence most forest certification procedures require evidence of public participation in the selection of management actions. This social requirement has been incorporated into the philosophy of a number of forest companies,

The challenge is to make sustainability evaluation for certification accessible to the public. This challenge is particularly acute in forestry, not only because of the degree of public involvement, but because of the large areas and long time periods involved. The use of well-documented models combined with visualization techniques to make the possible consequences of policy and management decisions more understandable to the general public can help to improve the communication of long-term sustainability issues.

Certification and adaptive management share many common aspects. The main goal of both methodologies is to promote sustainable management of forest resources. And as in the case of adaptive management, an important part of the process of certification is the continuous monitoring of the sustainability of forest activities. Companies who have achieved the certified label must undergo periodic revision by the certifying agencies to keep the label. The goal is to keep improving the quality of the operations through time. This implies a strong temporal component in the assessment of the sustainability of forest operations, which can be achieved only through monitoring of the ecosystem condition.

Monitoring as a component of certification, and its relationship to modelling

Traditional monitoring has several shortcomings that make it less than ideal for certification purposes (Kimmins et al, 2007). There is also the problem that certification agencies may not have the resources to analyse and interpret the monitoring data, thereby limiting feedback from monitoring to management. These and other shortcomings of conventional approaches to the monitoring of sustainability in forestry (discussed in Kimmins, 2007b) can be addressed by process-based monitoring at the ecosystem level (Figure 9.2). The emphasis of this type of monitoring is on the development and calibration (using monitoring and other data) of an ecosystem-level, process-based prediction system. With this approach, 'temporal fingerprints' of expected ecosystem change can be established that become the baseline against which data from context, compliance and other forms of monitoring are compared.

Process-based monitoring involves a combination of monitoring and ecosystem management modelling that reduces the long-term cost of monitoring and increases the utility of the data collected or assessing sustainability of multiple values. Process-based monitoring involves two distinct phases.

Phase I is conducted specifically to develop, calibrate and test new (or to calibrate and test existing) process-based or hybrid ecosystem-level models, and then to use them to explore which combinations of disturbance intensity and frequency are sustainable over various temporal and spatial scales. This first phase consists of two parts. In the first part, monitoring data collection is designed to calibrate the predictive relationships

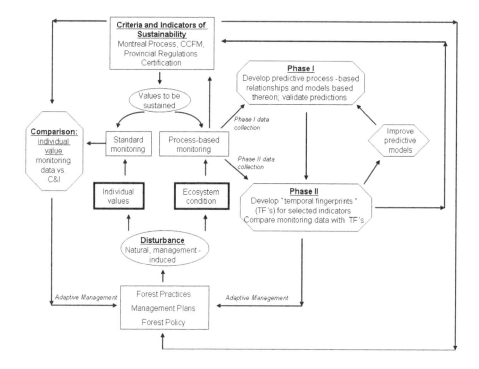

Figure 9.2 *Use of process-based, ecosystem-level monitoring in certification contrasted with conventional monitoring*

Note: conventional monitoring compares monitoring data that describe present conditions with selected indicators of sustainability. Process-based monitoring also collects data that describe present conditions, but in a manner that allows the development of causal and predictive process-based relationships between measurable ecosystem attributes (indicators) and values of interest. These relationships are used in Phase II to develop 'temporal fingerprints' of expected change in the values. Monitoring data are subsequently collected to compare with these temporal patterns of indicator values.

within the integrated ecosystem management modelling framework. In the second part, further data are collected to provide an independent test of these relationships and of the predictive power of the monitoring system. This establishes the level of accuracy and reliability of the decision-support system used to forecast growth, yield and other values under scrutiny in the certification process. After an acceptable level of reliability of the process-based monitoring system has been established, the sustainability of a proposed stand and landscape management strategy is assessed in Phase II.

Phase II includes the assessment conducted within the context of risks such as fire, insect outbreaks, wind damage and climate change, all of which should be addressable by the models used in certification. Once the monitoring support system has been established and verified, ongoing monitoring is reduced to periodic lower intensity, targeted data collection that provides a check on the efficacy of the management system, and to steadily increase the data upon which the predictive model is based. Alternative management regimes can then be considered to assess their relative sustainability in a series of scenario and value trade-off analyses (Morris et al, 1997; Wei et al, 2003; Blanco et al, 2005, 2007).

Although not very common yet, the use of ecosystem models in certification is increasing. One example is the work by Ranius et al (2003). These authors propose that strategies for preserving biodiversity in boreal forests should include the maintenance of coarse woody debris, because this is a key feature for the preservation of many threatened species, providing habitat for birds, insects, mammals, etc. For this reason, the amount of coarse woody debris present in a managed stand is one of the indicators used by the FSC's Forest Certification Standard to assess the sustainability of forest management. Ranius et al (2003) used a stand-level forest model to simulate the amount of coarse woody debris that can be found in the boreal forests of Norway spruce in central Scandinavia. They used the limits set by the FSC as a target for a certifiable management plan (for this indicator), and compared that 'ideal' plan with current practices in Fennoscandia. They found that those practices were removing too much coarse woody debris from the stands, and as a result they recommended that the current practices should be revised in order to fulfil the requirements of the certification scheme. Other examples of the use of models in certification can be found in Skreta (2006), who recommended the use of the model Forest Time Machine in the framework of the FSC certification scheme in Sweden.

Another example is the use of FORECAST to develop carbon curves to address the certification requirements of the Canadian Standards Association (Seely, 2005b). With public concerns over carbon emissions and their relationship to climate change, Canadian forest companies are under mounting pressure both nationally and internationally to account for the effects of their activities on terrestrial carbon stocks and greenhouse gas emissions. Because the capacity of forest ecosystems to sequester carbon has been increasingly recognized as an environmental value and thus a significant component of sustainable forest management plans, the Canadian Standards Association (together with other certification schemes) has identified the effects of management on global carbon cycles as a criterion for sustainable forest management. To successfully implement such a criterion within a certification system, it is necessary to utilize scientifically credible indicators and associated measures that can be valuated as part of a sustainable forest management plan (Kimmins et al, 2007).

Indicators based on correlations with volume and mean annual increments are usually insufficient. In Seely's (2005b) work, FORECAST was used as a tool to provide support for the Morice and Lakes IFPA certification (interior BC) under the Canada Standards Association forest certification initiative with respect to meeting the requirements related to measuring the impacts of forest management activities on global carbon cycles. In that project, the primary use of FORECAST was to generate a database of carbon curves showing the predicted evolution over time of different carbon pools in the forest ecosystem. A carbon curve was generated for different carbon analysis units in order to characterize the different stand types that are in the area. These curves were incorporated in the Tesera Scheduling Model (Tesera Systems, 2009), a forest planning tool that was used to generate landscape-level projections for a base case, a natural disturbance and a forest productivity scenario, allowing comparisons between

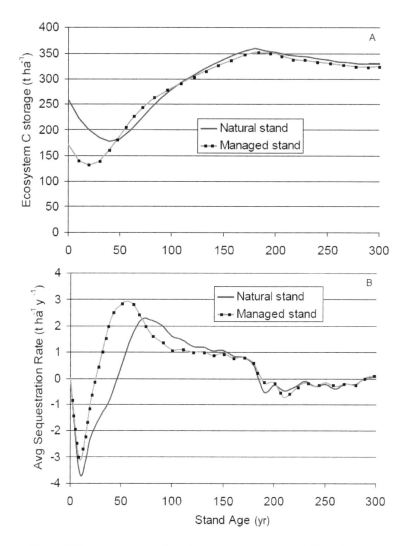

Figure 9.3 *A comparison of total ecosystem storage (panel A) and average sequestration rate (panel B) for a natural and an associated managed stand in Morice and Lakes timber supply areas (BC)*

the predicted effect of management on carbon sequestration rates and carbon storage in the ecosystem (Figure 9.3). This project demonstrated that the carbon curves generated using FORECAST can be used for the calculation of landscape-level indicators for use in support of certification applications. By incorporating the outputs of a hybrid model such as FORECAST as inputs for a second model, it was possible to estimate the effects of landscape-scale harvesting activities on the global carbon cycle. The full report of this project can be found in the related website of this book (www.forestry.ubc. ca/ecomodels).

Finally, hybrid models can be used to increase the participation of stakeholders in the certification process. This is a requirement of any certification scheme, such as in the case of the Pan European Forest Certification Council (PEFC, 2003). Sheppard and Meitner (2005) have successfully used FORECAST linked with FPS-ATLAS and 3D visualization techniques to increase the participation of communities in the management process and their understanding of the science behind the long-term model predictions used to support certifications procedures. Landscape visualizations offer a potential mechanism with which to address social implications of site-specific management actions or scenarios, such as impacts on scenic quality, recreation, spiritual/cultural values, general quality of life and property values. Furthermore, the general health of the forest is often judged by the public, and even by experts such as forest certification panels, in part by what they see on the ground (Sheppard and Salter, 2004). See Chapter 10 in this book for a more detailed discussion of this topic, and Chapter 2 for a discussion of the concept of forest health.

Ecosystem management models in reclamation planning

Reclamation[1] of ecosystems degraded by industrial activity is a problem common to many countries, including developed nations. For example, there are large areas of degraded land in Russia (Chertov et al, 1999c), the USA (Burger and Zipper, 2002) and Canada (Alberta Environment, 2006). Land degradation can occur from a broad range of activities, including the construction of transportation corridors and facilities, mining, poor forestry practices, peat excavation, and heavy metal toxicity. It varies in intensity from reductions in ecosystem productivity (due to acid deposition, for example) to the removal of entire ecosystems (as occurs in open-pit mining).

Typically, the goal of land reclamation is to achieve land capability equivalent to that which existed prior to disturbance (Alberta Environment, 2006). Field experiments can be a useful guide as to which reclamation practices are the most useful for achieving this goal. Basing reclamation protocols on field experiments alone, however, is impractical because the number of potential options for study is prohibitively high, and a given experiment has to run for many years before results can be considered as definitive (Chertov et al, 1999b). Ecosystem management models can play an important role in designing reclamation strategies and as a guide to empirical work. Models that include a representation of community dynamics along with the key drivers of ecosystem productivity allow for consideration of a very broad range of alternatives. The long-term outcome from each alternative can be compared and contrasted, from which the most promising are selected for further study and refinement. One example of this approach is in the design of land reclamation strategies following large-footprint disturbance, such as surface mining.

Land clearing prior to and during mine operation often results in the removal of entire ecosystems. In the oil sands region of Alberta, Canada, for example, a vast amount of bitumen material resides in relative close proximity to the surface. Access therefore requires that the existing surface vegetation (primarily forested communities) be removed along with the underlying overburden. The overburden piles as well as the tailings sand deposits created following bitumen extraction must then be reclaimed (Government of Alberta, 1999; Rowland et al, 2009). In some respects, this scenario

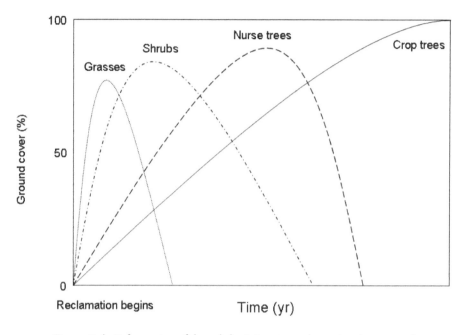

Figure 9.4 *Reforestation of degraded mining sites seeks to stimulate natural forest succession*

Note: all vegetation types are usually established during reclamation. As time passes, grasses and shrubs yield to fast-growing pioneer and nurse trees species, which are themselves overtopped by commercially valuable crop trees as the forest grows and matures.

resembles conditions that must have been present following the last glacial retreat (about 10,000 years before present). In principle, one option is to not engage in any active reclamation and simply let the vegetation re-establish at a 'natural' pace. Following disturbance, a plant community usually undergoes a series of seral stages, each stage resulting from the sequential replacement of the dominant species from the previous stage, until it regains its original composition (or, at least, something close to it) (Figure 9.4). This is the process of forest succession (Kimmins, 2004b). Typically, species such as grasses, shrubs and fast-growing hardwoods (*Populus, Alnus,* for example) establish first. Many of these colonizing species can tolerate fairly harsh abiotic conditions (acidity, low fertility, moisture deficits and temperature extremes) (Burger and Zipper, 2002). The pioneer community is eventually replaced by species with a narrower tolerance range, often hardwoods in this area, and then followed by shade-tolerant late-successional conifers. Due to their poor quality as a plant medium, oil sands spoil materials would probably require several hundred years (or more) before succession had produced a forest cover, if restoration was left entirely to natural processes. This time period is simply too long, either as an end goal or as a model system for reclamation. Firms operating in the oil sands have therefore invested considerable resources into developing more efficient and effective reclamation strategies. One approach to facilitate this process is using ecosystem models to explore and evaluate alternative options.

Organic matter and its decomposition are a key component for nutrient cycling and moisture storage in terrestrial ecosystems (Bradshaw and Chadwick, 1980; Sopper, 1992; Vetterlein and Hüttl, 1999). In oil sands reclamation, a surface soil amendment consisting of a peat-mineral mix consisting of 25 to 50 per cent by volume of mineral materials (Fung and Macyk, 2000) is applied to spoil materials, to a depth of between 20 and 50cm. This amendment constitutes the rooting zone for the vegetation community that is subsequently established on the reclaimed site. The peat supplies the organic component of these reconstructed soils, provides a limited supply of nutrients as it decomposes, and also serves to improve the water-holding capacity. Where lean oil sands (unprocessed oil sands with <6 per cent oil by mass) and saline overburden is being reclaimed (as opposed to tailings sand or non-saline overburden), an intermediate capping layer of up to 80cm of non-saline subsoil material is used to bury this material and protect it from water and plant root ingress (Oil Sands Vegetation Reclamation Committee, 1998; Stolte et al, 2000). A key question in oil sands reclamation is: what productivity can be expected from the soil prescriptions typically used to reclaim spoil materials? Related issues are how important to ecosystem productivity is the decomposition rate of the peat used in the capping material, and whether the quality of the underlying material used as subsoil (its organic carbon and nutrient content) has a significant impact upon ecosystem productivity.

To date, few models have the capability to represent the biogeochemical properties of these reclaimed systems and the relationship of these properties to vegetation development. One example is the work of Chertov et al (1999c) in Scots pine (*Pinus sylvestris* L.) plantations in Russia. They used the individual-based forest model EFIMOD (Chertov et al, 1999b) to simulate stand growth and soil organic matter accumulation on a humus-free mineral soil surface. Model results showed that the accumulation of soil organic matter was accelerated by nutrient additions through atmospheric nitrogen inputs and compost application. In the case of oil sands reclamation, a series of simulations have been conducted using the ecosystem simulation model FORECAST (Chapter 5). The overall objective of the modelling work was to evaluate current and potential reclamation practices and their impact on patterns of ecosystem development and productivity. Given that relatively little is known and understood about the determinants of ecosystem development within a reclamation context, model simulations were considered to be an important means of gaining an understanding of how various processes and management options interact and contribute to end land-use objectives. All simulations included two peat decomposition rates (termed slow and fast), and a range in tree seedling planting densities. The objective was to bracket a potential range in peat decomposition rates, from a highly conservative to a more optimistic value. Including a range in planting densities better reflects management options. Simulations also included the establishment of an understorey community comprised of one or more functional groups (tall shrubs, medium shrubs, forbs and grasses).

The simulations indicated that the peat decomposition rate was of over-riding importance in determining long-term productivity in a reclaimed plant community (Welham, 2005, 2006, 2009). If peat decomposition rates are 'fast', well-stocked reclaimed sites will be easily capable of achieving commercial forest capability. This would not be the case, however, if peat decomposes at a 'slow' rate. Model simulations also showed, however, that limitations in the quality of peat in the cap could be mitigated if the underlying material contained organic material (i.e. carbon) that could be mineralized to provide additional nitrogen (Welham, 2009). One option to mitigate nutrient limitation is simply to increase the thickness of the peat-based cap. Welham (2005) explored this option explicitly by simulating ecosystem productivity at peat-cap depths of 20, 50 and 100cm. Increasing the peat cap was indeed beneficial, though there may be diminishing returns as the depth increases, depending upon tree species.

Planting density is a variable that is easy to modify in the field. In all simulations, increasing the planting density always generated greater timber volume production. Adding to the tree population improved its ability to compete with the understorey, but the benefits were proportionately greater when peat decomposition was slower. For a given type of peat, there were often diminishing returns with increasing planting density, particularly when densities were greater than 2000 stems per hectare. Overall, the increase in cost associated with higher planting densities is probably small relative to the benefit of improving establishment success (by reducing the need for supplementary in-fill planting) and achieving long-term yield objectives. Higher planting densities also generate more snags in older stands (Welham, 2006). Snags are an important habitat element, serving as perching sites and a source of nesting cavities, for example. In general, more snags translate into better habitat.

A vigorous and diverse understorey of native species adds an important element of 'naturalness' to a reclaimed site and provides visual confirmation that key ecosystem processes have indeed been successfully established. Model simulations indicated, however, that the temporal dynamics of understorey vigour could not be evaluated independently of the associated tree population. For example, under nutrient-rich conditions, an understorey established during early stand development was not sustained when the stand entered canopy closure; light conditions declined severely due to vigorous tree growth, and litterfall was heavy, leading to the smothering of low-growing plants such as mosses and lichens (Welham, 2006). The highest and most sustained understorey populations, in contrast, occurred under relatively poor nutrient conditions when tree growth was insufficient to suppress light significantly. In this respect, model results indicated a clearly defined inverse relationship between overstorey and understorey development (Welham, 2009). Overstorey species composition also affected understorey vigour. Aspen (*Populus tremuloides*) canopies had higher light levels than mixed-species stands (aspen/spruce), which, in turn, had higher light levels than spruce-only stands. Understorey suppression was also exacerbated by higher planting densities (Welham, 2006).

Taken together, these results demonstrate the utility of employing an ecosystem-based approach to simulating reclamation practices, using a model that can link the key biogeochemical processes with the developing vegetation community. Community dynamics are complex and there are important interactions and feedbacks that are difficult to predict *a priori*. With the appropriate model, these relationships can be identified and used as guidelines to more effective and cost-efficient reclamation practices.

Take-home message

Ecosystem-based simulation models can be helpful tools in designing effective forest management plans in support of sustainable forest management initiatives and forest certification, and in designing reclamation plans for highly degraded industrial lands. Ultimately, however, model predictions cannot stand alone and must be tested and evaluated against field-based evidence. Forecasting and monitoring are thus inextricably linked (Duinker, 1989; Duinker and Trevisan, 2003).

Adaptive management in forestry requires a heavy component of computer-based forecasting because the long timescales involved preclude a simple empirical feedback mechanism for many important ecosystem attributes. Engaging appropriate computer tools in AM is also necessary to help non-technical stakeholders understand the complex nature of forest sustainability, an application that requires linkage with visualization and other communication techniques. The multiple values that must nowadays be the subject of adaptive management require the use of multi-value, process-driven, stand-level ecosystem model, linked with landscape models in meta-models where appropriate.

Certification processes have spawned extensive lists of indicators of the principles contained in certification standards. Many of these are snapshot measures of ecosystem conditions and do not relate well to the dynamics involved in ecosystem sustainability. Ecosystem-level models can help to identify a smaller set of measurable indicators that are good predictors of various aspects of sustainability, and by linking monitoring with process-based modelling can result over the long term in both a more effective and more economically efficient monitoring system.

Large-scale severe ecosystem disturbance often lies outside of our range of experience, and modelling can be an invaluable way of exploring the best way to re-establish functional ecosystems of desired character and values over acceptable timescales. We simply cannot wait for empirical evidence of the success of alternative strategies, nor will the public permit it.

The FORECAST family of models has been successfully used in these contexts, and their applications provide good examples of the utility of science-based ecological modelling in the context of competitive and sustainable forestry.

Additional material

Readers can access the complementary website (www.forestry.ubc.ca/ecomodels) to find full versions of the following additional material:

- Seely, B. (2005b) *Development of Carbon Curves for Addressing CSA Certification Requirements in the Morice and Lakes Timber Supply Areas*, FORRx Consulting, Belcarra, BC.
- Seebacher, T. (2005) 'Effect of capping depths and fertilization levels on white spruce seedling production and soil nutrient dynamics', report prepared for graduate course, University of British Columbia, Vancouver, BC.

Note

1   Ecosystem reclamation is defined as activities that result in a stable, self-sustaining ecosystem that may include exotic (as well as native) species and possess a structure and function that is similar but not necessarily identical to the original land (National Academy of Science, 1974). This is in subtle contrast to ecological restoration, the process of assisting the recovery of an ecosystem that has been degraded, damaged, or destroyed (SER, 2004).

Chapter 10

Future Perspectives in Hybrid Modelling[1]

Introduction

One sometimes encounters biologists, ecologists and environmentalists who assert that biophysical or ecological sciences are going to solve the environmental problems of the world. As comforting to its proponents as this simplistic view of the world is, it overlooks the realities of over 6.8 billion people and their social needs and economies, let alone the additional three to four billion people that some predictions suggest will be born. Issues in forestry are almost never simple, and are never simply biophysical in nature. Resource management must always consider the three major components of sustainability: environment, economy and society. Biologists and ecologists would insist that of these, the environment is the most fundamental since the other two are dependent on it; but without the other two, issues of sustainability, stewardship and intergenerational equity will not have been dealt with.

It has been asserted repeatedly in this book that decision-support tools in forestry should be multi-value and ecosystem-level, and either process-based or hybrid empirical-process. However, we also assert that unless such forest ecosystem-level models also address management, social and economic issues, they will fail to contribute to finding solutions to urgent issues in forestry. Finally, we assert that unless such combinations of modelling activities are linked to systems designed specifically for the communication of output – such as visualization – these models will remain largely the domain of academics and researchers.

This chapter gives examples of linking ecosystem-based, stand-level models to landscape models, to models addressing social values and to visualization systems. Opening with a discussion of meta-modelling, two examples are given of linkages involving the FORECAST and FPS-ATLAS models. This is followed by a discussion of the importance and role of linking forest ecosystem models to environmental issues of concern – climate-change impacts in this case – and linking environmental and forest management models to visualization systems of various types, including examples of applications in the city of Prince George, BC, and of the interactive 3D visualization CALP-Forester.

Linking forest management models that work at different scales

Forest management is highly complex because it involves multiple and often conflicting objectives, long time horizons, large geographical areas, and expertise in many subjects related to ecology, economics and social expectations. Forest management planning is therefore best conducted by a team that has the necessary expertise to address the broad scope of issues that are encountered.

Projecting a forest management scenario into the future to assess sustainability involves a large number of considerations as individual stands grow, are harvested or are impacted by natural disturbance. Fortunately, many decision-support systems have been developed to help forest planning teams incorporate this complexity into assessments of their management scenarios during the planning process. These decision-support systems require the linking of various models that operate at different scales and forecast or interpret data about different values into a common modelling framework known as a meta-model. Each component model of such a modelling framework has considerable complexity, and each requires an understanding of the underlying science, management implications and model limitations. Even if an individual could successfully program a single model that would perform all the functions of the individual models of a meta-model (i.e. produce a mega-model), it is unlikely she or he would possess sufficient scientific expertise in all the subject areas involved to adequately advise forest managers of the limitations of the model or propose alternative management strategies when necessary.

Basics of linking models (meta-modelling)

The basic design principle behind the meta-modelling approach is that each individual component model provides key outputs that are then incorporated as inputs to a subsequent model in the meta-model framework, and that this does not need to be done in real time (i.e. simultaneously). For example, an ecological model like FORECAST can be used to provide the volume–age tables (m^3 ha^{-1}) for individual stands that are subsequently used as an input to a landscape-level harvest-scheduling model like FPS-ATLAS (Nelson, 2003). The output from the donor model must be compatible with the input format of the recipient model, and this may require reformatting of the output of the stand-alone donor model. Simple text file output is usually the easiest way to export/import data between models, although spreadsheets and database programs can be utilized to streamline data formatting and facilitate transfer. The modelling team needs to carefully define what data are to be transferred between which models, and what modelling sequence will be so that the necessary inputs are available when needed. For example, before a visualization program can be run it needs stand-level information such as tree species, heights and stand density provided by a model such as FORECAST, plus the age, previous silviculture treatment, current stand type, etc. provided by a model like FPS-ATLAS. The meta-model framework eliminates the need to run the individual models simultaneously that occurs in a mega-model, thus providing greater scheduling flexibility for the modelling team.

Types of ecological inputs provided to forest-level models

Ecological models predict a suite of important indicators of vegetation/stand dynamics at the stand level that need to be incorporated into analyses conducted at the forest estate level. Forest estate planning models such as Woodstock (Remsoft, 2006) require libraries of merchantable volume per hectare by stand age to predict forest growth, harvest and growing stock over time. When stands are subjected to various silvicultural treatments or are affected by natural disturbance, the subsequent stand dynamics need to be known to accurately predict forest-scale (i.e. landscape-level) development. Forest estate models can use as inputs projections from ecological models of stand attributes, related to age and treatments, such as species composition, tree diameters, stand density and log size distributions, and track their changes over time and space. Forest habitat models also use inputs from ecological models, such as understorey (herb and shrub) composition, stand structure, snags (by size, species, age and state of decay) plus downed woody material (by size, species and decomposition stage). The current keen interest in carbon offsets (a financial instrument aimed at a reduction in greenhouse gas (GHG) emissions, using as the unit of currency the economic value of sequestering and storing one metric tonne of atmospheric CO_2, or avoiding the release of the same quantity of carbon from fossil fuels or ecosystem storage) has resulted in many forest estate models tracking carbon storage as part of the routine analysis of management options. Ecological models are essential for providing the carbon inputs to these forest estate models.

Examples of applications of links between FORECAST and FPS-ATLAS, and other models

Analysis of Lemon Creek, a watershed in south-eastern BC

Lemon Creek is a 40,000ha public forest located in the Arrow Lakes Timber Supply Area of south-eastern BC. The Arrow Lakes Innovative Forest Practice Agreement (IFPA) brought together industry and government to seek new ways to increase the harvest in the Arrow Lakes Timber Supply Area while conserving a variety of non-timber values. As part of this initiative, FORECAST and FPS-ATLAS were used to make long-term predictions (over 250 years) of harvest levels in Lemon Creek according to two levels of silviculture investment in reforestation (extensive and intensive) and two management scenarios (one based on forest regulations set by the BC Forest Practices Code, and the other a TRIAD zoning approach) (Nelson, 2006; Wells and Nelson, 2006). FORECAST was used to generate volume–age curves (m^3 ha^{-1}) for approximately 35 stand groups that covered species such as pine, larch, Douglas-fir, western red cedar, western hemlock, grand fir, sub-alpine fir and an aspen/birch mix under both the extensive and intensive silviculture treatments. These yield data were used as inputs to the FPS-ATLAS model that assigned approximately 13,000 stand polygons to these 35 stand groups. The combined analysis of these two models highlighted an important problem with management scenarios regulated by the Forest Practices Code (e.g. evenly distributed small patch cuts produced an 'unnatural' and socially unacceptable landscape pattern), and more areas could be reserved for OG and wildlife habitat values with less permanent road access while maintaining current harvest levels using the TRIAD zoning approach.

Analysis of Tree Farm Licence 48, a large management unit in north-eastern BC

Tree Farm Licence 48 (TFL 48) is an approximately 600,000-ha public forest located in north-eastern BC. TFL 48 is located primarily in the sub-boreal biogeoclimatic forest zone (Pojar et al, 1987) and contains species such as lodgepole pine, white and black spruce, sub-alpine fir, aspen, birch and cottonwood. A portion of TFL 48 was used to develop a comprehensive decision-support system (a meta-model) that included FORECAST, FPS-ATLAS, a fire disturbance model, the habitat model SIMFOR (Daust and Sutherland, 1997), a recreation opportunity model and a visualization model (see Figure 5.22, Seely et al, 2004). This decision-support system was used to examine the long-term consequences (over 250 years) of various levels of fire suppression and sizes of harvest opening. The forest-level analysis also included the potential carbon sequestering and storage in the forest according to each scenario provided by FORECAST.

As the mountain pine beetle (MPB) outbreak in central BC became a major epidemic in the 1990s and spread to threaten north-eastern BC over the past decade, an additional study was undertaken to determine how to mitigate the economic and ecological impacts over the entire area of TFL 48 (Seely et al, 2008). FORECAST was used to generate attributes of stand and understorey dynamics following three assumptions about the level of attack by the MPB (high, medium and low) and three management strategies (1, replant pine for high timber productivity; 2, replant other species to minimize future MPB-related economic risk; and 3, manage for high biodiversity). This involved the creation of approximately 130 stand groups that were ultimately assigned to approximately 220,000 polygons in the FPS-ATLAS forest estate model. FPS-ATLAS then predicted the forest level effects (e.g. harvest levels and growing stock) over 75 years using planning periods of 2–3 years for the first 15 years and then 10-year periods thereafter to capture the dynamics of the beetle attack and subsequent salvage operations. In addition, FORECAST provided data on tree mortality that were input to the estate model in the form of number of snags per hectare (by size and by coniferous and deciduous species). These stand-level data were scaled up to the landscape in FPS-ATLAS, and, together with additional data on habitat attributes provided by FORECAST, were subsequently used by the SIMFOR model to assess habitat supply for indicator species of cavity-nesting birds. This meta-model led to important insights into management strategies designed to mitigate the ecological impacts of the beetle attack, particularly the identification of strategies that protect mixed conifer and deciduous stands.

Using hybrid models and visualization to communicate with the general public

The potential utility of the hybrid and meta-modelling approaches extends well beyond forestry into many aspects of resource management and issues planning. This section provides an overview of the role of hybrid modelling, linked to visualization methods, in a variety of case histories. It describes applications in local climate planning and visioning, community-based land-use planning, and in assessing the sustainability of the supply of renewable biomass energy from a community forest.

A diversity of applications of modelling in planning and resource management

There are clearly many ways to incorporate hybrid modelling into sustainable forest management (SFM), and other applications, to achieve a balance of ecological, social and economic values. One of these is in public involvement processes. Despite the increasing demand for active public involvement in decision making, there are as yet few established and widely accepted methods for achieving this in a manner which addresses appropriate temporal and spatial scales and considers multiple values and their trade-offs. However, there is growing recognition that appropriate models can make an important contribution. For example, public interest in being involved in planning for climate-change issues such as accounting and management of GHG emissions, increasing fire risk, low-carbon and sustainable communities, local food production, and low-carbon energy generation requires integrated, model-driven, participatory planning processes. The issues are simply too complex and too interdependent for non-computer methods to be effective. Traditional methods have used the 'jigsaw puzzle' approach, in which the individual component issues are considered separately in relative isolation. Forest sustainability assessment, community planning, public participation, forest management and land use all require the application of integrated computer-based decision-support technology involving spatial modelling and visualization for enhanced decision making and policy development (Sheppard, 2005a).

Sheppard (2005a) documents numerous studies of public participation in natural resource management in the past, but the practical efficacy of the results of these studies has been disappointing (Duinker, 1998; De Marchi and Ravetz, 2001; Hamersley Chambers and Beckley, 2003). This is reflected in low public participation and satisfaction with forest planning processes (Forest Practices Board, 2000), or even stand-offs between environmental lobby groups and the timber industry (Moote et al, 1997; Martin et al, 2000). Where participation succeeded in reaching consensus at a strategic planning level (Daniels and Walker, 2001), the process may not have been carried through to implementation decisions at an on-the-ground tactical level (Mendoza and Prabhu, 2005), leaving considerable room for accusations by stakeholders of bias in final decision making (Hamersley Chambers and Beckley, 2003). Such may be the fate of planning processes that develop a static plan that lacks rigorous targets tied to processes of public participation. Despite the risks that such participation failures pose, scientists have continued to use expert-based decision-support systems and to develop sophisticated computer-modelling approaches to assist forest managers in making complex spatio-temporal decisions without public input (Kangas et al, 2000, 2001; de Steiguer et al, 2003).

The highly technical, expert-based, professional approach would benefit from a more nuanced, collaborative process that combines 'soft' qualitative participatory approaches and 'hard' quantitative modelling approaches (Mendoza and Prabhu, 2005). This would permit dove-tailing of expert-driven knowledge and well-informed recommendations with the immediate needs of decision makers, and could involve different forms of modelling. These could include numerical (e.g. projecting population growth), spatial (e.g. mapping land-use requirements) and 3D modelling (e.g. showing tree heights or rising sea levels on a shoreline). Such hybrid-modelling approaches and model-informed visualization processes may fail to meet the definition of a true meta-model, in which outputs of one model become the inputs for the next model. However, such techniques may be more effective in addressing the multiple issues raised by climate change while exploiting limited pre-existing datasets and local expertise. The use of realistic landscape visualizations requires the careful application of ethical procedures to reduce the risk posed by issues such as modelling uncertainty or non-visual phenomena (Sheppard, 2005c).

Climate-change planning and hybrid modelling

Underlining the utility of hybrid modelling as a driver of better land management is the emerging need to understand the complex spatio-temporal implications of climate change (Blanco et al, 2009a). This is an enormous challenge, as climate science is still struggling to downscale from global predictions of climate change and its possible impacts to local scales, where specific response measures need to occur (Nicholson-Cole, 2005). The timescale of predicted impacts is too long for most people to comprehend, complicating public involvement in decision making. Uncertainties in climate-change science and difficulties in modelling the feedback effects of future species adaptations and possible effects of mitigation measures complicate current decision making at the local community level (Schroth et al, 2009).

The approach of the Intergovernmental Panel on Climate Change (IPCC) to the problem of climate change has consistently emphasized three basic features: being global in focus, aiming at enhanced understanding, and being expert-driven (Shaw et al, 2009). This approach has not been successful in stimulating local and regional engagement of the public in the climate-change issue. However, planning processes that put climate-change impacts in the context of local and regional issues by using visualizations of familiar locations can make predictions of the possible impacts of climate change more meaningful to local communities (Leiserowitz, 2004; Balmford et al, 2004; Shaw et al, 2009; Burch et al, 2010). Model-driven visualization can provide a valuable method by which understanding of predicted climate-change impacts can be linked to the willingness and capacity of local communities to institute change (Nicholson-Cole, 2005; Sheppard, 2005a). The results of local visioning can effectively bridge between climate science and policy development through the shared generation of knowledge and predictions as part of the decision-making process, sometimes called 'local visioning'.

Climate change is a global problem with significant impacts on local landscapes. Rising snow lines, decreasing snow cover, reduced summer snowmelt and stream flow, changing precipitation, altered weather and fire regimes, and increases in insect and disease problems are just some of the potentially serious local or landscape effects (IPCC, 2007). Recent monitoring data confirm and sometimes exceed the most severe IPCC scenario trajectories (Richardson and Steffen, 2009), and it is a high environmental, social and political priority to drastically reduce GHG emissions, thereby avoiding the most severe potential climate-change impacts. Such mitigation efforts may include switching from fossil fuels to biomass fuels produced in community forests and woodlots. These and other energy-planning strategies for mitigation can benefit greatly from approaches based on ecosystem modelling (Flanders et al, 2009a, 2009b, 2009c). Some climate-change impacts are occurring already and others are probably unavoidable. Consequently, adaptation, such as planning for longer fire seasons (Schroth et al, 2009) and changing sea levels (Shaw et al, 2009), is now a reality for resource managers, landscape architects and city planners alike.

Climate change has only very recently been recognized in current planning practice as a driver of landscape change. Provided the ethical dilemmas of stimulating human behavioural change and the moral implications of advocacy are understood, generating model-based visualizations via transparent, participatory processes can be an effective way of visioning a more sustainable way of living (Sheppard, 2005b; Sheppard et al, 2008). A few of these holistic processes and strategies will be introduced here, and one will be explored in greater detail.

Local Climate Change Visioning Project in Vancouver, BC
Researchers in the Local Climate Change Visioning Project explored these concepts as part of a multi-year study with stakeholders within Metro Vancouver, BC (Shaw et al, 2009). Four development scenarios (referred to as World Views), projected out to 2100, integrated environmental, climatic, behavioural and socio-economic conditions in the rural coastal city of Delta (part of Metro Vancouver) with probabilistic risks of climate-change impacts occurring. These four 'worlds', entitled 'do nothing', 'adapt to risk', 'efficient development' and 'deep sustainability', represented four possible community futures for impact-oriented (highest GHG emissions), adaptation-oriented, mitigation- and adaptation-oriented, and intensive mitigation-oriented (lowest GHG emissions) development trajectories, respectively.

Unique storylines of impact causes, governmental policies, rates of change and local response strategies, along with corresponding GHG emissions, were generated by a team of scientists and community members and used to structure the four worlds. Local coastal vulnerability and ecosystem shift modelling were linked to quantitative, localized IPCC sea-level rise forecasts from Environment Canada (Hill, 2006), which, along with the regional storylines, structured visualizations of the different worlds. These showed impacts of varying levels of sea-level rise corresponding to the different atmospheric GHG emissions trajectories and corresponding CO_2 concentrations assumed to occur in each world. Forecasts of population, land use and economic growth were modelled with the regional socio-economic modelling program GB Quest (Robinson and Tansey, 2006) using assumptions that were consistent within the contexts of each of the four worlds. By mapping these storylines onto 3D GIS datasets of the local area, and inserting 3D computer-aided design (CAD) architectural elements into these simulations, the implications of varying degrees of climate impacts and action became clear. Narratives were drawn up depicting different urban forms, such as passive-solar housing; infrastructure modification, such as raising dykes; land-use changes, such as agricultural land conservation; and lifestyle changes, such as commuting less. This allowed participants and decision makers to explore the key drivers and outcomes associated with different sets of choices, with coherent, visual outcomes (Figure 10.1).

District of North Vancouver project
As part of this multi-locale study, the district of North Vancouver, a forested mountain slope within Metro Vancouver, was explored in a similar manner. This brought together a unique group of community stakeholders and experts to consider climate impact projections and suitable response options. In this community, climate-change impacts such as a decreasing snow pack, an ascending snow line, and shifting precipitation patterns were downscaled to the district scale from IPCC scenarios by researchers at

Figure 10.1 *A World 4 'efficient development' low-carbon scenario showing passive-solar, flood-resilient amphibious homes in high-risk coastal areas*

Note: A resident looking over an existing sea wall can be seen near the centre of the image. Off-shore storm surge barriers and a sea-wall breach are visualized in this time-step (2050) due to the inevitability of some risk of climate-change impacts despite the high level of proactive mitigation in this scenario.

Source: David Flanders, UBC CALP, rendered in Visual Nature Studio.

Environment Canada, and mapped into 3D GIS datasets enriched with buildings, rivers and trees. Changing risks of wild fires, wind storms, landslides, debris flows and pest infestation were made spatially explicit by formulating simple GIS algorithms that take into consideration current risk, slopes, the severity and extent of the dry season, dominant tree species and understorey vegetation cover, stand age and soil type. These were mapped onto the 3D geovisualizations in the form of GIS coverages. The damaging effects of changing hydrologies, rainfall patterns and reactive management regimes were shown in stream erosion, forest health effects and in the demands on the Capilano Dam municipal water supply. Suburban development patterns and future energy sources such as wind and micro-hydro were also explored (Figure 10.2).

Researchers have documented the efficacy of different types of model and model-driven visualization in communicating environmental change and planning scenarios to a broad spectrum of the public. Participants of the Local Climate-Change Visioning Project were shown a wide range of material, including scientific data and model output, maps, imagery, precedent photos, 3D visualizations and animations. The benefits of community visioning that merges complex climate and environmental modelling into geovisualization systems include increased comprehension of complex information, increased willingness to modify rankings of concern about environmental impacts,

Figure 10.2 *A World 1 'do nothing' high-carbon scenario on the 'north shore' of Metro Vancouver, showing high-energy suburban homes expanding into high-risk areas, fire scars, and forest species shifts following windstorm blow-downs in valleys*

Source: David Flanders, UBC CALP, rendered in Visual Nature Studio.

and increased willingness and capacity to make difficult decisions involving trade-offs between values (Lewis and Sheppard, 2006; Sheppard et al, 2008; Schroth et al, 2009; Salter et al, 2009; Burch et al, 2010).

The manner in which modelled information is presented can significantly affect its impact on the public, practitioners and planning professionals. Rural communities that do not currently have the complex spatial or sophisticated model calibration data required to run ecosystem and management models or visualization software can still benefit from their use by employing publicly available datasets and basic spatial and 3D modelling programs such as GIS and SketchUp. An example of this is a case study conducted in Kimberly, south-eastern BC (Schroth et al, 2009). The study employed multidimensional navigation and interactive visualizations using 'virtual globes' such as Google Earth and Biosphere 3D to enhance comprehension of multidimensional, climate-related development scenarios. Limited modelling sets included snow-depth projections and susceptibility to MPB infestation. As with the Local Climate-Change Visioning Project, a participant survey was conducted to quantify the effectiveness of the integrated approach. The study confirmed the benefits of a visioning package consisting of a variety of visualization formats, including posters, virtual globes, maps and oral presentations to communicate and demonstrate credible, reliable data. Since participants varied in the mode of communication they prefered, the diversity of presentation formats (e.g. posters alongside virtual globes) acted to complement each other. However, the perception of undesirable drama due to much movement and interactivity with the virtual globes led to the alienation of some users.

Landscape planning and design exercise in Tapalpa, Mexico

Other participatory planning studies in which much simpler, 2D photo-editing techniques were used without the aid of complex geovisualization software and modelling suggest that the effectiveness of a visioning exercise depends not only on the realism and complexity of visualizations, but also on the process by which the visions of the future were generated (Valencia-Sandoval et al, 2010). In the large rural district of Tapalpa, Mexico, a landscape planning and design exercise was conducted with limited budget, resources and data, and virtually no local experience of such processes. Researchers conducted a lengthy series of semi-structured interviews with residents and visitors, undertook a site analysis to identify major landscape units, identified environmental and socio-economic problems endemic to each unit, and then proposed design and planning recommendations to bring to the public. Stakeholders from across a range of cultural and socio-economic backgrounds came together for a one-day workshop in which participants self-organized into groups, one for each landscape unit, and, under the guidance of researchers and academic volunteers, assessed the suitability of the proposed recommendations, and modifications thereof when needed. Visualizations were generated to communicate the recommendations for each landscape unit, and these were presented to groups of community leaders within major towns in the region. The following year, Tapalpa's Strategic Plan for Sustainable Development, which incorporated the products of the public involvement process, was legislated. An exhibition in the town's plaza that concluded the project was the final example of facilitating the self-implementation of outcomes (Figure 10.3).

In summary, new modelling technologies and linked participatory processes can meet the changing contemporary challenges of resource management and landscape and community planning. Effective planning processes which incorporate hybrid modelling and visualization must demonstrate that they are founded on credible science, use ethically a variety of forms of communication such as a documented visioning package, encompass broad stakeholder input during the process, and a execute a dedicated, transparent public engagement process. Adaptable to unique environments, cultures and scopes, the model-driven planning and visioning processes described in this chapter will continue to be refined and applied to real-world issues. Ultimately, the effectiveness of these processes and the accuracy of the products they generate will be tested through comparison with reality as it unfolds in the landscape.

Modelling the forest biomass-energy supply for a community forest in Prince George, BC

Biomass is an alternative low-carbon energy source that could help reduce GHG emissions in cities like Prince George (the largest northern city in BC) as part of its strategy to plan for climate change. Current research on the potential for local bioenergy (energy derived from biomass) is informing the development of a sustainable 'Smart Growth'

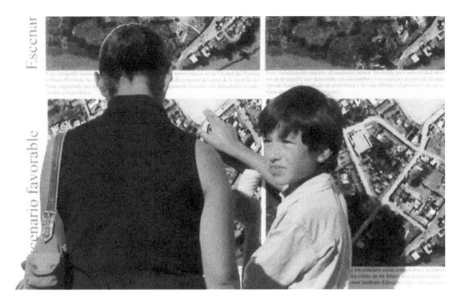

Figure 10.3 *Community members reacting to visualizations of two residential development scenarios displayed in the town plaza in Tapalpa, Mexico*

Note: the public exhibition played a major role in awareness-building and self-implementation of the plan in a region where the objectives of top–down planning are frequently unsuccessful at the local level.

concept plan for the downtown core in Prince George, which will include recommended provisions for the use of renewable energy. This concept plan is a key element in a multiphase-design *charrette* process (an intensive, multi-day design workshop) involving citizens, city staff and technical experts.

Prince George has grown to a population of over 80,000. Situated centrally in the interior of BC at the confluence of the Fraser and Nechako rivers, its principle industry has historically been forestry, and several pulp and saw mills have been established there. It has positioned itself as a centralized staging area for mining and exploration, government and regional transportation in northern and central BC. The city is surrounded by spruce and pine softwood forest stands, as well as hardwood and mixed stands with considerable volumes of aspen. The pine component has been heavily impacted by the MPB infestation.

The Prince George Community Forest (PGCF) occupies predominantly forested lands within municipal boundaries, encompassing 32,945ha, or almost 330km^2 of sub-boreal forest (Figure 10.4). Approximately one-fifth of the forested lands in the PGCF is either Crown-owned or municipal lands. 5000ha of these lands are under Crown ownership, which is historically the most viable form of land ownership for large-scale forest management for timber and non-timber forest products and values in Canada. Approximately 1000ha of the forested lands in the PGCF are municipal lands, which are subject to periodic thinning to manage forest fire risk.

Prince George - dominant biomass species

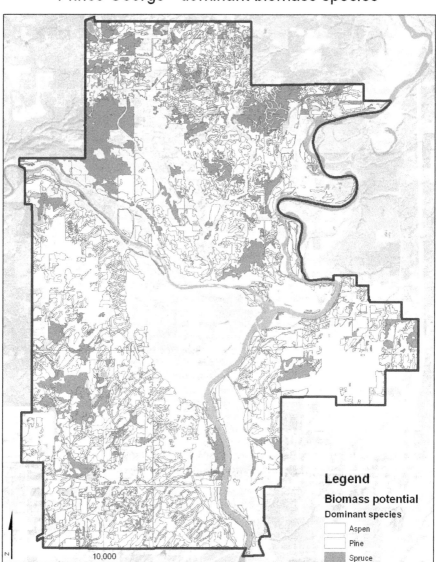

Figure 10.4 *The Prince George municipality and community forest*

Note: A colour version of this figure is shown in the plate section as Plate 5.

As a considerable potential source of local low-carbon bioenergy, the PGCF was the focus of a modelling-based assessment of bioenergy supply. Principal tree species suitable for biomass production include lodgepole pine, white spruce and aspen. Salvage of MPB-killed pine, and high-priority fuel-reduction activities in areas of high or very high

Figure 10.5 *Members of the community of stakeholders in Prince George creating a development plan in a charrette*

Source: Smart Growth on the Ground

wildfire risk in the Prince George region, together represent a significant potential source of biomass energy. However, the MPB-generated supply would only provide a short-term infusion of biomass and is therefore not considered a long-term bioenergy source (Stennes and McBeath, 2006). Although MPB-killed trees can be used for combustion up to seven years following mortality, current mortality rates are not expected to continue in the future (BC Ministry of Forests, 2009). All pine trees greater than a certain size will be dead.

The Smart Growth on the Ground (SGOG) Downtown Prince George Concept Plan is a vision of a more sustainable and vibrant downtown Prince George for the year 2035. A team of community stakeholders, experts and researchers created this vision during a four-day *charrette* in May 2009, where a diverse group of stakeholders created a development concept plan (Figure 10.5). The crafting of this vision was guided by eight SGOG principles of sustainable community planning (SGOG, 2009) to address challenges and opportunities that exist in the project area. Over the course of five public meetings, priorities and targets were crafted for each principle, which then guided the conversations, mapping, drawing and designing that took place during the charrette event (SGOG, 2009).

Unlike previous SGOG exercises in BC, the Prince George vision sought to address predicted climate-change impacts explicitly, as well as adaptation and mitigation responses. Climate scenarios looking as far forwards as 2100 were generated by researchers at the Pacific Climate Impacts Consortium (University of Victoria) as part of a parallel 'Adapting to Climate Change in Prince George' workshop with members of local government and the University of Northern BC (see www.sgog.bc.ca, under the 'Prince George partner community' section). This workshop identified a prioritized approach for developing a climate-change adaptation strategy for the city, including specific impacts to infrastructure, operations and planning, as well as allocating adaptation strategies to various local government departments.

Natural Resources Canada (NRCan) assisted in low-carbon energy planning as part of the climate-change mitigation efforts that would be integrated into the concept plan. Working in parallel to the SGOG process, researchers from the University of BC's Collaborative for Advanced Landscape Planning (CALP) coordinated with NRCan in a pilot study to assess local capacity to produce energy from forest resources under municipal jurisdiction. Prince George had already begun to explore the potential of a district energy system (DES) fuelled with biomass, but this had met with a mixed reaction from community representatives. Concerns that were voiced with regard to biomass harvesting, transportation, storage, and energy conversion include:

- loss of organic matter and woody debris at harvesting sites, with potential long-term reductions to ecosystem productivity;
- visual impacts of biomass harvesting in intensively managed, short rotation plantations, and associated effects on recreation, recreational properties, other land uses, etc.;
- air quality impacts of converting biomass to heat or electricity;
- traffic impacts on communities due to bulk deliveries of biomass; and
- cost effectiveness of the bioenergy production process.

Among the benefits of linking an energy planning study with a participatory community planning process is the opportunity to discuss options for a DES that can address community concerns. Key issues include distinguishing between a DES *per se* and its energy source; low-particulate combustion technologies such as gasification; energy plant siting and fuel sources; and net reductions in particulates and CO_2 emissions.

Potential co-benefits often cited for the use of bioenergy, in addition to the reduction in GHG emissions through switching from fossil fuels to biofuels, include:

- reduced community energy costs;
- increased local control over energy supplies and pricing, with reduced vulnerability to fluctuations in external fuel costs;
- additional value and employment from diversified forest products, beyond timber, fibre and pulp;
- reduced forest fire risk;
- opportunities for leadership in green energy and associated tourism; and
- community pride in the industry, ownership and self-reliance.

Concerns were expressed by some ecologists and stakeholders about the potential risk to other values of biomass harvesting in 'natural' forests, and the risk that intensive biomass removal could lead to a decline in forest productivity because of reductions in soil fertility over time (Kimmins, 2008b). These concerns required the use of an ecosystem-level, process-based ecosystem management model to explore the limits of harvesting that would ensure sustainable long-term biomass production and harvest. To address concerns over environmental impacts while building the case for locally supplied biomass as part of Prince George's energy mix, CALP relied on the FORECAST ecosystem modelling tool (see Chapter 5 in this volume and Kimmins,1997a; Kimmins et al, 1999 for a description of FORECAST applications for modelling sustainable bioenergy production, forest growth, and other resources and values). FORECAST provides estimates of total biomass produced and carbon storage, as well as exploration of various management/harvesting strategies and their effects on the sustainability of the ecosystem.

Using the BC Ministry of Forests and Range vegetation resource inventory (VRI), forest cover inventory, creek riparian area data, and BC biogeoclimatic zones (Pojar et al, 1987) as inputs, management scenarios for the Prince George Community Forest over 150 years were prepared (Table 10.1). Ten different analysis units (AUs) were delineated, each with a unique combination of dominant and secondary tree species, soil quality, site quality and biogeoclimatic zone. Trees are normally harvested at a younger age for energy than for traditional uses such as lumber, with implications for site nutrient depletion (Kimmins, 1977). There are several alternative silvicultural and harvest systems that can be used in bioenergy production, including clearcutting, thinning and a two-pass system described by Welham et al (2002).

In the two-pass harvest system, the fast-growing pioneer aspen is harvested at around 15 years, followed by harvest of commercially sized slower-growing coniferous species about 15–35 years later. Carefully applied, this system can enhance spruce volume and reduce the harvesting intervals through the protection of understorey spruce trees during overstorey removal (Welham et al, 2002). It is similar to stand thinning and other types of partial harvesting in that live standing biomass remains after a harvest, which serves to mitigate visual impacts that are related to social barriers to large-scale biomass harvesting. FORECAST was used to explore the effects of varying harvest intervals for aspen from 15 to 75 years, and for conifers (mainly lodgepole pine and black spruce) from 15 to 150 years. Aspen stands generally regenerate naturally, whereas spruce and pine stands may require replanting to ensure target stem densities. The two-pass strategy is only one of several sustainable harvest systems for these forests, requiring less replanting but more detailed harvest planning and execution.

For the FORECAST simulations of the Prince George Community Forest, soil organic matter and total ecosystem organic matter were used as an indicator of long-term forest sustainability (Morris et al, 1997). It is also a key measure of carbon storage. Organic carbon is beneficial for site productivity through its positive influence on many soil characteristics, and it facilitates key biological processes and nutrient release for tree growth (Seely et al, 2010). In an iterative process, the FORECAST model was run with variable stand rotation lengths (harvest intervals) to identify a management strategy that maintained no net change for these two variables after 150 years of management simulation (Figure 10.6). Initial modelling results suggested that in four of the more productive stand types, the overstorey and up to 50 per cent of understorey could be removed without reducing future harvest rates.

Figure 10.7 presents the sustainable biomass yield capacity for the areas of the PGCF that would be the main source of bioenergy, as well as for the high and very high wildfire risk areas outside of the biomass production stands, but excluding 30m riparian reserves for all creeks. Unclassified areas are mainly urban areas without significant tree canopy, or agricultural areas that lacked a dominant tree species. Biomass yields varied according to stand type, but ranged from 1 to 3BDT (bone dry tons) ha^{-1} y^{-1} with an overall average of 1.2BDT ha^{-1} y^{-1}.

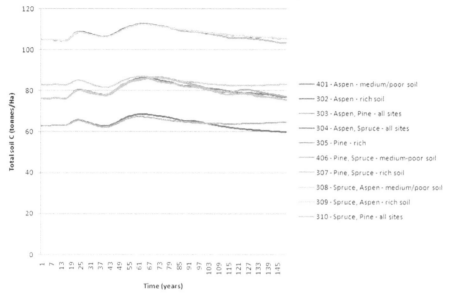

Figure 10.6 *FORECAST model output showing impact of harvesting rates on soil organic matter*

Note: A colour version of this figure is shown in the plate section as Plate 6.

Table 10.1 summarizes the sustainable harvest yields and resulting energy capacity from the PGCF. Total biomass yields, assuming all forested lands were to be harvested, results in potential heat energy generation, before losses due to system inefficiencies, of 419,700GJ y^{-1} given a conversion rate of 18GJ BDT^{-1} across hardwood and softwood species (Magelli et al, 2009). Crown and municipal lands, identified as the most viable option from an ownership/management perspective, would yield about one-third of that, at 138,500GJ y^{-1}. An additional 31,500GJ y^{-1} is available from other forested lands that require thinning for wildfire management purposes. However, approximately half of this biomass (12,000m^3 or 4500BDT) is currently removed as the annual allowable timber harvest (AAC). If the AAC continues to be used for timber production rather than energy use, the potential energy capacity would be reduced to 89,100GJ y^{-1}.

Eighty per cent of Prince George's currently planned community energy system's demand to serve downtown buildings is proposed to come from waste heat from existing industrial sites, leaving the remaining 20 per cent supplied by natural gas. According to this modelling exercise, the PGCF has enough biomass, harvested sustainably, to meet and exceed this remaining demand. The most efficient use of this biomass energy would be to provide for the base heating load; the relatively consistent heat demand that is required across all seasons, such as to supply hot water and night-time space

Prince George - biomass generation potential

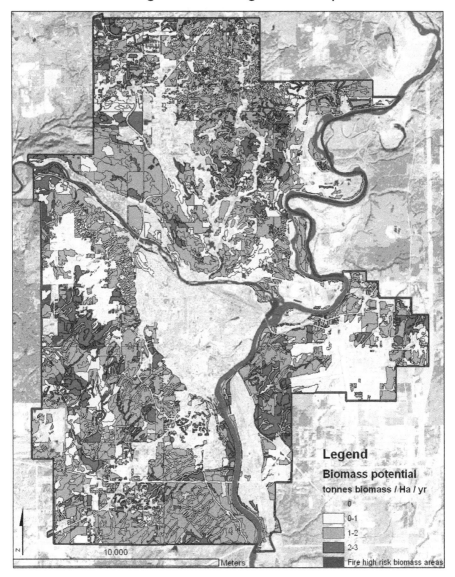

Figure 10.7 *Sustainable annual biomass harvests within the Prince George Community Forest*

Note: A colour version of this figure is shown in the plate section as Plate 7.

Table 10.1 *Summary of potential biomass and energy production for various management scales*

Sustainably harvested biomass from:	Biomass production	Potential heat energy capacity	Realized heat energy capacity	Number of dwelling units that could be supplied heat and hot water
	BDT y^{-1}	GJ y^{-1}	GJ y^{-1}	
All forest lands within PGCF	23,300	419,700	335,760	4330
Crown and municipal lands in the PGCF	7700	138,500	110,800	1430
Fire risk thinning on forested lands not in Crown and municipal lands in the PGCF	1750	31,500	25,200	325
Crown land, municipal land, fire risk thinning on forested lands in the PGCF minus the Crown Land AAC	4950	89,100	71,290	920
Approximate 20km forest buffer around Prince George	129,105	2,775,750	N/A	Approximates existing residential heating demand for Prince George, assuming 80% efficient systems

Source: for more detail, see Flanders (2009b).

heating, leaving short-duration but highly variable peak demand periods (such as for space heating during winter evenings) to be satisfied by natural gas.

If the heat energy generated from sustainably harvested biomass on Crown and municipal lands in the PGCF was applied across the entire residential sector of Prince George, it would satisfy about 5 per cent of the annual heating and hot water demand.[2] Using the simulated average sustainable harvest rate of around 1.2BDT ha^{-1}, 129,105ha would be required to satisfy 100 per cent of Prince George's 2002 residential heating demand. This is 25 times more energy production than could be extracted from PGCF Crown and municipal lands and a land base of approximately 20km around the city, as shown in Table 10.1.[3] More details of this study are in Flanders (2009b).

The results of this forest and land-use modelling exercise were circulated to community stakeholders before the Prince George charrette event. Amongst the topics presented were climate-change adaptation, rainwater management, residential energy characterization, urban trees and climate change, air quality, and GHG emissions and the business sector. This information was channelled into the series of workshops leading up to the *charrette*, and included indicator and target-setting workshops that corresponded to community priorities relating to the eight SGOG principles (SGOG, 2009). The research team's indicator relating to mitigation responses to climate change was 'percentage of Prince George heating demand to be supplied by local, low-carbon energy (such as biomass or solar)'. The target was set for 75 per cent. Although a range of low-carbon energy sources including solar thermal, solar air, and geoexchange would be required to satisfy this target, biomass was shown to be a significant component of this vision. Population growth estimates for Prince George showed only a moderate addition of residents, the majority of whom were to be located in the downtown core where they would have access to mixed-use and higher-density buildings that could be tied into the proposed DES. This being the case, the heating needs of 80 per cent of incoming downtown residents expected through to 2035 could be satisfied by sustainably harvested local biomass.

Products stemming from this research include a technical report (Flanders et al, 2009b), mapping of the local biomass supply and sustainable yield, and interactive, navigable Google Earth coverage (Figure 10.8). The 3D Google Earth mapping is a useful tool for exploring different scales of biomass supply (from neighbourhood woodlots to larger forest landscapes), and where these are located relative to the urban environment. These images were used when reporting back to a final public meeting in Prince George following the charrette exercise, and have been used in public presentations and educational seminars. Similar methods have since been used for the municipalities in the Northshore of Metro Vancouver, and will be integrated into their official community plan review process, as well as the extension of a DES in the district of North Vancouver and other low-carbon energy pilot projects envisaged for the region.

CALP-Forester: a user-friendly interface for landscape-level modelling with applications to engage stakeholders in modelling processes

The use of visualization in forestry has primarily been to communicate a plan or a limited set of alternatives to a wider audience. Visualization of forestry plans has become a normal part of forest practices in some areas, as forest companies seek approval for their harvesting in contentious areas or to meet visual impact assessment requirements. There are a variety of programs which focus on the production of highly detailed photo-realistic simulations. 3D Nature's World Construction Set (WCS), currently known as Visual Nature Studio (3D Nature, 2002) has become the *de facto* standard being used by private consultants worldwide to visualize the output from forestry models. These systems are GIS-based, and require the manual (or semi-automatic) entry of tree species, forest composition, tree heights and other stand and tree attributes in order to generate a simulation for a particular prescription, time-step or viewpoint. While the results can be visually impressive, the process is generally very time-consuming. WCS in particular requires a large number of poorly documented variables to be set for each rendering of a forest scene (Cavens, 2002).

Another widely used system is the US Forest Service suite of tools: Envision and the Stand Visualization System (SVS) (McGaughey, 1998). Envision and SVS were designed specifically to provide visualizations of forest stand and landscape data. As with WCS, the user must carefully populate the programs with forest data and specify camera and view

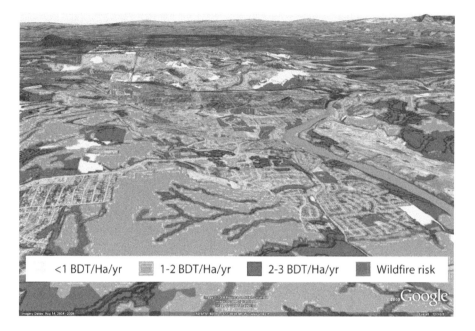

Figure 10.8 *Interactive 3D mapping of sustainable biomass yields modelled with FORECAST in Google Earth*

Note: A colour version of this figure is shown in the plate section as Plate 8.

parameters. These programs generate still images only, and depending on the complexity of the model, the images require anywhere from 15 seconds to 10 minutes to render. Although the user can query the final image to obtain basic information such as the location in 3D space of a given point of the visible terrain, the interface does not provide any ability for the user to interact directly with the forestry data being depicted.

In order to overcome this limitation, a new class of decision-support tools has been proposed: design-based decision support (DBDS) (Cavens, 2002). These tools combine the strengths of model-based decision support with the iterative sketching process traditionally used by spatial designers. A key goal of a DBDS tool is to allow the user to interactively test alternative management approaches and understand the implication of these approaches. The emphasis is on usability. The user should be able to focus on the task at hand, and not be distracted by excessive options. Obviously, this is a delicate balance. If the available user interface options are too limited, the user will not be able to accomplish their task. As a result, DBDS systems have to be domain-specific and not general-purpose modelling tools (such as a GIS or CAD system). In order to create accurate depictions of both present and future conditions, a DBDS must be tied to real data-driven models. These should be scientifically based, used to populate the visual interface, and should be integrated into the decision-support system in a bi-directional manner; the simulation should be able to visualize the output of the model, and the model inputs should be modifiable from within the simulation itself.

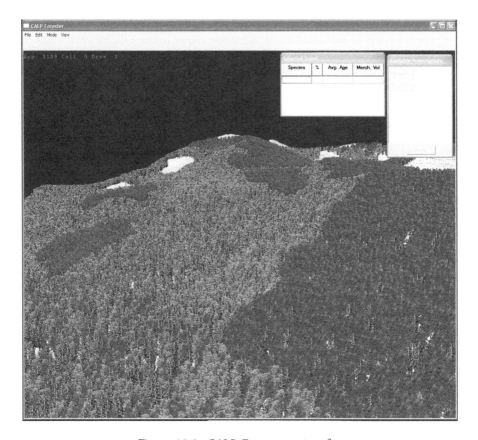

Figure 10.9 *CALP-Forester user interface*

Note: A colour version of this figure is shown in the plate section as Plate 9.

CALP-Forester has been developed by CALP (Faculty of Forestry, UBC) to test the ideas of a design-based decision-support system in a forestry setting. It is made up of two subsystems: the CALP-Forester program itself, which provides the graphical user interface that the end user interacts with; and the CALP Visualization System (CVS), which provides the link to the underlying forest models such as FORECAST and LLEMS.

CALP Forester is a program, written in C++, which provides the main DBDS interface to the user. It uses 3D rendering techniques to provide a perspective view of the landscape being investigated. The software relies on open source libraries to provide platform-independent 3D rendering and graphical user interface elements, respectively. The user interface is very simple. There are only a few windows that offer the user information and options (Figure 10.9). The interface displays a 3D view of the landscape, using 2D 'billboards' to represent individual tree stems, the characteristics of which (including species, stem height, height of crown) are derived from the underlying forest model(s). Other characteristics, such as timber volume, can also be specified and stored, but they have no impact on the user interface.

There are three main modes available to the user: movement, select by stand, and select by drawing. In movement mode, the mouse is used to move the viewer through the landscape. Different mouse buttons enact different kinds of motion, including glidding, zooming and rotating. In select by stand mode, the user indicates, with a mouse click, which stand to select. Summary data for the stand (in the current version, this includes stand composition, average height and stand age) are displayed in the 'Select Stand' window based on the underlying data from LLEMS (Chapter 6). The third mode, select by drawing, allows the user to draw directly on the screen to specify an area of interest, and, if necessary, define a new stand by applying a management prescription, selecting from the 'Available Prescriptions' window.

A drawback of other stand-based forestry modelling systems is that they force the user to work with the existing stands, and only allow the user to assign prescriptions to the existing spatial definitions of those stands. This is problematic, as the spatial extent and implied uniform characteristics within the stand polygons frequently do not correspond with real-world conditions or respond to new site-specific issues identified by the user. Stands are defined based on air-photo interpretation, and are often arbitrary delineations across a diffuse border. Working with previously defined stands is confining in forestry design, as one needs a much finer grain of control for purposes such as visual screening, hydrological and riparian buffering, block design or highly localized prescriptions such as seed tree locations.

The CALP-Forester interface allows the user to redefine the extent of a stand dynamically by drawing directly onto the 3D scene. The interface interprets the cursor position as representing a ray extending from the screen to the 3D terrain and can define an enclosed polygon which specifies the area of interest, which is then highlighted in red. The 3D polygons specified by the user are collapsed to 2D areas (as by definition they follow the terrain) and are compared with the existing stands (Figure 10.10). The system uses a variant of Vatti's polygon intersection test (Vatti, 1992) to find the geometric intersections of the polygon that the user has drawn within the existing stands. New stand polygons are generated from these intersections and re-triangulated in real time. Once a stand has been selected, the user can apply prescriptions to the particular stand. These can include different levels of harvesting or management operations such as thinning or fertilizing. The user selects a prescription from the 'Available Prescriptions' window, and this is applied to the currently selected stand (Figure 10.10).

The CALP-Forester visualization system was developed to integrate output from the ecosystem-level models FORECAST and LLEMS, previously described in this book. It was initially developed to speed up the modelling process for the Arrow Lakes IFPA project described above (Sheppard and Meitner, 2005), and to generate output from model runs that could be suitably displayed by WCS. It has been applied in other large-scale forest landscape visualization projects (Seely et al, 2004), and customized and extended for the CALP-Forester program. Preliminary tests with other researchers and practitioners indicate a high level of interest and perceived usability in both design and communication functions, with considerable potential as an engagement and education tool in forest design and modelling. Further testing of this interface with users is anticipated.

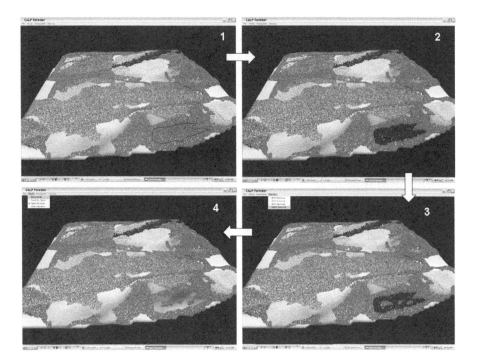

Figure 10.10 *Application of a variable retention prescription to a stand in CALP-Forester*

Note: 1 = drawing the area limits; 2 = selection of the stand, with the 'Selected Stand' window displaying the stand values for a set of variables as determined by LLEMS, whereas the 'Available Prescriptions' window shows the different prescriptions that can be applied to the selected stand; 3 = the area inside the stand that will not be harvested is selected by the same procedure as in the first step; 4 = the final stand, with the harvested areas shown as a clearcut and the areas not harvested looking like islands of trees inside the clearcut. A colour version of this figure is shown in the plate section as Plate 10.

Take-home message

With the end of timber-only forestry in sight, and the elusive goal of full ecosystem management beckoning us into the future, there must be a new set of decision-support, planning and communication tools available to policy makers, forest managers and other forest stakeholders. We believe that these should address multi-spatial scales, from stand to landscapes of various sizes. They must be capable of considering multiple values and their interactions and trade-offs: environmental, social and economic. But this is not enough. Unless forest policy and practice is successful in engaging public input and participation at the policy development stage dealing with principles, goals and objectives, and at the resource management planning stage dealing with ever-changing spatial landscape patterns of different ecosystem conditions and values, the desired social licence to manage public forests will remain elusive. Conflicts over forests will continue.

In this context, we see continued development of multi-value ecosystem management models that can address ecosystem processes and dynamics and address multiple values combined into meta-models. Complementing this, we see the need for communication tools based on various levels of visualization and public participation processes. Engagement of social scientists and other non-biologists in the process of agreeing on SFM strategies will be facilitated by interactive gaming tools in which visualization is the common language linking all involved. However, to be credible and more than a 'Potemkin village', such interactive visualization must be based on a credible, science-based representation of real ecosystems, their processes and dynamics.

Additional material

Readers can go to the companion website to this book (www.forestry.ubc.ca/ecomodels) to access the full versions of the following additional material:

- a movie clip visualization of the potential increase in the water level in Vancouver, BC;
- a movie clip showing the functionality of CALP-Forester;
- a copy of the Flanders et al (2009c) report.

Notes

1 This chapter was written in collaboration with researchers from the Department of Forest Resource Management (UBC, Faculty of Forestry), John Nelson (Forest Level Planning and Modelling Operations Research), Stephen Sheppard, David Flanders and Duncan Cavens (CALP).
2 Based on total residential demand of 2,775,750GJ y^{-1}, and 138,500GJ y^{-1} of PGCF biomass energy.
3 The 20km buffer does not include an assessment that these lands are forested and harvestable; further analysis using VRI GIS data and a FORECAST model simulation would be required to determine a more accurate sustainable yield buffer.

References

3D Nature (2002) *World Construction Set*, 3D Nature, LLC, Morrison, CO

Aber, J. D. (1997) 'Why don't we believe the models?', *Bulletin of the Ecological Society of America*, vol 78, pp232–233

Aber, J. D., and Melillo, J. M. (1982) *FORTNITE: A Computer Model of Organic Matter and Nitrogen Dynamics in Forest Ecosystems*, Research Bulletin R3130, University of Wisconsin, Madison, WI

Acevedo, M. F., Pamarti, S., Ablan, M., Urban, D. L. and Mikler, A. (2001) 'Modeling forest landscapes: parameter estimation from gap models over heterogeneous terrain', *Simulation*, vol 77, pp533–568

Adamowicz, W. L., White, W. and Phillips, W. E. (eds) (1993) *Forestry and the Environment: Economic Perspectives*, CAB International, Wallingford, UK

Aerts, R. and Chapin, F. S. (2000) 'The mineral nutrition of wild plants revisited: a re-evaluation of processes and patterns', *Advances in Ecological Research*, vol 30, pp1–67

Ågren, G. I. and Bosatta, E. (1996) *Theoretical Ecosystem Ecology: Understanding Element Cycles*, Cambridge University Press, Cambridge

Aitken, K. E. H., Weibe, K. L. and Martin, K. (2002) 'Nest-site reuse patterns for a cavity-nesting bird community in interior British Columbia', *The Auk*, vol 119, pp391–402

Akilli, G. K. (2007) 'Games and simulations: a new approach in education?', in D. Gibson, C. Aldrich and M. Prensky (eds) *Games and Simulations in Online Learning: Research and Development Frameworks*, Information Science Publishing, Hershey, PA

Alberta Environment (2006) *Land Capability Classification System for Forest Ecosystems in the Oil Sands, Volume 1: Field Manual for Land Capability Determination* (third edition), Goverment of Alberta, Edmonton, AB

Aldrich, C. (2005) *Learning By Doing: A Comprehensive Guide to Simulations, Computer Games, and Pedagogy in E-Learning and Other Educational Experiences*, Pfeiffer, San Francisco, CA

Alessi, S. M. and Trollip, S. R. (2001) *Multimedia for Learning: Methods and Development* (third edition), Ally and Bacon Publication, Boston, MA

Allen, G. M. and Gould, E. M. (1986) 'Complexity, wickedness, and public forests', *Journal of Forestry*, vol 84, pp20–23.

Amthor, J. S. (2000) 'The McCree–de Wit–Penning de Vries–Thornley respiration paradigms: 30 years later', *Annals of Botany*, vol 86, pp1–20

Andalo, C., Beaulieu, J. and Bouquet, J. (2005) 'The impact of climate change on growth of local white spruce populations in Québec, Canada', *Forest Ecology and Management*, vol 205, pp169–182

Anderson, P. W. (1972) 'More is different', *Science*, vol 177, pp393–396

Anderson, L. S. and Sinclair, F. L. (1993) 'Ecological interactions in agroforestry systems', *Agroforestry Abstracts*, vol 6, pp57–91

Anten, N. P. R. and Hirose, T. (1998) 'Biomass allocation and light partitioning among dominant and subordinate individuals in *Xanthium canadense* stands', *Annals of Botany*, vol 82, pp665–673

Arnott, J. T. and Beese, W. J. (1997) 'Alternatives to clearcutting in BC coastal montane forests', *The Forestry Chronicle*, vol 73, pp670

Assmann, E. (1970) *The Principles of Forest Yield Study*, Pergamon Press, New York, NY

Attiwill, P. M. (1994) 'The disturbance of forest ecosystems: the ecological basis for conservative management', *Forest Ecology and Management*, vol 63, pp247–300

Axelrod, M. C. and Kercher, J. R. (1981) 'User manual for SILVA: A computer code for estimating effects of pollution on the growth and succession of western coniferous forests', in *UCID-18594 Review*, Lawrence Livermore National Laboratory, Livermore, CA

Baker, W. L. (1994) 'Restoration of landscape structure altered by fire suppression', *Conservation Biology*, vol 8, pp763–769

Baker, W. L. (1999) 'Spatial simulation of the effects of human and natural disturbance regimes on landscape structure', in D. J. Mladenoff and W. L. Baker (eds) *Spatial Modelling of Forest Landscape Change: Approaches and Applications*, Cambridge University Press, Cambridge

Balandier, P., Bergez, J.-E. and Etienne, M. (2003) 'Use of the management-oriented silvopastoral model ALWAYS: calibration and evaluation', *Agroforestry Systems*, vol 57, pp159–171

Balmford, A., Manica, A., Airey, L., Birkin, L., Oliver, A. and Schleicher, J. (2004) 'Hollywood, climate change and the public', *Science*, vol 305, p1713

Barclay, H. J., Pang, P. C. and Pollard, D. F. W. (1986) 'Above-ground biomass distribution within trees and stands in thinned and fertilized Douglas-fir', *Canadian Journal of Forest Research*, vol 16, pp438–442

Bartelink, H. H. (2000) 'A growth model for mixed forest stands', *Forest Ecology and Management*, vol 134, pp29–43

Baskent, E. Z. and Keles, S. (2005) 'Spatial forest planning: a review', *Ecological Modelling*, vol 188, pp145–173

Baskerville, G. L. (1985) 'Adaptive management: wood availability and habitat availability', *The Forestry Chronicle*, vol 61, pp171–175

Battaglia, M., Sands, P., White, D. and Mummery, D. (2004) 'CABALA: a linked carbon, water and nitrogen model of forest growth for silvicultural decision support', *Forest Ecology and Management*, vol 193, pp251–282

Baudrillard, J. (1983) *Simulations*, Semiotext[e], New York, NY

BC Ministry of Forests (1996) *Tree Tales: Interactive Forest Science Computer Program – Teacher's Guide and CD*, British Columbia Ministry of Forests, Victoria, BC

BC Ministry of Forests (1999) *'Guidelines for Developing Stand Density Management Regimes*, BC Ministry of Forests, Forest Practice Branch, Victoria, BC

BC Ministry of Forests and Range (2006) *Preparing for Climate Change: Adapting to Impacts on British Columbia's Forest and Range Resources*, BC Ministry of Forests and Range, Victoria, BC

BC Ministry of Forests (2009) 'Mountain pine beetle: frequently asked questions', www.for.gov.bc.ca/hfp/mountain_pine_beetle/faq.htm, accessed October 2009

Beanlands, G. E. and Duinker, P. N. (1984) 'An ecological framework for environmental impact assessment', *Journal of Environmental Management*, vol 18, pp267–277

Beaudet, M., Messier, C. and Canham, C. D. (2002) 'Predictions of understorey light conditions in northern hardwood forests following parameterization, sensitivity analysis, and tests of the SORTIE light model', *Forest Ecology and Management*, vol 165, pp235–248

Beaumont, L. J., Hughes, L. and Poulsen, M. (2005) 'Predicting species distributions: use of climatic parameters in BIOCLIM and its impact on predictions of species' current and future distributions', *Ecological Modelling*, vol 186, pp250–269

Beck, J. and Wade, M. (2004) *Got Game: How the Gamer Generation is Reshaping Business Forever*, Harvard Business School Press, Cambridge, MA

Beese, W. (2000) 'Windthrow monitoring of alternative silvicultural systems in montane coastal forests', Windthrow Researchers Workshop, 31 January–1 February, Richmond, BC

Berger, U. and Hildenbrandt, H. (2000) 'A new approach to spatially explicit modelling of forest dynamics: spacing, ageing and neighbourhood competition of mangrove trees', *Ecological Modelling*, vol 132, pp287–302

Bergeron, Y. and Flannigan, M. D. (1995) 'Predicting the effects of climate change on fire frequency in the southeastern Canadian boreal forest', in M. J. Apps, D. T. Price and J. Wisniewski (eds) *Boreal Forests and Global Change: Proceedings*, Kluwer Academic Press, Dordrecht

Bergeron, Y., Harvey, B., Leduc, A. and Gauthier, S. (1999) 'Forest management guidelines based on natural disturbance dynamics: stand- and forest-level considerations', *The Forestry Chronicle*, vol 75, pp49–54

Bergeron, Y., Gauthier, S., Flannigan, M. and Kafka, V. (2004) 'Fire regimes at the transition between mixedwood and coniferous boreal forest in northwestern Quebec', *Ecology*, vol 85, pp1916–1932

Berninger, F. and Nikinmaa, E. (1997) 'Implications of varying pipe model relationships on Scots pine growth in different climates', *Functional Ecology*, vol 11, pp146–156

Beven, K. (2002) 'Towards a coherent philosophy for modelling the environment', *Proceedings of the Royal Society of London Series A*, vol 458, pp1–20

Bi, J., Blanco, J. A., Kimmins, J. P., Ding, Y., Seely, B. and Welham, C. (2007) 'Yield decline in Chinese Fir plantations: a simulation investigation with implications for model complexity', *Canadian Journal of Forest Research*, vol 37, pp1615–1630

Blackwood, J. S., Dresner, M. and Luh, H.-K. (2006) 'Using student generated qualitative ecological models', http://tiee.ecoed.net/vol/v4/experiments/ecological_models/abstract.html, accessed 20 June 2008

Blanco, J. A. (2007) 'The representation of allelopathy in ecosystem-level forest models', *Ecological Modelling*, vol 209, pp65–77

Blanco, J. A. (2010) 'The legacy of forest management: analyzing its influence on tree growth with ecosystem models', in W. P. Karam (ed.) *Tree Growth: Influences, Layers and Types*, Nova Science Publishers, New York, NY

Blanco, J. A. and González, E. (2010) 'Exploring the sustainability of current management prescriptions for *Pinus caribaea* plantations in Cuba: a modelling approach', *Journal of Tropical Forest Science*, vol 22, pp139–154

Blanco, J. A., Zavala, M. A., Imbert, J. B. and Castillo, F. J. (2005) 'Sustainability of forest management practices: evaluation through a simulation model of nutrient cycling', *Forest Ecology and Management*, vol 213, pp209–228

Blanco, J. A., Imbert, J. B. and Castillo, F. J. (2006) 'Effects of thinning on nutrient pools in two contrasting *Pinus sylvestris* L. forests in the western Pyrenees', *Scandinavian Journal of Forest Research*, vol 21, pp143–150

Blanco, J. A., Seely, B., Welham, C., Kimmins, J. P. and Seebacher, T. M. (2007) 'Testing the performance of a forest ecosystem model (FORECAST) against 29 years of field data in a *Pseudotsuga menziesii* plantation', *Canadian Journal of Forest Research*, vol 37, pp1808–1820

Blanco, J. A., Imbert, J. B. and Castillo, F. J. (2008) 'Nutrient return via litterfall in two contrasting *Pinus sylvestris* forests in the Pyrenees under different thinning intensities', *Forest Ecology and Management*, vol 256, pp1840–1852

Blanco, J. A., Welham, C., Kimmins, J. P., Seely, B. and Mailly, D. (2009a) 'Guidelines for modeling natural regeneration in boreal forests', *The Forestry Chronicle*, vol 85, pp427–439

Blanco, J. A., Imbert, J. B. and Castillo, F. J. (2009b) 'Nutrient retranslocation in two *Pinus sylvestris* forests in the western Pyrenees under different thinning intensities', *Ecological Applications*, vol 19, pp682–698

Boldor, M. (2007) 'A field and simulation study of the initiation phase in Douglas-fir plantations', MSc thesis, University of British Columbia, Vancouver, BC

Bonan, G. B. (1989) 'Environmental factors and ecological processes controlling vegetation patterns in boreal forests', *Landscape Ecology*, vol 3, pp111–130

Bossel, H. and Krieger, H. (1991) 'Simulation model of natural tropical forest dynamics', *Ecological Modelling*, vol 59, pp37–71

Bossel, H. and Krieger, H. (1994) 'Simulation of multi-species tropical forest dynamics using a vertically and horizontally structured model', *Forest Ecology and Management*, vol 69, pp123–144

Botkin, D. B. (1990) *Discordant Harmonies: A New Ecology for the Twenty-First Century*, Oxford University Press, New York, NY

Botkin, D. B. (1993a) *Forest Dynamics: An Ecological Model*, Oxford University Press, New York, NY

Botkin, D. B. (1993b) *JABOWA-II: A Computer Model of Forest Growth*, Oxford University Press, New York, NY

Botkin, D. B., Janak, J. F. and Wallis, J. R. (1972) 'Some ecological consequences of a computer model of forest growth', *Journal of Ecology*, vol 60, pp849–872

Bourgeois, W. W. (2008) 'Ecosystem-based management: its application to forest management in British Columbia', *BC Journal of Ecosystems and Management*, vol 9, pp1–11

Bowden-Dunham, M. T. (1998) *A Productivity Comparison of Clearcutting and Alternative Silviculture Systems in Coastal British Columbia* (Technical Report TR-122), Forest Engineering Research Institute of Canada, Vancouver, BC

Box, G. E. P. (1979) 'Robustness in scientific model building', in R. L. Launer and G. N. Wilkinson (eds) *Robustness in Statistics*, Academic Press, New York, NY

Bradshaw, A. D. and Chadwick, M. J. (1980) 'The restoration of land: the ecology and reclamation of derelict and degraded land', in *Studies in Ecology Volume 6*, Blackwell Scientific Publications, Oxford, pp1–317

Bransford, J. D., Brown, A. L. and Cocking, R. R. (eds) (2000) *How People Learn: Brain, Mind, Experience, and School*, Commission on Behavioural and Social Sciences and Education/National Research Council, National Academy Press, Washington, DC

Brix, H. (1983) 'Effects of thinning and nitrogen fertilization on growth of Douglas-fir: Relative contribution of foliage quantity efficiency', *Canadian Journal of Forest Research*, vol 13, pp167–175

Brugnach, M., Pahl-Wostl, C., Lindenschmidt, K. E., Janssen, J. A. E. B., Filatova, T., Mouton, A., Holrz, G., Van der Keur, P. and Gaber, N. (2008) 'Complexity and uncertainty: rethinking the modelling activity', in A. J. Jakeman, A. A. Voinov, A. E. Rizzoli and S. H. Chen (eds) *Environmental Modelling, Software and Decision Support: State of the Art and New Perspectives*, Elsevier, Amsterdam

Brunner, A. and Nigh, G. (2000) 'Light absorption and bole volume growth of individual Douglas-fir trees', *Tree Physiology*, vol 20, pp323–332

Bugmann, H. (1991) 'Development of a simplified forest gap model for workstations and personal computers' (internal report), Systems Ecology Group, ETHZ, Zürich

Bugmann, H. (1996) 'A simplified forest model to study species composition along climate gradients', *Ecology*, vol 77, pp2055–2074

Bugmann, H. (2001) 'A review of forest gap models', *Climatic Change*, vol 51, pp259–305

Bugmann, H., Grote, R., Lasch, P., Lindner, M. and Suckow, F. (1997) 'A new forest gap model to study the effects of environmental change on forest structure and functioning', in G. M. J. Mohren, K. Kramer and S. Sabaté (eds) *Global Change Impacts on Tree Physiology and Forest Ecosystems*, Kluwer Academic Publishers, Dordrecht, The Netherlands

Bunnell, F. L. and Boyland, M. (2003) 'Decision-support systems: it's the question not the model', *Journal of Natural Conservation*, vol 10, pp269–279

Burch, W. R. and Grove, M. J. (1999) 'Ecosystem management: some social and operational guidelines for practitioners', in W. T. Sexton, A. J. Malk, R. C. Szaro and N. C. Johnson (eds) *Ecological Stewardship: A Common Reference for Ecosystem Management*, Elsevier Science, Oxford

Burch, S., Sheppard, S. R. J., Shaw, A. and Flanders, D. (2010) 'Planning for climate change in a flood-prone community: municipal barriers to policy action and the use of visualizations as decision-support tools', *Journal of Flood Risk Management*, vol 3, pp126–139

Burger, J. A. and Zipper, C. E. (2002) *How to Restore Forests on Surface-Mined Land: Virginia Cooperative Extension*, Virginia State University, Blacksburg, VA

Burton, P. J. (1996) 'When is vegetation control needed?', in P. G. Comeau, G. J. Harper, M. E. Blache, J. O. Boateng and L. A. Gilkeson (eds) *Integrated Forest Vegetation Management: Options and Applications*, Canadian Forest Service, British Columbia Ministry of Forests, Victoria, BC

Burton, P. J. (2001) 'Windthrow patterns on cutblock edges and in retention patches in the SBSmc', Windthrow Researchers Workshop, 31 January–1 February, Richmond, BC

Burton, P. J. and Cumming, S. G. (1991) *ZELIG BC: User's Guide to the Prototype of a Forest Succession Simulator for the Evaluation of Partial Cutting Options in British Columbia*, University of British Columbia, Vancouver, BC

Burton, P., Kneeshaw, D. and Coates, D. (1999) 'Managing forest harvesting to maintain old growth in boreal and sub-boreal forests', *The Forestry Chronicle*, vol 75, pp623–631

Busing, R. T. (1991) 'A spatial model of forest dynamics', *Vegetatio*, vol 92, pp167–179

Busing, R. T. and Clebsch, E. E. C. (1987) 'Application of a spruce-fir forest canopy gap model', *Forest Ecology and Management*, vol 20, pp151–169

Busing, R. T. and Mailly, D. (2004) 'Advances in spatial, individual-based modelling of forest dynamics', *Journal of Vegetation Science*, vol 15, pp831–842

Cai, T. and Dang, Q.-L. (2002) 'Effects of soil temperature on parameters of a coupled photosynthesis–stomatal conductance model', *Tree Physiology*, vol 22, pp819–827

Cabanettes, A., Auclair, D. and Imam, W. (1999) 'Diameter and height growth curves for widely-spaced trees in European agroforestry', *Agroforestry Systems*, vol 43, pp169–181

Cabarle, B., Cashwell, J., Coulombe, M., Mater, J., Stuart, W., Winterhalter, D. and Hill, L. (1995) 'Forestry certification: an SAF study group report', *Journal of Forestry*, vol 93, pp6–10

Camiré, C., Trofymow, J. A., Duschene, L., Moore, T. R., Kozak, L., Titus, B., Kranabetter, M., Prescott, C., Visser, S., Morrison, I., Siltanen, M., Smith, S., Fyles, J. and Wein, R. (2002) 'Rates of litter decomposition over six years in Canadian forests: influence of litter quality and climate', *Canadian Journal of Forest Research*, vol 32, pp789–804

Canham, C. D., Finzi, A. C., Pacala, S. W. and Burbank, D. H. (1994) 'Causes and consequences of resource heterogeneity in forests: interspecific variation in light transmission by canopy trees', *Canadian Journal of Forest Research*, vol 24, pp337–349

Canham, C. D., Coates, K. D., Bartemucci, P. and Quaglia, S. (1999) 'Measurement and modelling of spatially explicit variation in light transmission through cedar-hemlock forests of British Columbia', *Canadian Journal of Forest Research*, vol 29, pp1775–1783

Canham, C. D., Cole, J. J. and Lauenroth, W. K. (eds) (2003) *Models in Ecosystem Science*, Princeton University Press, Princeton, NJ

Carey, A. B. (2009) 'Maintaining biodiversity in managed forests', in T. A. Spies and S. L. Duncan (eds) *Old Growth in a New World: A Pacific Northwest Icon Revisited*, Island Press, New York, NY, pp58–69

Cary, G. J., Keane, R. E., Gardner, R. H., Lavorel, S., Flannigan, M. D., Davies, I. D., Li, C., Lenihan, J. M., Rupp, T. S. and Mouillot, F. (2006) 'Comparison of the sensitivity of landscape-fire-succession models to variation in terrain, fuel pattern, climate and weather', *Landscape Ecology*, vol 21, pp121–137.

Cashore, B., van Kooten, G. C., Vertinsky, I., Auld, G. and Affolderback, J. (2005) 'Private or self-regulation? A comparative study of forest certification choices in Canada, the United States and Germany', *Forest Policy and Economics*, vol 7, pp53–69

Carstens, A. and Beck, J. (2005) 'Get ready for the gamer generation', *TechTrends*, vol 49, pp22–25

Case, M. J. and Peterson, D. L. (2005) 'Fine-scale variability in growth–climate relationship of Douglas-fir, North Cascade Range, Washington', *Canadian Journal of Forest Research*, vol 35, pp2743–2755

Case, M. J. and Peterson, D. L. (2007) 'Growth–climate relations of lodgepole pine in the North Cascades National Park, Washington', *Northwest Science*, vol 81, pp62–75

Cavens, D. (2002) 'A semi-immersive visualisation system for model-based participatory forest design and decision support', MSc thesis, University of British Columbia at Vancouver, BC

Chave, J. (1999) 'Study of structural, successional and spatial patterns in tropical rain forests using TROLL, a spatially explicit forest model', *Ecological Modelling*, vol 124, pp233–254

Chen, R. and Twilley, R. R. (1998) 'A gap dynamic model of mangrove forest development along gradients of soil salinity and nutrient resources', *Journal of Ecology*, vol 86, pp37–51

Chertov, O. G., Komarov, A. S. and Karev, G. P. (1999a) *Modern Approaches in Forest Ecosystem Modelling*, European Forest Institute, Leiden, The Netherlands

Chertov, O. G., Komarov, A. S. and Tsiplianovsky, A. M. (1999b) 'A combined simulation model of Scots pine, Norway spruce and Silver birch ecosystems in the European boreal zone', *Forest Ecology and Management*, vol 116, pp189–206

Chertov, O. G., Komarov, A. S. and Tsiplianovsky, A. M. (1999c) 'The simulation of soil organic matter and nitrogen accumulation in Scots pine plantations on bare parent material using the combined forest model EFIMOD', *Plant and Soil*, vol 213, pp31–41

Chertov, O., Komarov, A., Loukianov, A., Mikhailov, A., Nadporozhskaya, M. and Zubkova, E. (2005) 'The use of forest ecosystem model EFIMOD for research and practical implementation at forest stand, local and regional levels', *Ecological Modelling*, vol 194, pp227–232

Chew, J. D., Stalling, C. and Moeller, K. (2004) 'Integrating knowledge for simulating vegetation change at landscape scales', *Western Journal of Applied Foresters*, vol 19, pp102–108

Choi, J., Lorimer, C. G. and Vanderwerker, J. M. (2007) 'A simulation of the development and restoration of old-growth structural features in northern hardwoods', *Forest Ecology and Management*, vol 30, pp204–220

Coates, K. D., Canham, C. D., Beaudet, M., Sachs, D. L. and Messier, C. (2003) 'Use of a spatially explicit individual-tree model (SORTIE/BC) to explore the implications of patchiness in structurally complex forests', *Forest Ecology and Management*, vol 186, pp297–310

Coffin, D. P. and Lauenroth, W. K. (1989) 'A gap dynamics simulation model of succession in a semiarid grassland', *Ecological Modelling*, vol 49, pp229–266

Comeau, P. G. (1996) 'Why mixedwoods?', in P. G. Comeau and K. D. Thomas (eds) *Silviculture of Temperate and Boreal Broadleaf–Conifer Mixtures* (Land Management Handbook 36), BC Ministry of Forests, Victoria, BC

Comeau, P. G., Biring, B. S. and Harper, G. J. (2000) *Conifer Response to Brushing Treatments: A Summary of BC Data*, BC Ministry of Forests, Victoria, BC

Comins, H. N. and McMurtrie, R. E. (1993) 'Long-term response of nutrient-limited forests to CO_2-enrichment: Equilibrium behaviour of plant-soil models', *Ecological Applications*, vol 3, pp666–681

Conklin, J. (2005) *Dialogue Mapping: Building Shared Understanding of Wicked Problems*, John Wiley & Sons, Chichester, NY

Costanza, R., Norton, B. G. and Haskell, B. D. (eds) (1992) *Ecosystem Health*, Island Press, Washington, DC

Council for Regulatory Environmental Modeling (CREM) (2009) *Guidance on the Development, Evaluation, and Application of Environmental Models*, US Environmental Protection Agency, Washington, DC

Courbaud, B. (2000) 'Comparing light interception with stand basal area for predicting tree growth', *Tree Physiology*, vol 20, pp407–414

Covington, W. W. and DeBano, L. F. (eds) (1994) *Sustainable Ecological Systems: Implementing an Ecological Approach to Land Management*, USDA, Fort Collins, CO

Crookston, N. L. and Dixon, G. E. (2005) 'The forest vegetation simulator: a review of its structure, content and applications', *Computers and Electronics in Agriculture*, vol 49, pp60–80

Crout, N., Kokkonen, T., Jakeman, A. I., Norton, J. P., Newham, L. T. H., Anderson, R., Assaf, H., Croke, B. F. W., Gaber, N., Gibbons, J., Holzworth, D., Mysiak, J., Reichl, J., Seppelt, R., Wagener, T. and Whitfield, P. (2008) 'Good modelling practice', in A. J. Jakeman, A. A. Voinov, A. E. Rizzoli and S. H. Chen (eds) *Environmental Modelling, Software and Decision Support: State of the Art and New Perspectives*, Elsevier, Amsterdam

Crown, M. and Brett, C. P. (1975) *Fertilization and Thinning Effects on a Douglas-fir Ecosystem at Shawnigan Lake: An Establishment Report*, Canadian Forest Service, Pacific Forest Research Centre, Victoria, BC.

Dale, V. H. and Hemstrom, M. A. (1984) *CLIMACS: A Computer Model of Forest Stand Development for Western Oregon and Washington*, USDA/Forest Service, Portland, OR

Daniels, S. E. and Walker, G. B. (2001) *Working through Environmental Conflict: The Collaborative Learning Approach*, Praeger Publishers, Westport, CT

Davis, A. J., Jenkinson, L. S., Lawton, J. H., Shorrocks, B. and Wood, S. (1998) 'Making mistakes when predicting shifts in species range in response to global warming', *Nature*, vol 391, pp783–786

Daubenmire, R. F. (1974) *Plants and Environment: A Textbook of Plant Autecology* (third edition), John Wiley & Sons, New York, NY

Daust, D. K. and Sutherland, G. D. (1997) 'SIMFOR: software for simulating forest management and assessing effects on biodiversity', in I. D. Thompson (ed.) *The Status of Forestry/Wildlife Decision Support Systems in Canada: Proceedings of a Symposium, Toronto, Canada, 1994*, Natural Resources Canada, Canadian Forestry Service, Sault St Marie, ON

DeAngelis, D. M. and Mooij, W. M. (2005) 'Individual-based models of ecological and evolutionary processes', *Annual Review of Ecology, Evolution and Systematics*, vol 36, pp147–168

De Leo, G. A. and Levin, S. (1997) 'The multifaceted aspects of ecosystem integrity', *Conservation Ecology*, vol 1, p3, www.consecol.org/vol1/issi/art3/

De Marchi, B. and Ravetz, J. R. (2001) 'Participatory approaches to environmental policy', *Environmental Evaluation in Europe* (Policy Research Brief No. 10), Cambridge Research for the Environment, Cambridge, UK

De Reffye, P., Fourcaud, T., Blaise, F., Barthélémy, D. and Houllier, F. (1997) 'A functional model of tree growth and tree architecture', *Silva Fennica*, vol 31, pp297–311

Del Río, M. and Montero, G. (2001) 'Modelo de simulación de claras en masas de *Pinus sylvestris* L.', *Monografías INIA Forestal vol 3*, Madrid, Spain

Desanker, P. V. and Prentice, I. C. (1994) 'MIOMBO: a vegetation dynamics model for the miombo woodlands of Zambezian Africa', *Forest Ecology and Management*, vol 69, pp87–95

De Steiguer, J. E., Liberti, L., Schuler, A. and Hansen, B. (2003) *Multi-criteria Decision Models for Forestry and Natural Resource Management: An Annotated Bibliography* (General Technical Report NE-307), USDA Forest Service, Newton Square, PA

Deutschman, D. H., Levin, S. A. and Pacala, S. W. (1999) 'Error propagation in a forest succession model: the role of fine-scale heterogeneity in light', *Ecology*, vol 80, pp1927–1943

Develice, R. L. (1988) 'Test of a forest dynamics simulator in New Zealand', *New Zealand Journal of Botany*, vol 26, pp387–392

Dicus, C. A. and Delfino, K. (2003) *A Comparison of California Forest Practice: Rules and Two Forest Certification Systems* (Report to California Forest Products Commission), California Polytechnic State University, San Luis Obispo, CA

Didion, M., Kupferschmid, A. D., Zingg, A., Fahse, L. and Bugmann, H. (2009) 'Gaining local accuracy while not losing generality – extending the range of gap model applications', *Canadian Journal of Forest Research*, vol 39, pp1092–1107

Di Lucca, C. M. (1999) 'TASS/SYLVER/TIPSY: systems for predicting the impact of silvicultural practices on yield, lumber value, economic return and other benefits', in C. R. Bamsey (ed.) *Proceedings: Stand Density Management Conference: Using the Planning Tools*, Clear Lake, Edmonton, AL

Di Lucca, C. M., Goudie, J. W. and Stearns-Smith, S. (2009) 'TASS III: A new generation growth and yield prediction model for complex stands in British Columbia', in *Proceedings of the XIII World Forestry Congress*, Buenos Aires, Argentina, 18–23 October 2009

Ditzer, T., Glauner, R., Förster, M., Köhler, P. and Huth, A. (2000) 'The process-based stand growth model Formix 3-Q applied in a GIS environment for growth and yield analysis in a tropical rain forest', *Tree Physiology*, vol 20, pp367–381

Dixon, G. E. (ed.) (2002) *Essential FVS: A User's Guide to the Forest Vegetation Simulator*, US Department of Agriculture, Forest Service, Forest Management Service Center, Fort Collins, CO

Doyle, T. W. (1981) 'The role of disturbance in the gap dynamics of a montane rain forest: an application of a tropical forest succession model', in D. C. West, H. H. Shugart and D. B. Botkin (eds) *Forest Succession: Concepts and Application*, Springer Verlag, New York, NY

Dresner, M. and Elser, M. (2009) 'Enhancing science teachers' understanding of ecosystem interactions with qualitative conceptual models', http://tiee.ecoed.net/vol/v6/research/dresner/abstract.html, accessed 20 June 2009

Drouet, J. L. and Pages, L. (2007) 'GRAAL-CN: a model of GRowth, Architecture and ALlocation of Carbon and Nitrogen dynamics within whole plants formalised at the organ level', *Ecological Modelling*, vol 206, pp231–249

Duinker, P. N. (1989) 'Ecological effects monitoring in environmental impact assessment: what can it accomplish?', *Environmental Management,* vol 13, pp797–805

Duinker, P. N. (1998) 'Public participation's promising progress: advances in forest decision-making in Canada', *Commonwealth Forest Review*, vol 77, pp107–112

Duinker, P. N. and Trevisan, L. M. (2003) 'Adaptive management: progress and prospects for Canadian forests', in P. J. Burton, C. Messier, D. W. Smith and W. L. Adamowicz (eds) *Towards Sustainable Management of the Boreal Forest*, NRC Research Press, Ottawa, ON

Dunning, J. B., Stewart, D. J., Danielsom, B. J., Noon, B. R., Root, T. L., Lumberson, R. H. and Stevens, E. E. (1995) 'Spatially explicit population models: current forms and future users', *Ecological Applications*, vol 5, pp3–11

Dunnivant, F., Danowski, D., Timmens-Haroldson, A. and Newman, M. (2002) 'EnviroLand: a simple computer program for quantitative stream assessment', *The American Biology Teacher*, vol 64, pp589–595

Dykstra, D. P. and Monserud, R. A. (eds) (2009) *Forest Growth and Timber Quality: Crown Models and Simulation Methods for Sustainable Forest Management*, US Department of Agriculture, Forest Service, Pacific Northwest Research Station, Portland, OR

Ebermayer, E. (1876) *Die gesamte Lehre der Waldstreu mit Rücksicht auf die chemische Statik des Waldbaues*, Springer, Berlin

Ecosystem Marketplace (2010) *State of the Forest Carbon Markets 2009: Taking Root & Branching Out*, Ecosystem Marketplace, Washington, DC

Egler, F. E. (1954) 'Vegetation science concepts: 1. Initial floristic composition, a factor in old-field vegetation development', *Vegetatio*, vol 4, pp412–417

Ehrlich, P. R. (1968) *The Population Bomb*, Buccaneer Books, Cutchoque, NY

Ek, A. R. and Monserud, R. A. (1979) 'Performance and comparison of stand growth models based on individual tree and diameter-class growth', *Canadian Journal of Forest Research*, vol 9, pp231–244

El Bayoumi, M. A., Shugart, H. H. and Wein, R. W. (1984) 'Modeling succession of the eastern Canadian mixed-wood forest', *Ecological Modelling*, vol 21, pp175–198

Ellis, E. A., Bentrup, G. and Schoeneberger, M. M. (2004) 'Computer-based tools for decision support in agroforestry: current state and future needs', *Agroforestry Systems*, vol 61, pp401–421

Elton, C. (1924) 'Periodic fluctuations in the numbers of animals: their causes and effects', *British Journal of Experimental Biology*, vol 2, pp19–163

Elton, C. (1927) *Animal Ecology*, Sedgwick and Jackson, London

Eng, M. (2004) *Forest Stewardship in the Context of Large-Scale Salvage Operations*, BC Ministry of Forests, Victoria, BC

Enquist, B. J. (2002) 'Universal scaling in tree and vascular plant allometry: toward a general quantitative theory linking plant form and function from cells to ecosystems', *Tree Physiology*, vol 22, pp1045–1064

Everett, R. L. (ed.) (1994) *Eastside Forest Ecosystem Health Assessment, Volume IV: Restoration of Stressed Sites, and Processes*, USDA, Portland, OR

FAO (2009) *State of the World's Forests 2009*, Food and Agriculture Organization, United Nations, Rome

Finney, M. A. (1999) 'Mechanistic modelling of landscape fire patterns', in D. J. Mladenoff and W. L. Baker (eds) *Spatial Modelling of Forest Landscape Change: Approaches and Applications*, Cambridge University Press, Cambridge

Fisher, F. and Binkley, D. (2000) *Ecology and Management of Forest Soils*, John Wiley & Sons, New York, NY

Flanders, D., Pond, E. and Sheppard, S. R. J. (2009a) *A Preliminary Assessment of Renewable Energy Capacity in the Northshore, Metro Vancouver, BC*, University of British Columbia, Vancouver, BC

Flanders, D., Sheppard, S. R. J. and Blanco, J. A. (2009b) *The Potential for Local Bioenergy in Low-Carbon Community Planning: Smarth Growth on the Ground* (Foundation Research Bulletin #4), Smart Growth BC, Vancouver, BC

Flanders, D., Salter, J., Tatebe, K., Pond, E. and Sheppard, S. R. J. (2009c) *A Preliminary Assessment of Renewable Energy Capacity in Prince George, BC* (report prepared for Natural Resources Canada), University of British Columbia, Vancouver, BC

Ford, A. (1999) *Modeling the Environment*, Island Press, Washington, DC

Forest Practices Board (2000), 'A review of the forest development planning process in BC', www.fpb.gov.bc.ca/content.aspx?id=4248, accessed January 2009

Forscher, B. K. (1963) 'Chaos in the brickyard', *Science*, vol 142, p3590

Fourcaud, T., Zhang, X., Stokes, A., Lambers, H. and Körner, C. (2008) 'Plant growth modelling and applications: the increasing importance of plant architecture in growth models', *Annals of Botany*, vol 101, pp1053–1063

Franc, A. and Picard, N. (1997) *Quelques remarques sur la description et la modélisation des forêts homogènes et hétérogènes. Applications aux 'gap models'*, Programme GIP Forêts Hétérogènes, Seignosse, France

Franklin, J. F. and Spies, T. A. (1991) 'Ecological definitions of old-growth Douglas-fir forests', in L. F. Ruggiero, K. B. Aubry, A. B. Carey and M. H. Huff, *Wildlife and Vegetation of Unmanaged Douglas-fir Forests*, USDA, Portland, OR

Franklin, J. F., Berg, D. R., Thornburgh, D. A. and Tappeiner, J. C. (1997) 'Alternative silvicultural approaches to timber harvesting: variable retention harvest systems', in K. A. Kohn and J. F.

Franklin (eds) *Creating a Forestry for the 21st Century: The Science of Ecosystem Management*, Island Press, Washington, DC

Franklin, J. F., Spies, T. A., Van Pelt, R., Carey, A. B., Thornburgh, D. A., Berg, D. R., Lindenmayer, D. B., Harmon, M. E., Keeton, W. S., Shaw, D. C., Bible, K. and Chen, J. (2002) 'Disturbances and structural development of natural forest ecosystems with silvicultural implications, using Douglas-fir forests as an example', *Forest Ecology and Management*, vol 155, pp399–423

Freese, F. (1960) 'Testing accuracy', *Forest Science*, vol 6, pp139–145

Frelich, L. E. (2002) *Forest Dynamics and Disturbance Regimes: Studies from Temperate and Evergreen-Deciduous Forests*, Cambridge University Press, Cambridge

Friedlingstein, P., Joel, G., Field, C. B. and Fung, I. Y. (1999) 'Toward an allocation scheme for global terrestrial carbon models', *Global Change Biology*, vol 5, pp755–770

Friend, A. D., Shugart, H. H. and Running, S. W. (1993) 'A physiology-based gap model of forest dynamics', *Ecology*, vol 74, pp792–797

Friend, A. D., Stevens, A. K., Knox, R. G. and Cannell, M. G. R. (1997) 'A process-based, terrestrial biosphere model of ecosystem dynamics (Hybrid v3.0)', *Ecological Modelling*, vol 95, pp249–287

Fry, G. J. and Lungu, C. (1996) 'Assessing the benefits of interventions to improve soil moisture conditions using the PARCH Crop Environment Model: in land degradation – a challenge to agricultural production', *Proceedings of the Fifth Annual Scientific Conference*, SADC Land and Water Management Research Programme, Harare, Zimbabwe

Fulton, M. R. (1991) 'A computationally efficient forest succession model: design and initial tests', *Forest Ecology and Management*, vol 42, pp23–34

Fung, M. Y. P. and Macyk, T. M. (2000) 'Reclamation of oil sands mining areas', in R. I. Barmhisel, R. G. Darmondy and W. L. Daniels (eds) *Reclamation of Drastically Disturbed Lands* (Agronomy Monographs 41), ASA, CSSA and SSSA, Madison, WI

Galindo-Leal, C. and Bunnell, F. L. (1995) 'Ecosystem management: implications and opportunities of a new paradigm', *The Forestry Chronicle*, vol 71, pp601-606

Gardner, R. H. and Urban, D. L. (2003) 'Model validation and testing: past lessons, present concerns, future prospects', in C. D. Canham, J. J. Cole and W. K. Lauenroth (eds) *Models in Ecosystem Science*, Princeton University Press, Princeton, NJ

Génard, M., Dauzat, J., Frnack, N., Lescourret, F., Moitrier, N., Vaast, P. and Vercamber, G. (2008) 'Carbon allocation in fruit trees: from theory to modelling', *Trees*, vol 22, pp269–282

Gerzon, M. (2009) 'Modelling the recovery of Old-Growth attributes in Coastal Western Hemlock forests following management and natural disturbances', MSc thesis, University of British Columbia, Vancouver, BC, Canada

Gerzon, M. and Seely, B. (in review) 'Simulating the development of old-growth structural attributes in second-growth coastal western hemlock forests: evaluation of the mechanistic forest growth model FORECAST', *Ecological Modelling*

Giske, J. (1998) 'Evolutionary models for fisheries management', in T. Pitcher, D. Pauly and P. J. B. Hart (eds) *Re-inventing Fisheries Management*, Chapman & Hall, New York, NY

Giske, J., Skjoldal, H. R. and Aksnes, D. L. (1992) 'A conceptual model of distribution of capelin in the Barents Sea', *Sarsia*, vol 77, pp147–156

Gleeson, S. K. (1993) 'Optimization of tissue nitrogen and root/shoot allocation', *Annals of Botany*, vol 71, pp23–31

Godfrey, K. (1983) *Compartmental Models and their Applications*, Academic Press, New York, NY

González, R., Wright, J. and Saloni, D. (2009) 'Filling a need: forest plantations for bioenergy in the south', *Ethanol Produce Magazine*, www.ethanolproducer.com/article.jsp?article_id=6025&q=&page=all, accessed November 2009

Gordon, S. N. (2006) 'Decision support systems for forest biodiversity management: a review of tools and an analytical-deliberative framework for understanding their successful application', PhD thesis, Oregon State University, Portland, OR

Government of Alberta (1999) *Conservation and Reclamation Information Letter: Guidelines for Reclamation to Forest Vegetation in the Athabasca Oil Sands Region*, Goverment of Alberta, Edmonton, AB

Government of British Columbia (1994) *Forest Practices Code Act of British Columbia*, Government of British Columbia, Victoria, BC

Grace, J., Berninger, F. and Nagy, L. (2002) 'Impacts of climate change on tree line', *Annals of Botany*, vol 90, pp537–544

Grant, T. and Littlejohn, G. (eds) (2004) *Teaching Green: The Middle Years, Hands-On Learning in Grades 6–8*, New Society Publishers, Toronto, ON

Grant, R. F., Araina, A., Arora, V., Barr, A., Black, T. A., Chen, J., Wang, S., Yuan, F. and Zhang, Y. (2005) 'Intercomparison of techniques to model high temperature effects on CO_2 and energy exchange in temperate and boreal coniferous forests', *Ecological Modelling*, vol 188, pp217–252

Grant, R. F., Zhang, Y., Yuan, F., Wang, S., Hanson, P. J., Gaumont-Guay, D., Chen, J., Black, T. A., Barr, A., Baldocchi, D. D. and Arain, A. (2006) 'Intercomparison of techniques to model water stress effects on CO_2 and energy exchange in temperate and boreal deciduous forests', *Ecological Modelling*, vol 196, pp289–312

Grant, W. E., Pedersen, E. K. and Martin, S. L. (1997) *Ecology and Natural Resources Management: System Analysis and Simulation*, Wiley & Sons, New York, NY

Gredler, M. E. (1992) *Designing and Evaluating Games and Simulations: A Process Approach*, Kogan Page, London

Gredler, M. E. (1996) 'Educational games and simulations: a technology in search of a research paradigm', in Jonassen, D. H. (ed.) *Handbook of Research for Educational Communications and Technology*, MacMillan, New York, pp521–539

Green, E. J., MacFarlane, D. W., Valentine, H. T. and Strawderman, W. E. (1999) 'Assessing uncertainty in a stand growth model by Bayesian synthesis', *Forest Sciences*, vol 45, pp528–538

Greenblat, C. S. (1975) 'Basic concepts and linkages', in C. S. Greenblat and R. D. Duke (eds) *Gaming-Simulation: Rationale, Design and Applications – A Text with Parallel Readings for Social Scientists, Educators, and Community Workers*, Wiley & Sons, Toronto, ON

Grimm, V., Schmidt, E. and Wissel, C. (1992) 'On the application of stability concepts in ecology', *Ecological Modelling*, vol 63, pp143–161

Groot, A. (2004) 'A model to estimate light interception by tree crowns, applied to black spruce', *Canadian Journal of Forest Research*, vol 34, pp788–799

Grover, B. E. and Greenway, K. J. (1999) 'The ecological and economic basis for mixedwood management', in T. S. Veeman, D. W. Smith, B. G. Purdy, F. J. Salkle and G. A. Larkin, *Proceedings Sustainable Forest Management Network Conference: Science and Practice – Sustaining the Boreal Forest*, University of Alberta, Edmonton, AB

Gunderson, L., Allen, C. and Holling, C. (2009) *Foundations of Ecological Resilience*, Island Press, Washington, DC

Gustafson, E. J. and Crow, T. R. (1999) 'HARVEST: linking timber harvesting strategies to landscape patterns', in D. J. Mladenoff and W. L. Baker (eds) *Spatial Modelling of Forest Landscape Change: Approaches and Applications*, Cambridge University Press, Cambridge

Gustafson, E. J., Lytle, D. E., Swaty, R. and Loehle, C. (2007) 'Simulating the cumulative effects of multiple forest management strategies on landscape measures of forest sustainability', *Landscape Ecology*, vol 22, pp141–156

Haener, M. K. and Luckert, M. K. (1998) 'Forest certification: economic issues and welfare implications', *Canadian Public Policy*, vol 25, ppS83–S94

Håkanson, L. (2003) 'Propagation and analysis of uncertainty in ecosystem models', in C. D. Canham, J. J. Cole and W. K. Lauenroth (eds) *Models in Ecosystem Science*, Princeton University Press, Princeton, NJ

Hall, G. M. J. and Hollinger, D. Y. (2000) 'Simulating New Zealand forest dynamics with a generalised temperate forest gap model', *Ecological Applications*, vol 10, pp115–130

Hamann, A. and Wang, T. (2006) 'Potential effects of climate change on ecosystem and tree species distribution in British Columbia', *Ecology*, vol 87, pp2773–2786

Hamersley Chambers, F. and Beckley, T. (2003) 'Public involvement in sustainable boreal forest management', in P. J. Burton, C. Messier, D. W. Smith and W. L. Adamowicz (eds) *Towards Sustainable Management of the Boreal Forest*, NRC Research Press, Ottawa, ON

Hanks, J. and Ritchie, J. T. (eds) (1991) *Modeling Plant and Soil Systems*, American Society of Agronomics/Crop Science Society of America/Soil Science Society of America, Madison, WI

Harringon, E. A. and Shugart, H. H. (1990) 'Evaluating performance on an Appalachian oak forest dynamics model', *Vegetatio*, vol 86, pp1–13

Harvey, B., Leduc, A. and Bergeron, Y. (1995) 'Post-harvest succession in relation to site type in the southern boreal forest', *Canadian Journal of Forest Research*, vol 25, pp1658–1672

Hasenauer, H. (1994) 'A single tree simulator for uneven-aged mixed-species stands', in M. E. Pinto da Costa and T. Preuhsler (eds) *Mixed Stands: Research Plots, Measurements and Results, Models*, Instituto Superior de Agronomia, Lisboa, Portugal

Hasenauer, H. (ed.) (2006) *Sustainable Forest Management: Growth Models for Europe*, Springer, Berlin

He, H. S. (2008) 'Forest landscape models: definitions, characterization, and classification', *Forest Ecology and Management*, vol 254, pp484–498

He, H. S., Mladenoff, D. J. and Boeder, J. (1999) 'An object-oriented forest landscape model and its representation of tree species', *Ecological Modelling*, vol 119, pp1–19

He, H. S., Li, W., Sturtevant, B. R., Yang, J., Shang, Z. B., Gustafson, E. J. and Mladenoff, D. J. (2005) *LANDIS: A Spatially Explicit Model of Forest Landscape Disturbance, Management, and Succession – LANDIS 4.0 User's Guide*, USDA Forest Service North Central Research Station, Newton Square, PA

Hertel, J. P. and Millis, B. J. (2002) *Using Simulations to Promote Learning in Higher Education: An Introduction*, Stylus Publishing, Sterling, VA

Hilborn, R. and Mangel, M. (1997) *The Ecological Detective: Confronting Models with Data*, Princeton University Press, Princeton, NJ

Hill, P. R. (ed.) (2006) *Biophysical Impacts of Sea Level Rise and Changing Storm Conditions on Roberts Bank* (draft assessment report), Geological Survey of Canada, Ottawa, ON

Holling, C. S. (1973) 'Resilience and stability of ecological systems', *Annual Review of Ecology and Systematics*, vol 4, pp1–23

Holling, C. S. (ed.) (1978) *Adaptive Environmental Assessment and Management*, John Wiley and Sons, Toronto, ON

Holling, C. S. and Gunderson, L. H. (2002) 'Resilience and adaptive cycles', in L. H. Gunderson and C. S. Holling (eds) *Panarchy*, Island Press, Washington, DC

Huang, S. and Titus, S. J. (1994) 'An age-independent individual tree height prediction model for boreal spruce–aspen stands in Alberta', *Canadian Journal of Forest Research*, vol 24, pp1295–1301

Huffaker, C. B. (1958) 'Experimental studies on predation: dispersion factors and predator–prey oscillations', *Hilgardia*, vol 27, 795–835

Huffaker, C. B. and Messenger, P. S. (1964) 'The concept and significance of natural control', in P. DeBach (ed.) *Biological Control of Insect Pests and Weeds*, Reinhold Publishing, New York, NY

Hurtley, B. (1991) 'How plants respond to climate change: migration rates, individualism and the consequences for plant communities', *Annals of Botany*, vol 67, pp15–22

Huston, M. A. and Smith, T. M. (1987) 'Plant succession: life history and competition', *The American Naturalist*, vol 130, pp168–198

Huth, A., Ditzer, T. and Bossel, H. (1998) *The Rain Forest Growth Model FORMIX3 – Model Description and Analysis of Forest Growth and Logging Scenarios for the Deramakot Forest Reserve (Malaysia)*, Verlag Erich Goltze, Göttingen, Germany

Huth, N. I., Snow, V. O. and Keating, B. A. (2001) 'Integrating a forest modelling capability into an agricultural production systems modelling environment: current applications and future possibilities', *Proceedings of the International Congress on Modelling and Simulation*, Australian National University, Canberra, Australia

Huth, A., Drechsler, M. and Kohler, P. (2005) 'Using multicriteria decision analysis and a forest growth model to assess impacts of tree harvesting in Dipterocarp lowland rain forests', *Forest Ecology and Management*, vol 207, pp215–232

Innes, J. L. (1993) *Forest health: its assessment and status*, CAB International, Wallingford

IPCC (Intergovernmental Panel on Climate Change) (2007) *Climate Change 2007: Fourth Assessment Report (Synthesis Report)*, Cambridge University Press, Cambridge

ISEE Systems (2009) *STELLA: Systems Thinking for Education and Research*, Lebanon, NH

IUCN (2008) *Forests and Conflict* (IUCN Forest Conservation Programme Newsletter), issue 38, www.iucn.org/forest/av, accessed October 2009

Ives, A. R. and Carpenter, S. R. (2007) 'Stability and diversity in ecosystems', *Science*, vol 317, pp58–62

Jacobs, J. W. and Dempsey, J. V. (1993) 'Simulation and gaming: fidelity, feedback and motivation', in J. V. Dempsey and G. C. Sales (eds) *Interactive Instruction and Feedback*, Educational Technology Publications, Engelwood Cliffs, NJ

Jakeman, A. J., Chen, S. H., Rizzoli, A. E. and Voinov, A. A. (2008) 'Modelling and software as instruments for advancing sustainability', in A. J. Jakeman, A. A. Voinov, A. E. Rizzoli and S. H. Chen (eds) *Environmental Modelling, Software and Decision Support: State of the Art and New Perspectives*, Elsevier, Amsterdam

Johnsen, K., Samuelson, L., Teskey, R., McNulty, S. and Fox, T. (2001) 'Process models as tools in forestry and management', *Forest Science*, vol 47, pp2–8

Johnson, D. H. (2001) 'Validating and evaluating models', in T. M. Shenk and A. B. Franklin (eds) *Modeling in Natural Resource Management: Development, Interpretation, and Application*, Island Press, Washington, DC

Johnson, E. A. and Miyanishi, K. (2008) 'Creating new landscapes and ecosystems: the Alberta oil sands', *Annals of the New York Academy of Sciences*, vol 1134, pp120–145

Johnson, M. and Williamson, T. (2005) 'Climate change implication for stand yields and soil expectation values: a northern Saskatchewan case study', *The Forestry Chronicle*, vol 81, pp683–690

Jones, K. (1980) *Simulations: A Handbook for Teachers*, Kogan Page, London

Jones, K. (1988) 'Why gamesters die in space', in D. Crookall, J. H. G. Klabbers, A. Coote, D. Saunders, A. Cecchini and A. Delle Piane (eds) *Simulation-Gaming in Education and Training: Proceedings of the International Simulation and Gaming Association's 18th International Conference*, Pergamon Press,Toronto, ON

Jørgensen, S. E., Fath, B. D., Grant, W. and Nielsen, S. N. (2006) 'The editorial policy of ecological modelling', *Ecological Modelling*, vol 199, pp1–3

Jorritsma, I. T. M., van Hees, A. F. M. and Mohren, G. M. J. (1999) 'Forest development in relation to ungulate grazing: a modeling approach', *Forest Ecology and Management*, vol 120, pp23–34

Joseph, P., Keith, T., Kline, L., Schanke, J., Kanaskie, A. and Overhulser, D. (1991) 'Restoring forest health in the Blue Mountains: a 10-year strategic plan', *Forest Log*, vol 61, pp3–12

Jourdan, C. and Rey, H. (1997) 'Modelling and simulation of the architecture and development of the oil-palm (*Elaeis guineensis* Jacq) root system: 1. The model', *Plant and Soil*, vol 190, pp217–233

Jungst, S. E., Thompson, J. R. and Atchison, G. J. (2003) 'Academic controversy: fostering constructive conflict in natural resources education', *Journal of Natural Resources and Life Sciences Education*, vol 32, pp36–42

Kay, J. J. (1993) 'On the nature of ecological integrity: some closing comments', in S. Woodley, J. Kay and G. Francis (eds) *Ecological Integrity and the Management of Ecosystems*, St Lucie Press, Delray Beach, FL

Kangas, J., Store, R., Leskinen, P. and Mehtatalo, L. (2000) 'Improving the quality of landscape ecological forest planning by utilizing advanced decision-support tools', *Forest Ecology and Management*, vol 132, pp157–171

Kangas, J., Kangas, A., Leskinen, P. and Pykalainen, J. (2001) 'MCDM methods in strategic planning of forestry on state-owned lands in Finland', *Journal of Multi-Criteria Decision Analysis*, vol 10, pp257–271

Keane, R. E., Arno, S. F. and Brown, J. K. (1989) *FIRESUM: An Ecological Process Model for Fire Succession in Western Conifer Forests*, USDA Forest Service Intermountain Research Station, Ogden, UT

Keane, R. E., Morgan, P. and Running, S. W. (1996) *Fire-BGC: A Mechanistic Ecological Process Model for Simulating Fire Succession on Coniferous Forest Landscapes of the Northern Rocky Mountains*, USDA Forest Service, Intermountain Research Station, Ogden, UT

Keane, R. E., Cary, G. J., Davies, I. D., Flannigan, M. D., Gardner, R. H., Lavorel, S., Lenihan, J. M., Li, C. and Rupp, T. S. (2004) 'A classification of landscape fire succession models: spatial simulations of fire and vegetation dynamics', *Ecological Modelling*, vol 179, pp3–27

Keenan, R. (1993) 'Structure and function of western red cedar and western hemlock forests on Northern Vancouver Island', PhD thesis, University of British Columbia, Vancouver, BC

Kellomäki, S. and Strandman, H. (1995) 'A model for the structural growth of young Scots pine crowns based on light interception by shoots', *Ecological Modelling*, vol 80, pp237–250

Kellomäki, S. and Väisänen, H. (1991) 'Application of a gap model for the simulation of forest ground vegetation in boreal conditions', *Forest Ecology and Management*, vol 42, pp35–38

Kellomäki, S., Väisänen, H., Hänninen, H., Kolström, T., Lauhanen, R., Mattila, U. and Pajari, B. (1992) 'SIMA: a model for forest succession based on the carbon and nitrogen cycles with application to silvicultural management of the forest ecosystem', *Silva Carelica*, vol 22, pp91–99

Kercher, J. R. and Axelrod, M. X. (1984) 'A process model of fire ecology and succession in a mixed-conifer forest', *Ecology*, vol 65, pp1725–1742

Kessell, S. R. (1979) 'Gradient modelling: A new approach to fire modelling and wilderness resource management', *Environmental Management*, vol 1, pp39–48

Kienast, F. and Kuhn, N. (1989) 'Simulating forest succession along ecological gradients in southern Central Europe', *Vegetatio*, vol 79, pp7–20

Kimmins, J. P. (1977) 'Evaluation of the consequences for future tree productivity of the loss of nutrients in whole-tree harvesting', *Forest Ecology and Management*, vol 1, pp169–183

Kimmins, J. P. (1988) 'Community organization: methods of study and prediction of the productivity and yield of forest ecosystems', *Canadian Journal of Botany*, vol 66, pp2654–2672

Kimmins, J. P. (1990a) 'Monitoring the condition of the Canadian forest environment: the relevance of the concept of "ecological indicators"', *Environmental Monitoring and Assessment*, vol 15, pp231–240

Kimmins, J. P. (1990b) 'Workgroup issue paper: indicators and assessments of the state of forests', *Environmental Monitoring and Assessment*, vol 15, pp297–299

Kimmins, J. P. (1990c) 'Modelling the sustainability of forest production and yield for a changing and uncertain future', *The Forestry Chronicle*, vol 66, pp271–280

Kimmins, J. P. (1993a) 'Ecology, environmentalism and green religion', *The Forestry Chronicle*, vol 69, pp285–289

Kimmins, J. P. (1993b) *Scientific Foundations for the Simulation of Ecosystem Function and Management in FORCYTE-11*, Forestry Canada, Northwest Region, Northern Forest Centre, Edmonton, AB

Kimmins, J. P. (1996a) 'Importance of soil and the role of ecosystem disturbance for sustained productivity of cool temperate and boreal forests', *Soil Science Society of America Journal*, vol 60, pp1643–1654

Kimmins, J. P. (1996b) 'The health and integrity of forest ecosystems: are they threatened by forestry?', *Ecosystem Health*, vol 2, pp5–18

Kimmins, J. P. (1997a) 'Predicting sustainability of forest bioenergy production in the face of changing paradigms', *Biomass and Bioenergy*, vol 13, pp201–212

Kimmins, J. P. (1997b) 'Biodiversity and its relationship to ecosystem health and integrity', *The Forestry Chronicle*, vol 73, pp229–232

Kimmins, J. P. (1999a) 'Biodiversity, beauty and the beast: are beautiful forests sustainable, are sustainable forests beautiful, and is "small" always ecologically desirable?', *The Forestry Chronicle*, vol 75, pp955–960

Kimmins, J. P. (1999b) *Balancing Act: Environmental Issues in Forestry*, UBC Press, Vancouver, BC

Kimmins, J. P. (2000a) 'Visible and non-visible indicators of forest sustainability: beauty, beholders and belief systems', in S. R. J. Sheppard and H. W. Harshaw (eds) *Forests and Landscapes: Linking Ecology, Sustainability and Aesthetics*, CABI Publishing, London

Kimmins, J. P. (2000b) 'Respect for nature: an essential foundation for sustainable forest management', in R. G. D'Eon, J. F. Johnson and E. A. Ferguson (eds) *Ecosystem Management of Forested Landscapes: Directions and Implementations*, UBC Press, Vancouver, BC

Kimmins, J. P. (2002) 'Future shock in forestry. Where have we come from; where are we going; is there a "right way" to manage forests? Lessons from Thoreau, Leopold, Toffler, Botkin and nature', *The Forestry Chronicle*, vol 78, pp263–271

Kimmins, J. P. (2003) 'Old growth forest: an ancient and stable sylvan equilibrium or a relatively transitory ecosystem condition that offers people a visual and emotional feast? Answer: it depends', *The Forestry Chronicle*, vol 79, pp429–440

Kimmins, J. P. (2004a) 'Environmental issues in forestry', in *Forest Ecology: A Foundation for Sustainable Management and Environmental Ethics in Forestry*, Pearson/Prentice Hall, Upper Saddle River, NJ

Kimmins, J. P. (2004b) 'Ecological sucession: processes of change in ecosystems', in *Forest Ecology: A Foundation for Sustainable Management and Environmental Ethics in Forestry*, Pearson/Prentice Hall, Upper Saddle River, NJ

Kimmins, J. P. (2004c) 'Emulation of natural forest disturbance: what does this mean?', in A. H. Perrera, L. J. Buse and M. Weber (eds) *Emulating Natural Forest Landscape Disturbances: Concepts and Application*, Columbia University Press, New York, NY

Kimmins, J. P. (2004d) *Forest Ecology* (third edition), Prentice Hall, Upper Saddle River, NJ

Kimmins, J. P. (2007a) 'Forest ecosystem management: miracle or mirage?', in T. B. Harrington and G. E. Nicholas (eds) *Managing for Wildlife Habitat in Westside Production Forests*, USDA Forest Service, PNW Research Station, Portland, OR

Kimmins, J. P. (2007b) 'Sustainability: a focus on forests and forestry', in P. N. Nemetz (ed.) *Sustainable Resource Management*, Edward Elgar, Camberley

Kimmins, J. P. (2008a) 'From science to stewardship: harnessing forest ecology in the service of society', *Forest Ecology and Management*, vol 256, pp1625–1635

Kimmins, J. P. (2008b) 'Carbon and nutrient cycling: is forest bioenergy sustainable?', *Canadian Silviculture*, spring, pp12–16

Kimmins, J. P. and Scoullar, K. A. (1979) *FORCYTE (FORest Cycling Trend Evaluator): A Computer Simulation Model to Examine the Long-Term Consequences for Site Nutrient Capital and Productivity of Intensive Forest Harvesting*, Department of Environment, Canadian Forestry Service

Kimmins, J. P., Scoullar, K. A. and Feller, M. C. (1981) 'FORCYTE-9: a computer simulation approach to evaluating the effect of whole tree harvesting on nutrient budgets and future forest productivity', *Mitteilungen der Forstilichen Bundesversuchsanstalt*, vol 140, pp189–205

Kimmins, J. P., Scoullar, K. A. and Apps, M. J. (1990) *FORCYTE-11 User's Manual for the Benchmark Version*, Forestry Canada, Northwest Region, Northern Forest Centre, Edmonton, AB

Kimmins, J. P., Scoullar, K. A., Kremsater, L., Thauberger, R. and Waldie, W. C. (1997) *FORTOON Forest Management Program, Version 2.7*, Science Programming, Naramata, BC

Kimmins, J. P., Mailly, D. and Seely, B. (1999) 'Modelling forest ecosystem net primary production: the hybrid simulation approach used in FORECAST', *Ecological Modelling*, vol 122, pp195–224

Kimmins, J. P., Welham, C., Seely, B., Meitner, M., Rempel, R. and Sullivan, T. (2005) 'Science in forestry: why does it sometimes disappoint or even fail us?', *The Forestry Chronicle*, vol 81, pp723–734

Kimmins, J. P., Rempel, R., Welham, C., Seely, B. and van Rees, K. (2007) 'Biophysical sustainability, process-based monitoring and forest ecosystem management decision support systems', *The Forestry Chronicle*, vol 83, pp502–514

Kimmins, J. P., Blanco, J. A., Seely, B. and Welham, C. (2008a) 'Complexity in modeling forest ecosystems: how much is enough?', *Forest Ecology and Management*, vol 256, pp1646–1658

Kimmins, J. P., Welham, C., Fuliang, C., Wangpakapattanawong, P. and Christanty, L. (2008b) 'The role of ecosystem-level models in the design of agroforestry systems for future environmental conditions and social needs', in S. Jose and A. M. Gordon (eds) *Towards Agroforestry Design: An Ecological Approach*, Springer, New York, NY

Kirschbaum, M. U. F. (1999) 'CenW, a forest growth model with linked carbon, energy, nutrient and water cycles', *Ecological Modelling*, vol 118, pp17–59

Kitching, R. L. (1983) *System Ecology: An Introduction to Ecological Modelling*, University of Queensland Press, Saint Lucia, Australia

Kneeshaw, D. D., Leduc, A., Drapeau, P., Gauthier, S., Paré, D., Carigan, R., Doucet, R., Bouthiller, L. and Messier, C. (2000) 'Development of integrated ecological standards of sustainable forest management at the operational scale', *The Forestry Chronicle*, vol 76, pp481–493

Kobe, R. K. and Coates, K. D. (1997) 'Models of sapling mortality as a function of growth to characterize interspecific variation in shade tolerance of eight tree species of northwestern British Columbia', *Canadian Journal of Forest Research*, vol 27, pp227–236

Köhler, P. and Huth, A. (1998) 'The effects of tree species grouping in tropical rainforest modelling: simulations with the individual-based model FORMIND', *Ecological Modelling*, vol 109, pp301–321

Kokkila, T., Mäkelä, A.,and Nikinmaa, E. (2002) 'A method for generating stand structures using Gibbs marked point process', *Silva Fennica*, vol 36, pp265–277

Kolb, T. E., Wagner, W. R. and Covington, W. W. (1994) 'Concepts of forest health: utilitarian and ecosystem perspectives', *Journal of Forestry*, vol 92, pp10–15

Kolström, T. (1993) 'Modelling the development of uneven aged stand of *Picea abies*', *Scandinavian Journal of Forest Research*, vol 8, pp373–383

Kolström, M. (1998) 'Ecological simulation model for studying diversity of stand structure in boreal forests', *Ecological Modelling*, vol 111, pp17–36

Komarov, A., Chertov, O., Zudin, S., Nadporozhskaya, M., Mikhailov, A., Bykhovets, S., Zudina, E. and Zoubkova, E. (2003) 'EFIMOD 2: a model of growth and cycling of elements in boreal forest ecosystems', *Ecological Modelling*, vol 170, pp373–392

Korzukhin, M. D., Ter-Mikaelian, M. T. and Wagner, R. G. (1996) 'Process versus empirical models: which approach for forest management?', *Canadian Journal of Forest Research*, vol 26, pp879–887

Kräuchi, N. and Kienast, F. (1995) 'Modelling subalpine forest dynamics as influenced by a changing environment', *Water, Air and Soil Pollution*, vol 68, pp185–197

Kurz, W. A. and Apps, M. J. (1999) 'A 70-year retrospective analysis of carbon fluxes in the Canadian forest sector', *Ecological Application*, vol 9, pp526–547

Kurz, W. A. and Apps, M. J. (2006) 'Developing Canada's national forest carbon monitoring, accounting and reporting system to meet the reporting requirement of the Kyoto protocol', *Mitigation and Adaptation Strategies for Global Change*, vol 11, pp33–43

Kurz, W. A., Stinson, G. and Rampley, G. (2008) 'Could increased boreal forest ecosystem productivity offset carbon losses from increased disturbances?', *Philosophical Transactions of the Royal Society, B – Biological Sciences*, vol 363, pp2259–2268

Lafon, C. W. (2004) 'Ice-storm disturbance and long-term forest dynamics in the Adirondack Mountains', *Vegetation Science*, vol 15, pp267–276

Landsberg, J. (2003a) 'Physiology in forest models: history and the future', *Forest Biometry, Modelling and Information Sciences*, vol 1, pp49–63

Landsberg, J. (2003b) 'Modelling forest ecosystems: state of the art, challenges, and future directions', *Canadian Journal of Forest Research*, vol 33, pp385–397

Landsberg, J. J. and Waring, R. H. (1997) 'A generalised model of forest productivity using simplified concepts of radiation-use efficiency, carbon balance, and partitioning', *Forest Ecology and Management*, vol 95, pp209–228

Landsberg, J. J., Waring, R. H. and Coops, N. C. (2003) 'Performance of the forest productivity model 3-PG applied to a wide range of forest types', *Forest Ecology and Management*, vol 172, pp199–214

Larocque, G. R., Archambault, L. and Delisle, C. (2006) 'Modelling forest succession in two southeastern Canadian mixedwood ecosystem types using the ZELIG model', *Ecological Modelling*, vol 199, pp350–362

Larocque, G. R., Bhatti, J. S., Boutin, R. and Chertov, O. (2008) 'Uncertainty analysis in carbon cycle models of forest ecosystems: research needs and development of a theoretical framework to estimate error propagation', *Ecological Modelling*, vol 219, pp400–412

Lasch, P., Badeck, F.-W., Suckow, F., Lindner, M. and Mohr, P. (2005) 'Model-based analysis of management alternatives at stand and regional level in Brandenburg (Germany)', *Forest Ecology and Management*, vol 207, pp59–74

Lathman II, L. G. and Scully, E. P. (2008) 'Critters! A realistic simulation for teaching evolutionary biology', *The American Biology Teacher*, vol 70, pp30–33

Lauer, T. E. (2003) 'Conceptualizing ecology: a learning cycle approach', *The American Biology Teacher*, vol 65, pp518–522

Lawlor, D. J. (2000) *Photosynthesis*, Bios Scientific Publishers, Oxford

Lawson, G. J., Crout, N. M. J., Levy, P. E., Mobbs, D. C., Wallace, J. S., Cannell, M. G. R., Bradley, R. G. and Sinclair, F. (1995) 'The tree–crop interface: representation by coupling of forest and crop process-models', *Agroforestry Systems*, vol 30, pp199–221

Lederman, L. C. (1984) 'Debriefing: a critical re-examination of the post-experience analytic process with implications for its effective use', *Simulation & Games*, vol 15, pp415–431

Leemans, R. and Prentice, I. C. (1987) 'Description and simulation of tree-layer composition and size distributions in a primaeval Picea-Pinus forest', *Vegetatio*, vol 69, pp147–156

Leemans, R. and Prentice, I. C. (1989) *FORSKA, A General Forest Succession Model*, Institute of Ecological Botany, Uppsala, Sweden

Leiserowitz, A. (2004) 'Before and after *The Day After Tomorrow*: A US study of climate risk perception', *Environment*, vol 46, pp22–37

Leopold, A. (1949) 'The land ethic', in *A Sand County Almanac*, Oxford University Press, New York, NY

Le Roux, X., Lacointea, A., Escobar-Gutiérrez, A. and Le Dizèsa, S. (2001) 'Carbon-based models of individual tree growth: a critical appraisal', *Annals of Forest Sciences*, vol 58, pp469–506

Lett, C., Silber, C. and Barret, N. (1999) 'Comparison of a cellular automata network and an individual-based model for the simulation of forest dynamics', *Ecological Modelling*, vol 121, pp277–293

Levin, S. A. (1999a) *Fragile Dominion: Complexity and the Commons*, Perseus Books, Reading, MA

Levin, S. A. (1999b) 'Towards a science of ecological management', *Conservation Ecology*, vol 3, pp6–9

Levins, R. (1966) 'The strategy of model building in population biology', in E. Sober (ed.) *Conceptual Issues in Evolutionary Biology* (first edition), MIT Press, Cambridge, MA

Lewis, J. L. and Sheppard, S. R. J. (2006) 'Culture and communication: can landscape visualization improve forest management consultation with indigenous communities?', *Landscape and Urban Planning*, vol 77, pp291–313

Lexer, M. J. and Hönninger, K. (1998) 'Simulated effects of bark beetle infestations on stand dynamics in *Picea abies* stands: coupling a patch model and a stand risk model', in M. Beniston and J. L. Innes (eds) *The Impacts of Climate Variability on Forests*, Lecture Notes in Earth Sciences 74, Springer Verlag, Berlin

Li, C., Barclay, H., Liu, J. W. and Campbell, D. (2005) 'Simulation of historical and current fire regimes in central Saskatchewan', *Forest Ecology and Management*, vol 208, pp319–329

Li, H., Franklin, J. F., Swanson, F. J. and Spies, T. A. (1993) 'Developing alternative forest cutting patterns: A simulation approach', *Landscape Ecology*, vol 8, pp63–75

Li, H., Gartner, D. I., Mou, P. and Trettin, C. C. (2000) 'A landscape model (LEEMATH) to evaluate effects of management impacts on timber and wildlife habitat', *Computers and Electronics in Agriculture*, vol 27, pp263–292

Lieffers, V. J., Macmillan, R. B., MacPherson, D., Branter, K. and Stewart, J. D. (1996) 'Semi-natural and intensive silvicultural systems for the boreal mixedwood forest', *The Forestry Chronicle*, vol 72, pp286–292

Lin, Y. and Tanaka, S. (2006) 'Ethanol fermentation from biomass resources: current state and prospects', *Applied Microbiology and Biotechnology*, vol 69, pp627–642

Lindenmayer, D. B., Margules, C. R. and Botkin, D. B. (2000) 'Indicators of biodiversity for ecologically sustainable forest management', *Conservation Biology*, vol 14, pp941–950

Linder, M., Sievänen, R. and Pretzsch, H. (1997) 'Improving the simulation of stand structure in a forest gap model', *Forest Ecology and Management*, vol 95, pp183–195

Lindgren, P. M. F., Sullivan, T. P., Sullivan, D. S., Brockley, R. P. and Winter, R. (2007) 'Growth response of young lodgepole pine to thinning and repeated fertilization treatments: 10-year results', *Forestry*, vol 80, pp187–211

Lischke, H., Zimmermann, N. E., Bolliger, J., Rickebusch, S. and Löffler, T. J. (2006) 'TreeMig: a forest-landscape model for simulating spatio-temporal patterns from stand to landscape scales', *Ecological Modelling*, vol 199, pp409–420

Liu, J. (1993) 'ECOLECON: a spatially-explicit model for ECOLogical-ECONomics of species conservation in complex forest landscapes', *Ecological Modelling*, vol 70, pp63–87

Liu, J. and Ashton, P. S. (1998) 'FORMOSAIC: an individual-based spatially explicit model for simulating forest dynamics in landscape mosaics', *Ecological Modelling*, vol 106, pp177–200

Liu, J. and Ashton, P. S. (1999) 'Simulating effects of landscape context and timber harvest on tree species diversity', *Ecological Applications*, vol 9, pp186–201

Liu, J., Ickes, K., Ashton, P. S., LaFrankie, J. V. and Monakaran, N. (1999) 'Spatial and temporal impacts of adjacent areas on the dynamics of species diversity in a primary forest', in D. J.

Mladenoff and W. L. Baker (eds) *Spatial Modelling of Forest Landscape Change: Approaches and Applications*, Cambridge University Press, Cambridge

Lo, Y.-H. (2009) 'Relationships between climate and annual radial growth in three coniferous species in interior British Columbia, Canada', PhD thesis, University of British Columbia, Vancouver, BC

Lo, Y.-H., Blanco, J. A. and Kimmins, J. P. (2010a) 'A word of caution when projecting future shifts of tree species ranges', *The Forestry Chronicle*, vol 86, pp312–316

Lo, Y.-H., Blanco, J. A., Seely, B., Welham, C. and Kimmins, J. P. (2010b) 'Relationships between climate and tree radial growth in interior British Columbia, Canada', *Forest Ecology and Management*, vol 259, pp932–942

Lockwood, C. and Moore, T. (1993) 'Harvest scheduling with spatial constraints: a simulated annealing approach', *Canadian Journal of Forest Research*, vol 23, 467–478

Löffler, T. J. and Lischke, H. (2001) 'Incorporation and influence of variability in an aggregated forest model', *Natural Resource Modeling*, vol 14, pp103–137

Lovelock, J. E. (1979) *Gaia: A New Look at Life on Earth*, Oxford University Press, Oxford, UK

Luan, J., Muetzelfeldt, R. I. and Grace, J. (1996) 'Hierarchical approach to forest ecosystem simulation', *Ecological Modelling*, vol 86, pp37–50

Lucas, N. S., Curran, P. J., Plummer, S. E. and Danson, F. M. (2000) 'Estimating the stem carbon production of a coniferous forest using an ecosystem model driven by the remotely sensed red edge', *International Journal of Remote Sensing*, vol 2, pp619–631

McCown, R. L., Hammer, G. L., Hargreaves, J. N. G., Holzworth, D. P. and Freebairn, D. M. (1996) 'APSIM: A novel software system for model development, model testing, and simulation in agricultural systems research', *Agricultural Systems*, vol 50, pp255–271

McCown, R. L., Carberry, P. S., Foale, M. A., Hochman, Z., Coutts, J. A. and Dalgliesh, N. P. (1998) 'The FARMSCAPE approach to farming systems research', *Proceedings of the Ninth Australian Society of Agronomy Conference*, Australian Society of Agronomy, Wagga Wagga, NSW, Australia

MacDicken, K. G. and Vergara, N. T. (1990) *Agroforestry: Classification and Management*, John Wiley and Sons, New York, NY

McGarigal, K., Cushman, S. A., Neel, M. C. and Ene, E. (2002) *FRAGSTATS: Spatial Pattern Analysis Program for Categorical Maps*, University of Massachusetts, Amherst, MS

McGaughey, R. J. (1997) 'Visualizing forest stand dynamics using the stand visualization system', in *Proceedings of the 1997 ACSM/ASPRS Annual Convention and Exposition Volume 4*, American Society for Photogrammetry and Remote Sensing, Seattle, WA

McGaughey, R. J. (1998) 'Techniques for visualizing the appearance of forest operations', *Journal of Forestry*, vol 96, pp9–14

McIntire, E. J. B., Duchesneau, R. and Kimmins, J. P. (2005) 'Seed and bud legacies interact with varying fire regimes to drive long-term dynamics of boreal forest communities', *Canadian Journal of Forest Research*, vol 35, pp2765–2773

McIntosh, B. S., Giupponi, C., Voinov, A. A., Smith, C., Matthews, K. B., Monticino, M., Kolkman, M. J., Crossman, N., van Ittersum, M., Haase, D., Haase, A., Mysiak, J., Groot, J. C. J., Sieber, S., Verweij, P., Quinn, N., Waeger, P., Gaber, N., Hepting, D., Scholten, H., Sulis, A., van Delden, H., Gaddis, E. and Assaf, H. (2008) 'Bridging the gaps between design and use: developing tools to support environmental management policy', in A. J. Jakeman, A. A. Voinov, A. E. Rizzoli and S. H. Chen (eds) *Environmental Modelling, Software and Decision Support: State of the Art and New Perspectives*, Elsevier, Amsterdam

McKeachie, W. J. (1994) 'Teaching tips: strategies, research, and theory for college and university teachers', D. C. Heath, Lexington, MA

MacKenzie, A. H. (2005) 'The brain, the biology classroom and kids with video games', *The American Biology Teacher*, vol 67, pp517–518

McMahan, A. J., Milner, K. S. and Smith, E. L. (2002) *FVS-BGC: User's Guide to Version 1.1*, USDA Forest Service, State & Private Forestry, Forest Health Protection, Forest Health Technology Enterprise Team, Fort Collins, CO

McWilliams, E. R. G. and Thérien, G. (1997) *Fertilization and Thinning Effects on a Douglas-Fir Ecosystem at Shawnigan Lake: 24-Year Growth Response*, BC Ministry of Forests, Victoria, BC

Magee, M. (2006) 'State of the field review: simulation in education, final report', Alberta Online Learning Consortium, Calgary, AB

Magelli, F., Boucher, K., Bi, H. T., Melin, S. and Bonoli, A. (2009) 'An environmental impact assessment of exported wood pellets from Canada to Europe', *Biomass and Bioenergy*, vol 33, pp434–441

Maier, H. R., Ascough II, J. C., Wattenbach, M., Renschler, C. S., Labiosa, W. B. and Ravalico, J. K. (2008) 'Uncertainty in environmental decision making: issues, challenges and future directions', in A. J. Jakeman, A. A. Voinov, A. E. Rizzoli and S. H. Chen (eds) *Environmental Modelling, Software and Decision Support: State of the Art and New Perspectives*, Elsevier, Amsterdam

Mailly, D., Kimmins, J. P. and Busing, R. T. (2000) 'Disturbance and succession in a coniferous forest of northwestern North America: simulations with DRYADES, a spatial gap model', *Ecological Modelling*, vol 127, pp183–205

Mäkelä, A. (1988) 'Performance analysis of a process-based stand growth model using Monte Carlo techniques', *Scandinavian Journal of Forest Research*, vol 3, pp315–331

Mäkelä, A. (2003) 'Process-based modelling of tree and stand growth: towards a hierarchical treatment of multiscale processes', *Canadian Journal of Forest Research*, vol 33, pp298–409

Mäkelä, A. (2009) 'Hybrid models of forest stand growth and production', in D. P. Dykstra and R. A. Monserud (eds) *Forest Growth and Timber Quality: Crown Models and Simulation Methods for Sustainable Forest Management*, US Department of Agriculture, Forest Service, Pacific Northwest Research Station, Portland, OR

Mäkelä, A., Landsberg, J., Ek, A. R., Burk, T. E., Ter-Mikaelian, M., Ågren, G. I., Oliver, C. D. and Puttonen, P. (2000) 'Process-based models for forest ecosystem management: current state of the art and challenges for practical implementation', *Tree Physiology*, vol 20, pp289–298

Malanson, G. P. (1996) 'Effects of dispersal and mortality on diversity in a forest stand model', *Ecological Modelling*, vol 87, pp103–110

Malkönen, E. (1974) 'Annual primary production and nutrient cycle in some Scots pine stands', *Communications of Finnish Forest Research Institute*, vol 84, pp1–87

Man, C. D., Comeau, P. G. and Pitt, D. G. (2008) 'Competitive effects of woody and herbaceous vegetations in a young boreal mixedwood stand', *Canadian Journal of Forest Research*, vol 38, pp1817–1828

Mangel, M., Fiksen, Ø. and Gisk, J. (2001) 'Theoretical and statistical models in natural resource management and research', in T. M. Shenk and A. B. Franklin (eds) *Modeling in Natural Resource Management: Development, Interpretation, and Application*, Island Press, Washington, DC

Marcot, B. G., Raphael, M. G. and Berry, K. H. (1983) 'Monitoring wildlife habitat and validation of wildlife-habitat relationships models', *North American Wildlife and Natural Resources Conference*, vol 48, pp315–329

Marshall, P., Parysow, P. and Akindele, S. (2008) 'Evaluating growth models: a case study using PROGNOSIS[BC]', in R. N. Havis and N. L. Crookston (eds) *Proceedings: Third Forest Vegetation Simulator Conference*, US Department of Agriculture, Forest Service, Rocky Mountain Research Station RMRS-P-54, Fort Collins, CO

Martin, P. (1992) 'EXE: a climatically sensitive model to study climate change and CO_2 enhancement effects on forests', *Australian Journal of Botany*, vol 40, pp717–735

Martin, W. E., Wise Bender, H. and Shields, D. J. (2000) 'Stakeholder objectives for public lands: rankings of forest management alternatives', *Journal of Environmental Management*, vol 58, pp21–32

Martin, W. L., Bradley, R. L. and Kimmins, J. P. (2002) 'Post-clearcutting chronosequence in the BC coastal western hemlock zone: I. changes in forest floor mass and N storage', *Journal of Sustainable Forestry*, vol 14, pp1–22

Mayer, D. G. and Butler, D. G. (1993) 'Statistical validation', *Ecological Modelling*, vol 68, pp21–32

Meadows, D. H., Meadows, D. L., Randers, J. and Behrens III, W. W. (1972) *The Limits to Growth: A Report for the Club of Rome's Project on the Predicament of Mankind*, Pan Books, London

Meadows, D. H., Randers, J. and Meadows, D. L. (2004) *The Limits to Growth: The 30-Year Update*, Earthscan, London

Medawar, P. (1984) *Pluto's Republic*, Oxford University Press, New York, NY

Meir, E. (1999) *EcoBeaker 2.0: Laboratory Guide and Manual*, BeakerWare, Ithaca, NY

Menaut, J., Gignoux, C., Prado, J. and Clobert, J. (1990) 'Tree community dynamics in a humid savanna of the Côte d'Ivoire: modelling the effects of fire and competition with grass and neighbours', *Journal of Biogeography*, vol 17, pp471–481

Mendoza, G. A. and Prabhu, R. (2000) 'Development of a methodology for selecting criteria and indicators of sustainable forest management: a case study on participatory assessment', *Environmental Management*, vol 26, pp659–673

Mendoza, G. A. and Prabhu, R. (2005) 'Combining participatory modelling and multi-criteria analysis for community-based forest management', *Forest Ecology and Management*, vol 207, 145–156

Merganicová, K., Pietsch, S. A. and Hasenauer, H. (2005) 'Testing mechanistic modeling to assess impacts of biomass removal', *Forest Ecology and Management*, vol 207, pp37–57

Messier, C. and Kneeshaw, D. D. (1999) 'Thinking and acting differently for sustainable management of the boreal forest', *The Forestry Chronicle*, vol 75, pp929–938

Messier, C., Fortin, M.-J., Schmiegelow, F., Doyon, F., Cumming, S. G., Kimmins, J. P., Seely, B., Welham, C. and Nelson, J. (2003) 'Modelling tools to assess the sustainability of forest management scenarios', in P. J. Burton, C. Messier, D. W. Smith and W. L. Adamowicz (eds) *Towards Sustainable Management of the Boreal Forest*, NRC Research Press, Ottawa, ON

Mielke, D. L., Shugart, H. H. and West, D. C. (1978) *A Stand Model for Upland Forests of Southern Arkansas*, Oak Ridge National Laboratory, Oak Ridge, TN

Miller, C. and Urban, D. L. (1999) 'A model of surface fire, climate and forest pattern in the Sierra Nevada, California', *Ecological Modelling*, vol 114, pp113–135

Milner, K. S., Coble, D. W., McMahan, A. J. and Smith, E. L. (2003) 'FVSBGC: a hybrid of the physiological model STAND-BGC and the forest vegetation simulator', *Canadian Journal of Forest Research*, vol 33, pp466–479

Miner, C. L., Walters, N. R. and Belli, M. L. (1988) *A Guide to the TWIGS Program for the North Central United States* (General Technical Report NC-125), USDA Forest Service, North Central Forest Experiment Station, St Paul, MN

Ministerio de Medio Ambiente (2003) *Deposición atmosférica en la estación de Burguete (Navarra)*, Servicio de Protección Contra Agentes Nocivos en los Montes, Madrid

Mitchell, K. J. (1969) 'Simulation of the growth of even-aged stands of white spruce', *Yale University School of Forestry Bulletin*, pp1–48

Mitchell, K. J. (1975) 'Dynamics and simulated yield of Douglas-fir', *Forest Science Monograph* (supplement to *Forest Science*, vol 21), p17

Mitchell, S. J. and Beese, W. J. (2002) 'The retention system: reconciling variable retention with the principles of silvicultural systems', *The Forestry Chronicle*, vol 78, pp397–403

Mitchell, A. K., Barclay, H. J., Brix, H., Pollard, D. F. W., Benton, R. and deJong, R. (1996) 'Biomass and nutrient element dynamics in Douglas-fir: effects of thinning and nitrogen fertilization over 18 years', *Canadian Journal of Forest Research*, vol 26, pp376–388

Mladenoff, D. J. (2004) 'LANDIS and forest landscape models', *Ecological Modelling*, vol 180, pp7–19

Mladenoff, D. J. and Baker, W. L. (1999) 'Development of forest and landscape modelling approaches', in D. J. Mladenoff and W. L. Baker (eds) *Spatial Modelling of Forest Landscape Change: Approaches and Applications*, Cambridge University Press, Cambridge

Mladenoff, D. J. and DeZonia, B. (2004) *APACK 2.23 Analysis Software User's Guide*, Forest Landscape Ecology, University of Wisconsin, Madison, WI

Mladenoff, D. J. and He, H. S. (1999) 'Design, behaviour and applications of LANDIS, an object-oriented model of forest landscape disturbance and succession', in D. J. Mladenoff and W. L. Baker (eds) *Spatial Modelling of Forest Landscape Change: Approaches and Applications*, Cambridge University Press, Cambridge

Mobbs, D. C., Cannell, M. G. R., Crout, N. M. J., Lawson, G. J., Friend, A. D. and Arah, J. (1998) 'Complementarity of light and water use in tropical agroforests: I. Theoretical model outline, performance and sensitivity', *Forest Ecology and Management*, vol 102, pp259–274

Mohren, G. M. J., van Hees, A. F. M. and Bartelink, H. H. (1991) 'Succession models as an aid for forest management in mixed stands in The Netherlands', *Forest Ecology and Management*, vol 42, pp111–127

Monning, E. and Byler, J. (1992) *Forest Health and Ecological Integrity in the Northern Rockies*, USDA Forest Service, Missoula, MT

Monserud, R. A. (2003) 'Evaluating forest models in a sustainable forest management context', *FBMIS*, vol 1, pp35–47

Monserud, R. A. and Sterba, H. (1996) 'A basal area increment for even- and uneven-aged forest stands in Austria', *Forest Ecology and Management*, vol 80, pp57–80

Moorcroft, P. R., Hurtt, G. C. and Pacala, S. W. (2001) 'A method for scaling vegetation dynamics: the ecosystem demography model (ED)', *Ecological Monographs*, vol 71, pp557–586

Moote, M. A., Mitchel, M. P. and Chickering, D. K. (1997) 'Theory in practice: applying participatory democracy theory to public land planning', *Journal of Environmental Management*, vol 21, pp877–889

Morgan, P., Aplet, G. H., Haufler, J. B., Humphries, H. C., Moore, M. M. and Wilson, W. D. (1994) 'Historical range of variability: a useful tool for evaluating ecosystem change', *Journal of Sustainable Forestry*, vol 2, pp87–111

Morris, D. M., Kimmins, J. P. and Duckert, D. R. (1997) 'The use of soil organic matter as a criterion of the relative sustainability of forest management alternatives: a modelling approach using FORECAST', *Forest Ecology and Management*, vol 94, pp61–78

Morton, A. (1990) 'Mathematical modelling and contrastive explanation', *Canadian Journal of Philosophy*, vol 16, pp251–270

Munro, D. D. (1974) 'Forest growth models: a prognosis', in J. Fries (ed.) *Growth Models for Tree and Stand Simulation*, Royal College of Forestry, Stockholm

Nagel, J. (1997) *BWIN: Program for Standard Analysis and Prognosis. User's Manual for Version 3.0*, Niedersächsische Forstliche Versuchsanstalt, Göttingen, The Netherlands

National Academy of Science (1974) *Rehabilitation Potential of Western Coal Lands*, Ballinger, Cambridge, MA

Nelson, J. (2003) *Forest Planning Studio (FPS)-ATLAS Program: Reference Manual*, University of British Columbia, Vancouver, BC

Nelson, J. (2006) 'Criterion 4: timber economic benefits', Arrow IFPA Series: Note 7 of 8, *British Columbia Journal of Ecosystems and Management*, vol 7, pp92–98

Nemetz, P. N. (ed.) (2004) 'Sustainable resource management: reality or illusion?', *Journal of Business Administration and Policy Analysis*. Faculty of Commerce and Business Administration, University of British Columbia, Vancouver, BC

Nichols, J. D. (2001) 'Using models in the conduct of science and management of natural resources', in T. M. Shenk and A. B. Franklin (eds) *Modeling in Natural Resource Management: Development, Interpretation, and Application*, Island Press, Washington, DC

Nicholson-Cole, S. A. (2005) 'Representing climate change futures: a critique on the use of images for visual communication', *Computers, Environment and Urban Systems*, vol 29, pp255–273

Nitschke, C. R. and Innes, J. L. (2008) 'A tree and climate assessment tool for modelling ecosystem response to climate change', *Ecological Modelling*, vol 210, pp263–277

Noble, I. R. and Slatyer, R. D. (1980) 'The use of vital attributes to predict successional changes in plant communities subject to recurrent disturbances', *Vegetatio*, vol 43, pp5–21

Noon, B. R. (2009) 'Old growth forests as wildlife habitat', in T. A. Spies and S. L. Duncan (eds) *Old Growth in a New World: A Pacific Northwest Icon Revisited*, Island Press, New York, NY

Oil Sands Vegetation Reclamation Committee (OSVRC) (1998) *Guidelines for Reclamation to Forest Vegetation in the Alberta Oil Sands Region* (Alberta Environment Report ESD/LM/99-1), Alberta Environmental Protection, Edmonton, AB

Olarieta, J. R., Domingo, F. and Usón, A. (2000) 'FORTOON: a useful teaching tool in land evaluation for forest management', *The Land*, vol 4, pp29–44

Oliver, C. D. and Larson, B. C. (1990) *Forest Stand Dynamics*, McGraw-Hill, New York, NY

Oliver, C. D., Kimmins, J. P., Harshaw, H. W. and Sheppard, S. R. J. (2000) 'Criteria and indicators of sustainable forestry: a system approach', in S. R. J. Sheppard and H. W. Harshaw (eds) *Forest and Landscapes: Linking Ecology, Sustainability and Aesthetics* (IUFRO Research Series no 6), CABI Publishing, Wallingford, UK

Ollinger, S. V., Aber, J. D. and Federer, C. A. (1998) 'Estimating regional forest productivity and water yield using an ecosystem model linked to a GIS', *Landscape Ecology*, vol 13, pp323–334

O'Neill, R. V., DeAngelis, D. L., Waide, J. B. and Allen, T. F. H. (1986) *A Hierarchical Concept of the Ecosystem*, Princeton University Press, Princeton, NJ

Oreskes, N. (2003) 'The role of quantitative models in science', in C. D. Canham, J. J. Cole and W. K. Lauenroth (eds) *Models in Ecosystem Science*, Princeton University Press, Princeton, NJ, pp13–31

Oreskes, N., Shrader-Frechette, K. and Belitz, K. (1994) 'Verification, validation and confirmation of numerical models in the earth sciences', *Science*, vol 263, pp641–646

Orians, G. H. (1975) 'Diversity, stability and maturity in natural ecosystems', in W. H. van Dobben and R. H. Lowe-McConnell (eds) *Unifying Concepts in Ecology*, BV Publishers, The Hague

Ortiz-Barney, E., Stromberg, J. C. and Beauchamp, V. B. (2005) 'The floristic relay game: a board game to teach plant community succession and disturbance dynamics', http://tiee.ecoed.net/vol/v3/experiments/floristic/abstract.html, accessed 20 June 2008

Osborne, C. P. (2004) 'Modelling the ecology of plants', in J. Wainwright and M. Mulligan (eds) *Environmental Modelling: Finding Simplicity in Complexity*, John Wiley and Sons, New York, NY

Osborne, C. P. and Beerling, D. J. (2002) 'Sensitivity of tree growth to a high CO_2 environment: consequences for interpreting the characteristics of fossil woods from ancient "greenhouse" worlds', *Palaeogeography, Palaeoclimatology, Palaeoecology*, vol 182, pp15–29

Overdevest, C. and Rickenbach, M. G. (2006) 'Forest certification and institutional governance: an empirical study of forest stewardship council certificate holders in the United States', *Forest Policy and Economics*, vol 9, pp93–102

Pacala, S. W. and Deutschman, D. H. (1995) 'Details that matter: the spatial distribution of individual trees maintains forest ecosystem function', *Oikos*, vol 74, pp357–365

Pacala, S. W., Canham, C. D. and Silander, J. A. (1993) 'Forest models defined by field measurements: I. The design of a northeastern forest simulator', *Canadian Journal of Forest Research*, vol 23, pp1980–1988

Pacala, S. W., Canham, C. D., Saponara, J., Silander, J. A., Kobe, R. K. and Ribbens, E. (1996) 'Forest models defined by field measurements: estimation, error analysis and dynamics', *Ecological Monographs*, vol 66, pp1–43

Pages, L., Vercambre, G., Drouet, J. L., Lecompte, F., Collet, C. and Le Bot, J. (2004) 'Root Typ: a generic model to depict and analyse the root system architecture', *Plant and Soil*, vol 258, pp103–119

PEFC (Pan European Forest Certification Council) (2003) 'Annex 6: certification and accreditation procedures', www.pefc.org/index.php/standards/technical-documentation/pefc-international-standards/item/416-annex-6-certification-and-accreditation-procedures, accessed May 2009

Pastor, J. and Post, W. M. (1986) 'Influence of climate, soil moisture, and succession on forest carbon and nitrogen cycles', *Biogeochemistry*, vol 2, pp3–27

Pausas, J. G. (1999) 'Response of plant functional types to changes in the fire regime in Mediterranean ecosystems: a simulation approach', *Journal of Vegetation Science*, vol 10, pp717–722

Pausas, J. G., Austin, M. P. and Noble, I. R. (1997) 'A forest simulation model for predicting eucalypt dynamics and habitat quality for arboreal marsupials', *Ecological Applications*, vol 7, pp921–933

Pearlstine, L., McKellar, H. and Kitchens, W. (1985) 'Modeling the impacts of river diversion on bottomland forest communities in the Santee River Floodplain, South Carolina', *Ecological Modelling*, vol 29, pp283–302

Pearson, R. G. and Dawson, P. T. (2003) 'Predicting the impacts of climate change on the distribution of species: are climate envelope models useful?', *Global Ecology and Biogeography*, vol 12, pp361–371

Pelz, D. R. (1978) 'Computer-assisted instruction in forestry', *Journal of Forestry*, vol 76, pp570–573

Peng, C., Liu, J., Dang, Q., Apps, M. J. and Jiang, H. (2002) 'TRIPLEX: a generic hybrid model for predicting forest growth and carbon and nitrogen dynamics', *Ecological Modelling*, vol 153, pp109–130

Pennanen, J. and Kuuluvainen, T. (2002) 'A spatial simulation approach to natural forest landscape dynamics in boreal Fennoscandia', *Forest Ecology and Management*, vol 164, pp157–175

Pennanen, J., Greene, D. F., Fortin, M. J. and Messier, C. (2004) 'Spatially explicit simulation of long-term boreal forest landscape dynamics: incorporating prediction of quantitative stand attributes', *Ecological Modelling*, vol 180, pp195–209

Perera, A. H., Buse, L. J. and Weber, M. (eds) (2004) *Emulating Natural Forest Landscape Disturbances: Concepts and Application*, Columbia University Press, New York, NY

Perttunen, J., Sievanen, R. and Nikinmaa, E. (1998) 'LIGNUM: a model combining the structure and the functioning of trees', *Ecological Modelling*, vol 108, pp189–198

Peterken, G. F. (1996) *Natural Woodland: Ecology and Conservation in Northern Temperate Regions*, Cambridge University Press, Cambridge

Peters, R. H. (1991) *A Critique for Ecology*, Cambridge University Press, Cambridge, UK

Peterson, G., Allen, C. R. and Holling, C. S. (1998) 'Ecological resilience, biodiversity, and scale', *Ecosystems*, vol 1, pp6–18

Petranek, C., Corey, F. and Black, R. (1992) 'Three levels of learning in simulations: participating, debriefing, and journal writing', *Simulation & Gaming*, vol 23, pp174–185

Phipps, R. L. (1979) 'Simulation of wetlands forest dynamics', *Ecological Modelling*, vol 7, pp257–288

Pimentel, D. and Patzek, T. W. (2005) 'Ethanol production using corn, switchgrass, and wood; biodiesel production using soybean and sunflower', *Natural Resources Journal*, vol 14, pp65–76

Pojar, J., Klinka, K. and Meidinger, D. V. (1987) 'Biogeoclimatic ecosystem classification in British Columbia', *Forest Ecology and Management*, vol 22, pp119–154

Popper, K. R. (1963) 'Conjectures and refutations', Routledge and Kegan Paul, London

Portney, K. E. and Cohen, S. (2006) 'Practical contexts and theoretical frameworks for teaching complexity with digital role-play simulations', in S. Cohen, K. E. Portney, D. Rehberger and C. Thorsen (eds) *Virtual Decisions*, Lawrence Erlbaum, Mahwah, NJ

Posch, P. (1991) 'The educational perspective', in Centre for Education Research and Innovation (ed.) *Environment, Schools and Active Learning*, OCDE, Paris

Power, M. (1993) 'The predictive validation of ecological and environment models', *Ecological Modelling*, vol 68, pp33–50

Prescott, C. E., Maynard, D. G. and Laiho, R. (2000) 'Humus in northern forests: friend or foe?', *Forest Ecology and Management*, vol 133, pp23–36

Prescott, C. E., Blavins, L. L. and Staley, C. (2004) 'Litter decomposition in BC forests: controlling factors and influences of forestry activities', *Journal of Ecosystems and Management*, vol 5, pp30–43

Pretzsch, H. (2001) *Modellierung des Waldwachstums*, Blackwell, Berlin

Pretzsch, H. (2009) *Forest dynamics, growth and yield*, Springer, Berlin, Heidelberg

Pretzsch, H. and Kahn, M. (1995) 'Modelling growth of Bavarian mixed stands in a changing environment', in *Proceedings of the 20th IUFRO World Congress – Caring for the Forest: Research in a Changing World*, Tempere, Finland

Pretzsch, H., Biber, P., Dursky, J., von Gadow, K., Hasenauer, H., Kändler, G., Kenk, G., Kublin, E., Nagel, J., Pukkala, T., Skovsgaard, J. P., Sodtke, R. and Sterba, H. (2006) 'Standardizing and categorizing tree growth models', in H. Hasenauer (ed.) *Growth Models for Forest Management in Europe*, Springer-Verlag, Heidelberg, Germany

Pretzsch, H., Grote, R., Reineking, B., Rötzer, T. H. and Seifert, S. T. (2008) 'Models for forest ecosystem management: a European perspective', *Annals of Botany*, vol 101, pp1065–1087

Proulx, G. (2004) 'Integrating scientific method and critical thinking in classroom debates on environmental issues', *The American Biology Teacher*, vol 66, pp26–33

Prusinkiewicz, P. and Rolland-Lagan, A. G. (2006) 'Modeling plant morphogenesis', *Current Opinion in Plant Biology*, vol 9, pp83–88

Puettmann, K. J., Coates, K. D. and Messier, C. (2009) *A Critique of Silviculture: Managing for Complexity*, Island Press, Washington, DC

Quinn, C. (2005) *Engaging Learning: Designing e-Learning Simulation Games*, Pfeiffer, San Francisco, CA

Randers, J. (1980) 'Guidelines for model conceptualization', in J. Randers (ed.) *Elements of System Dynamics Method*, Pegasus Communications, Walthan, MA

Ranius, T., Kindvall, O., Kruys, N. and Jonsson, B. G. (2003) 'Modelling dead wood in Norway spruce stands subject to different management regimes', *Forest Ecology and Management*, vol 182, pp13–29

Rauscher, H. M. (1999) 'Ecosystem management decision support for federal forests in the United States: a review', *Forest Ecology and Management*, vol 114, pp173–197

Rauscher, H. M., Lloyd, F. T., Loftis, D. L. and Twery, M. J. (2000) 'A practical decision-analysis process for ecosystem management', *Computers and Electronics in Agriculture*, vol 27, pp195–226

Redmond, K. T. (2007) 'Climate variability and change as a backdrop for western resource management', in L. Joyce, R. Haynes, R. White and R. J. Barbour (eds) *Proceedings of the Workshop Bringing Climate Change into Natural Resource Management*, USDA, Portland, OR

Rees, W. E. (1992) 'Ecological footprints and appropriated carrying capacity: what urban economics leaves out', *Environment and Urbanization*, vol 4, pp121–130

Reeve, D. and Petch, J. (1999) *GIS, Organisations and People: A Socio-technical Approach*, Taylor and Francis, London

Refsgaard, J. C., Henriksen, H. J., Harrar, W. G., Sholten, H. and Kassahun, A. (2005) 'Quality assurance in model based water management: review of existing practice and outline of new approaches', *Environmental Modelling and Software*, vol 20, pp1201–1215

Rempel, R. S., Donnelly, M., van Damme, L., Gluck, M., Kushneriuk, R. and Moore, T. (2006) 'Spatial landscape assessments models: a meta-modelling framework for biodiversity conservation planning', in R. Lafortezza and G. Sanesi (eds) *Patterns and Processes in Forest Landscapes: Consequences of Human Management*, Locorotondo, Bari, Italy

Remsoft (2006) *Woodstock User Guide*, Remsoft Inc., Fredericton, NB, Canada

Rennie, P. J. (1955) 'The uptake of nutrients by mature forest growth', *Plant and Soil*, vol 7, pp49–95

Reynolds, M. R. (1984) 'Estimating the error in model predictions', *Forest Science*, vol 30, pp454–469

Richardson, G. and Pugh, A. (1981) *Introduction to System Dynamics Modelling with Dynamo*, Pegasus Communications, Walthan, MA

Richardson, J. R. and Berish, C. W. (2003) 'Data and information issues in modelling for resource management decision making: communication is the key', in V. Dale (ed.) *Ecological Modelling for Resource Management*, Springer-Verlag, New York, NY

Richardson, K. and Steffen, W. (2009) 'The IARU International Scientific Congress on Climate Change: global risks, challenges and decisions (12 March, Copenhagen, Denmark)', *IOP Conference Series: Earth and Environmental Science*, vol 6, section 001002, pp1–3

Ripley, B. D. (1996) *Pattern Recognition and Neural Networks*, Cambridge University Press, Cambridge, UK

Rittel, H. J. and Webber, M. M. (1973) 'Dilemmas in a general theory of planning', *Policy Science*, vol 4, pp155–169

Roberts, D. W. and Betz, D. W. (1999) 'Simulating landscape vegetation dynamics of Bryce Canyon National Park with the vital attributes/fuzzy system model VAFS/LANDSIM', in D. J. Mladenoff and W. L. Baker (eds) *Spatial Modelling of Forest Landscape Change: Approaches and Applications*, Cambridge University Press, Cambridge

Robinson, A. P. and Froese, R. E. (2004) 'Model validation using equivalence tests', *Ecological Modelling*, vol 176, pp349–358

Robinson, D. C. E. (1996) *GIZELA: Model Description, User's Guide and Tutorial*, ESSA Technologies, Vancouver, BC

Robinson, J. and Tansey, J. (2006) 'Co-production, emergent properties and strong interactive social research: the Georgia Basin Futures Project', *Science and Public Policy*, vol 33, pp151–160

Rockström, J., Steffen, W., Noone, K., Persson, A., Chapin III, F. S., Lambin, E. F., Lenton, T. M., Scheffer, M., Folke, C., Schellnhuber, H. J., Nykvist, B., de Wit, C. A., Hughes, T., van der Leeuw, S., Rodhe, H., Sörlin, S., Snyder, P. K., Costanza, R., Svedin, U., Falkenmark, M., Karlberg, L., Corell, R. W., Fabry, V. J., Hansen, J., Walker, B., Liverman, D., Richardson, K., Crutzen, P. and Foley, J. A. (2009) 'A safe operating space for humanity', *Nature*, vol 461, pp472–475

Rowe, J. S. (1961) 'The level-of-integration concept and ecology', *Ecology*, vol 42, pp420–427

Rowland, S. M., Prescott, C. E., Grayston, S. J., Quideau, S. A. and Bradfield, G. E. (2009) 'Recreating a functioning forest soil in reclaimed oil sands in northern Alberta: an approach for measuring success in ecological restoration', *Journal of Environmental Quality*, vol 38, pp1580–1590

Ruben, B. D. (1999) 'Simulations, games, and experience-based learning: the quest for a new paradigm for teaching and learning', *Simulation and Gaming*, vol 30, pp498–505

Rykiel, E. J. (1996) 'Testing ecological models: the meaning of validation', *Ecological Modelling*, vol 90, pp229–244

Sachs, D. and Trofymow, J. A. (1991) *Testing the Performance of FORCYTE-11 Against Results from the Shawnigan Lake Thinning and Fertilization Trials on Douglas-Fir*, Forestry Canada, Pacific and Yukon Region, Pacific Forestry Centre, Victoria, BC

Sainsbury, K. J., Punt, A. E. and Smith, A. D. M. (2000) 'Design of operational management strategies for achieving fishery ecosystem objectives', *ICES Journal of Marine Science*, vol 57, pp731–741

Salisbury, F. B. and Ross, C. W. (1992) *Plant Physiology* (fourth edition), Wadsworth Publishing Company, Belmont, CA

Salt, G. W. (1979) 'A comment on the use of the term emergent properties', *The American Naturalist*, vol 113, pp145–161

Salter, J., Campbell, C., Journeay, M. and Sheppard, S. R. J. (2009) 'The digital workshop: exploring the use of interactive and immersive visualisation tools in participatory planning', *Journal of Environmental Management*, vol 90, pp2090–2101

Salwasser, H. (2002) 'Navigating through the wicked messiness of natural resource problems: roles for science, coping strategies, and decision analysis', *Proceedings of the Sierra Science Summit*, USDA Forest Service, Kings Beach, CA

Sampson, R. N. and Adams, D. L. (eds) (1994) *Assessing Forest Ecosystem Health in the Inland West*, Haworth Press, New York, NY

Sayer, J. and Maginnis, S. (eds) (2007) *Forests in Landscapes: Ecosystem Approaches to Sustainability*, Earthscan, London

Scheller, R. M., Domingo, J. B., Sturtevan, B. R., Williams, J. S., Rudy, A., Gustafson, E. J. and Mladenoff, D. J. (2007) 'Design, development, and application of LANDIS-II, a spatial landscape simulation model with flexible temporal and spatial resolution', *Ecological Modelling*, vol 201, pp409–419

Schreiber, E. S. G., Bearlin, A. G., Nicol, S. J. and Todd, C. R. (2004) 'Adaptive management: a synthesis of current understanding and effective application', *Ecological Management and Restoration*, vol 5, pp177–182

Schroth, O., Pond, E., Muir-Owen, S., Campbell, C. and Sheppard, S. R. J. (2009) *Tools for the Understanding of Spatio-Temporal Climate Scenarios in Local Planning: Kimberley (BC) Case Study*, Centre for Advance Landscape Planning, UBC, Vancouver, BC

Schumaker, N. H. (1998) *A User's Guide to the PATCH Model* (Environmental Protection Agency report EPA/600/R-98/135), Environmental Research Laboratory, Corvallis, OR

Seagle, S. W. and Liang, S.-Y. (2001) 'Application of a forest gap model for prediction of browsing effects on riparian forest succession', *Ecological Modelling*, vol 144, pp213–229

Seely, B. (2005a) *South Coast LLEMS Project Phase 3: Application and Evaluation*, FORRx Consulting, Belcarra, BC

Seely, B. (2005b) *Development of Carbon Curves for Addressing CSA Certification Requirements in the Morice and Lakes Timber Supply Areas*, FORRx Consulting, Belcarra, BC

Seely, B. and Welham, C. (2006) *Towards the Application of SOM as a Measure of Ecosystem Productivity in the Quesnel Forest District: Deriving Thresholds, Determining Effective Sampling Regimes, and Evaluating Practices*, British Columbia Ministry of Forests Forest Sciences Program, Victoria, BC

Seely, B., Arp, P. and Kimmins, J. P. (1997) 'A forest hydrology submodel for simulating the effect of management and climate change on stand water stress', in A. Amaro and M. Tomé (eds) *Proceedings of Empirical and Process-Based Models for Forest, Tree and Stand Growth Simulation*, Edições Salamandra, Lisboa

Seely, B., Kimmins, J. P., Welham, C. and Scoullar, K. A. (1999) 'Ecosystem management models: defining stand-level sustainability, exploring stand-level stewardship', *Journal of Forestry*, vol 97, pp4–10

Seely, S., Welham, C. and Kimmins, H. (2002) 'Carbon sequestration in a boreal forest ecosystem: results from the ecosystem simulation model, FORECAST', *Forest Ecology and Management*, vol 169, pp123–135

Seely, B., Nelson, J., Wells, R., Peter, B., Meitner, M., Anderson, A., Harshaw, H., Sheppard, S., Bunnell, F. L., Kimmins, H. and Harrison, D. (2004) 'The application of a hierarchical, decision-support system to evaluate multi-objective forest management strategies: a case study in northeastern British Columbia, Canada', *Forest Ecology and Management*, vol 199, pp283–305

Seely, B., Nelson, J., Vernier, P., Wells, R. and Moy, A. (2007a) *Exploring Opportunities for Mitigating the Ecological Impacts of Current and Future Mountain Pine Beetle Outbreaks Through Improved Planning: A Focus on Northeastern BC* (Mountain Pine Beetle Initiative working paper 8.27), Natural Resources Canada, Victoria, BC

Seely, B., Liu, G. and Makitalo, A. (2007b) *Moving Towards a Desirable Future: Developing and Evaluating Alternative MPB Salvage Strategies in the Prince George Forest District*, BC Forest Science Program, BC Ministry of Forests, Victoria, BC

Seely, B., Hawkins, C., Blanco, J. A., Welham, C. and Kimmins, J. P. (2008) 'Evaluation of a mechanistic approach to mixedwood modelling', *The Forestry Chronicle*, vol 84, pp181–193

Seely, B., Welham, C. and Blanco, J. A. (2010) 'Towards the application of soil organic matter as an indicator of ecosystem productivity: deriving thresholds, developing monitoring systems, and evaluating practices', *Ecological Indicators*, vol 10, pp999–1008

SER (Society for Ecological Restoration International Science & Policy Working Group) (2004) *The SER International Primer on Ecological Restoration*, Society for Ecological Restoration International, Tuscon, AZ

Sessions, J., Johnson, N. K., Franklin, J. F. and Gabriel, J. T. (1999) 'Achieving sustainable forest structures on fire-prone landscapes while pursuing multiple goals', in D. J. Mladenoff and W. L. Baker (eds) *Spatial Modelling of Forest Landscape Change: Approaches and Applications*, Cambridge University Press, Cambridge

SGOG (Smart Growth on the Ground) (2009) 'Smart Growth on the Ground: Prince George concept plan', www.sgog.bc.ca/content.asp?contentID=144, accessed October 2009

Shao, G., Schall, P. and Weishampel, J. F. (1994) 'Dynamic simulations of mixed broadleaved-*Pinus koraiensis* forests in the Changbaishan Biosphere Reserve of China', *Forest Ecology and Management*, vol 70, pp169–181

Shaw, A., Sheppard, S., Burch, S., Flanders, D., Weik, A., Carmichael, J., Robinson, J. and Cohen, S. (2009) 'How futures matter: synthesizing, downscaling, and visualizing climate change scenarios for participatory capacity building', *Journal of Global Environmental Change*, vol 19, pp447–463

Shenk, T. M. and Franklin, A. B. (eds) (2001) 'Modeling in natural resource management: development, interpretation, and application', Island Press, Washington, DC

Sheppard, S. R. J. (2005a) 'Participatory decision support for sustainable forest management: a framework for planning with local communities at the landscape level in Canada', *Canadian Journal of Forest Research*, vol 35, pp1515–1526

Sheppard, S. R. J. (2005b) 'Landscape visualisation and climate change: the potential for influencing perceptions and behaviour', *Environmental Science & Policy*, vol 8, pp637–654

Sheppard, S. R. J. (2005c) 'Validity, reliability, and ethics in visualization', in I. Bishop and E. Lange (eds) *Visualization in Landscape and Environmental Planning: Technology and Applications*, Taylor and Francis, London

Sheppard, S. R. J. and Harshaw, H. W. (2000) 'Conclusions: towards a research agenda for forest landscape management', in S. R. J. Sheppard and H. W. Harshaw (eds) *Forests and Landscapes: Linking Ecology, Sustainability, and Aesthetics*, CAB International, Wallingford, UK

Sheppard, S. R. J. and Harshaw, H. W. (eds) (2001) *Forests and Landscapes: Linking Ecology, Sustainability and Aesthetics*, CABI Publishing, London, UK

Sheppard, S. R. J. and Meitner, M. (2005) 'Using multi-criteria analysis and visualisation for sustainable forest management planning with stakeholder groups', *Forest Ecology and Management*, vol 207, pp171–187

Sheppard, S. R. J. and Salter, J. (2004) 'The role of visualization in forest planning', in J. Burley, J. Evans and J. Youngquist (eds) *Encyclopaedia of Forest Sciences*, Academic Press/Elsevier, Oxford

Sheppard, S. R. J., Shaw, A., Flanders, D. and Burch, S. (2008) 'Can visualization save the world? Lessons for landscape architects from visualizing local climate change' (paper delivered to Digital Design in Landscape Architecture 2008; Ninth International Conference on IT in Landscape Architecture, Anhalt University of Applied Sciences, Dessau/Bernburg, Germany)

Shindler, B. and Cramer, L. A. (1999) 'Changing public values: consequences for Pacific northwest forestry', *Western Journal of Applied Forestry*, vol 13, pp28–34

Shore, T. L. and Safranyik, L. (1992) *Susceptibility and Risk Rating Systems for the Mountain Pine Beetle in Lodgepole Pine Stands*, Information Report BC-X-336, Canadian Forest Service, Pacific Forestry Centre, Victoria, BC

Shugart, H. H. (1998) *Terrestrial Ecosystems in Changing Environments*, Cambridge University Press, Cambridge

Shugart, H. H. and Noble, I. R. (1981) 'A computer model of succession and fire response of the high altitude eucalyptus forest of the Brindabella Range, Australian Capital Territory', *Australian Journal of Ecology*, vol 6, pp149–164

Shugart, H. H. and West, D. C. (1977) 'Development and application of an Appalachian deciduous forest succession model', *Environmental Management*, vol 5, pp161–179

Shugart, H. H., Hopkins, M. S., Burgess, I. P. and Mortlock, A. T. (1980) 'The development of a succession model for subtropical rain forest and its application to assess the effects of timber harvest at Wiangaree State Forest, New South Wales', *Environmental Management*, vol 11, pp243–265

Silverstone, S. (2004) 'Managing the gamer generation', http://hbswk.hbs.edu/item.jhtml?id=4429&t=innovation, accessed 20 June 2008

Simard, S. W., Heineman, J. L., Mather, W. J., Sachs, D. L. and Vyse, A. (2001) *Effects of Operational Brushing on Conifers and Plant Communities in the Southern Interior of British Columbia: Results from PROBE 1991–2000*, Land Management Handbook no 48, British Columbia Ministry of Forests, Victoria, BC

SimBiotic Software for Teaching and Research (2009) *EcoBeaker 2.5*, SimBio, Missoula, MN

Sinclair, A. R. E., Krebs, C. J., Fryxell, J. M., Turkington, R., Boutin, S., Boonstra, R., Lundberg, P. and Oksanen, L. (2000) 'Testing hypotheses of trophic level interactions using experimental perturbations of a boreal forest ecosystem', *Oikos*, vol 89, pp313–328

Sinoquet, H. and Le Roux, X. (2000) 'Short-term interactions between tree foliage and the aerial environment: an overview of modelling approaches available for tree structure-function models', *Annals of Forest Sciences*, vol 57, pp477–496

Sirois, I., Bonan, G. B. and Shugart, H. H. (1994) 'Development of a simulation model of the forest–tundra transition zone of north-eastern Canada', *Canadian Journal of Forest Research*, vol 24, pp697–706

Skreta, M. (2006) 'Applying Swedish programmes projecting forest development to Polish forestry conditions', final PhD thesis, Southern Swedish Forest Research Centre, SLU, Alnarp, Sweden

Smith, B., Prentice, I. C. and Sykes, M. T. (2001) 'Representation of vegetation dynamics in modelling of terrestrial ecosystems: comparing two contrasting approaches within European climate space', *Global Ecology Biogeography*, vol 10, pp621–637

Smith, T. M. and Urban, D. L. (1988) 'Scale and the resolution of forest structural pattern', *Vegetatio*, vol 74, pp143–150

Soares, P., Tomé, M., Skovsgaard, J. P. and Vanclay, J. K. (1995) 'Evaluating a growth model for forest management using continuous forest inventory data', *Forest Ecology and Management*, vol 71, pp251–265

Sober, E. (1981) 'The principle of parsimony', *British Journal of Philosophy of Science*, vol 32, pp145–156

Society of American Foresters (1993) *Task Force Report on Sustaining Long-Term Forest Health and Productivity*, Society of American Foresters, Bethesda, MD

Solomon, A. M. (1986) 'Transient response of forests to CO_2-induced climate change: simulation experiments in eastern North America', *Oecologia*, vol 68, pp567–679

Solomon, A. M., West, D. C. and Solomon, J. A. (1981) 'Simulating the role of climate change and species immigration in forest succession', in D. C. West, H. H. Shugart and D. B. Botkin (eds) *Forest Succession: Concepts and Application*, Springer-Verlag, New York, NY

Solomon, A. M. and Shugart, H. H. (1984) 'Integrating forest stand simulations with paleoecological records to examine the long-term forest dynamics', in G. I. Ågren (ed.) *State and Change of Forest Ecosystems: Indicators in Current Research*, Swedish University of Agricultural Science, Uppsala, Sweden

Sopper, W. E. (1992) 'Reclamation of mine land using municipal sludge', *Advances in Soil Science*, vol 17, pp351–431

Sougavinski, S. and Doyon, F. (2002) *Variable Retention: Research Findings, Trial Implementation and Operational Issues*, Sustainable Forest Management Network, Edmonton, AB

Spies, T. A. (2009) 'Science of old growth', in T. A. Spies and S. L. Duncan (eds) *Old Growth in a New World: A Pacific Northwest Icon Revisited*, Island Press, New York, NY

Spies, T. A. and Duncan, S. L. (eds) (2009) *Old Growth in a New World: A Pacific Northwest Icon Revisited*, Island Press, New York, NY

Standish, J. T., Manning, G. H. and Demaershalk, J. P. (1985) *Development of Biomass Equations for British Columbia Tree Species*, Canadian Forest Service, Victoria, BC

Starfield, A. M. and Bleloch, A. L. (1991) *Building Models for Conservation and Wildlife Management* (second edition), Interaction Book Company, Edina, MN

Stennes, B. and McBeath, A. (2006) *Bioenergy Options for Woody Feedstock: Are Trees Killed by Mountain Pine Beetle in British Columbia a Viable Bioenergy Resource?* Canadian Forest Service, Pacific Forestry Centre, Victoria, BC

Sterba, H. and Monserud, R. A. (1997) 'Applicability of the forest stand growth simulator PROGNAUS for the Austrian part of the Bohemian Massif', *Ecological Modelling*, vol 98, pp23–34

Sterman, J. (2002) 'All models are wrong: reflections on becoming a systems scientist', *System Dynamics Review*, vol 18, pp501–531

Stolte, W. J., Barbour, S. L. and Boese, C. D. (2000) 'Reclamation of saline-sodic waste dumps associated with the oilsands industry', in A. Etmanski (ed.) *Global Land Reclamation/ Remediation 2000 and Beyond: Proceedings of the Canadian Land Reclamation Association 25th Annual Meeting*, Edmonton, AB, Canada, 17–20 September, Canadian Land Reclamation Association, Edmonton, AB

Straka, T. J. and Childers, C. J. (2006) 'Consulting foresters' view of professional forestry education', *Journal of Natural Resources & Life Sciences Education*, vol 35, pp48–52

Stratton, R. D. (2004) 'Assessing the effectiveness of landscape fuel treatments on fire growth and behavior', *Journal of Forestry*, Oct/Nov, pp32–40

Sullivan, T. P., Sullivan, D. S., Lindgren, P. M. F. and Ransome, D. B. (2006) 'Long-term responses of ecosystem components to stand thinning in young lodgepole pine forest: III. Growth of crop trees and coniferous stand structure', *Forest Ecology and Management*, vol 228, pp69–81

Swedish Environmental Advisory Council (2002) *Resilience and Sustainable Development: Building Adaptive Capacity in a World of Transformations*, Swedish Environmental Advisory Council, Stockholm

Tansley, A. G. (1935) 'The use and abuse of vegetational concepts and terms', *Ecology*, vol 16, pp284–307

Tesera Systems (2009) *Tesera Scheduling Model*, Cochrane, AB

Tharp, M. L. (1978) 'Modeling major perturbations on a forest ecosystem', MSc thesis, University of Tennessee, Knoxville, TN

Theil, H. (1966) *Applied Econometric Forecasting*, North-Holland, Amsterdam

Thiagarajan, S. and Stolovitch, H. D. (1978) *Instructional Simulation Games*, Educational Technology Publications, Englewood Cliffs, NJ

Thomas, J. W. (1995) *The Forest Service Program for Forest and Rangeland Resources: A Long-Term Strategic Plan*, RPA Program, USDA Forest Service, Washington, DC

Thuiller, W., Albert, C., Araújo, M. B., Berry, P. M., Cabeza, M., Guisan, A., Hickler, T., Midgley, G. F., Paterson, J., Schurr, F. M., Sykes, M. T. and Zimmermann, N. E. (2008) 'Predicting global change impacts on plant species' distributions: future challenges', *Perspectives in Plant Ecology*, vol 9, pp137–152

Thysell, D. R. and Carey, A. B. (2000) *Effects of Forest Management on Understory and Overstory Vegetation: A Retrospective Study* (Technical Report PNW-GTR-488), USDA Forest Service, Portland, OR

Tomé, M. J. and Soares, P. (2009) 'Is there a niche for hybrid models?', in D. P. Dykstra and R. A. Monserud (eds) *Forest Growth and Timber Quality: Crown Models and Simulation Methods for Sustainable Forest Management*, US Department of Agriculture, Forest Service, Pacific Northwest Research Station, Portland, OR

Tonu, O. (1983) 'Metsa suktsessiooni ja tasandilise struktuuri imiteerimisest', *Yearbook of the Estonian Naturalist Society*, vol 69, pp110–117

Trofymow, J. A., Barclay, H. J. and McCullough, K. M. (1991) 'Annual rates and elemental concentrations of litterfall in thinned and fertilized Douglas-fir', *Canadian Journal of Forest Research*, vol 21, pp1601–1615

Trofymow, J. A., Moore, T. R., Titus, B., Prescott, C., Morrison, I., Siltanen, M., Smith, S., Fyles, J., Wein, R., Camire, C., Duschene, L., Kozak, L., Kranabetter, M. and Visser, M. (2002) 'Rates of litter decomposition over six years in Canadian forests: influence of litter quality and climate', *Canadian Journal of Forest Research*, vol 32, pp 789–804

Trofymow, J. A., Addison, J., Blackwell, B. A., He, F., Preston, C. A. and Marshall, V. G. (2003) 'Attributes and indicators of old growth and successional Douglas-fir forests on Vancouver Island', *Environmental Review*, vol 11, pp187–204

Tscharntke, T. and Hawkins, B. A. (eds) (2002) *Multitrophic Level Interactions*, Cambridge University Press, Cambridge, UK

Turner, M. G. (1988) 'A spatial simulation model of land use changes in a Piedmont county in Georgia', *Applied Mathematics and Computation*, vol 27, pp39–51

Twery, M. (2004) 'Modelling in forest management', in J. Wainwright and M. Mulligan (eds) *Environmental Modelling: Finding Simplicity in Complexity*, John Wiley and Sons, New York, NY

Tyrrell, L. E. and Crow, T. R. (1994) 'Structural characteristics of old-growth hemlock-hardwood forests in relation to age', *Ecology*, vol 75, p370

UNCED (United Nations Conference on Environment and Development) (1993) *Agenda 21: Earth Summit: The United Nations Programme of Action from Rio*, United Nations, Rome

United Nations (2004) *World Population to 2300*, United Nations, Department of Economic and Social Affairs, New York, NY

Urban, D. L. (1990) *A Versatile Model to Simulate Forest Pattern: A User's Guide to ZELIG Version 1.0*, Department of Environmental Sciences, University of Viginia, Charlottesville, VA

Urban, L. (2005) 'Modeling ecological processes across scales', *Ecology*, vol 86, pp1996–2006

Urban, D. L. and Shugart, H. H. (1992) 'Individual based models of forest succession', in D. C. Glenn-Lewin, R. K. Peet and T. T. Veblen (eds) *Plant Succession: Theory and Prediction*, Chapman & Hall, London

Urban, D. L., Harmon, M. E. and Halpern, C. B. (1993) 'Potential response of pacific north-western forests to climatic change, effects of stand age and initial composition', *Climatic Change*, vol 23, pp247–266

Urban, L., Acevedo, M. F. and Garman, S. L. (1999) 'Scaling fine-scale processes to large-scale patterns using models derived from models: meta-models', in D. J. Mladenoff and W. L. Baker (eds) *Spatial Modelling of Forest Landscape Change: Approaches and Applications*, Cambridge University Press, Cambridge

USDA Forest Service (1993) *Healthy Forests for America's Future: A Strategic Plan*, USDA Forest Service, Washington, DC

Valencia-Sandoval, C., Flanders, D. N. and Kozak, R. A. (2010) 'Participatory landscape planning and sustainable community development: methodological observations from a case study in rural Mexico', *Journal of Landscape and Urban Planning*, vol 94, pp63–70

Valentine, H. T., Gregoire, T. G., Burkhart, H. E. and Hollinger, D. Y. (1997) 'A stand-level model of carbon allocation and growth, calibrated for loblolly pine', *Canadian Journal of Forest Research*, vol 27, pp817–830

Vanclay, J. K. (1994) *Modelling Forest Growth and Yield: Applications to Mixed Tropical Forests*, CAB International, Wallingford, UK

Vanclay, J. K. and Skovsgaard, J. P. (1997) 'Evaluating forest growth models', *Ecological Modelling*, vol 98, pp1–12

Van Daalen, I. C. and Shugart, H. H. (1989) 'OUTENIQUA: a computer model to simulate succession in the mixed evergreen forests of the southern Cape, South Africa', *Landscape Ecology*, vol 24, pp255–267

Van Horn, R. (1969) 'Validation', in T. H. Naylor (ed.) *The Design of Computer Simulation Experiments*, Duke University Press, Durham, NC

Van Mantgem, P. J., Stephenson, N. L., Bryne, J. C., Daniels, L. D., Franklin, J. F., Fulé, P. Z., Harmon, M. E., Larson, A. J., Smith, J. M., Taylor, A. H. and Veblen, T. T. (2009) 'Widespread increase of tree mortality rates in the Western United States', *Science*, vol 323, pp521–524

Van Noordwijk, M. (2002) 'Scaling trade-offs between crop productivity, carbon stocks and biodiversity in shifting cultivation landscape mosaics: the FALLOW model', *Ecological Modelling*, vol 149, pp113–126

Van Noordwijk, M. and Lusiana, B. (1999) 'WaNuLCAS, a model of water, nutrient and light capture in agroforestry systems', in D. Auclair and C. Dupraz (eds) *Agroforestry for Sustainable Land Use: Fundamental Research and Modelling with Emphasis on Temperate and Mediterranean Applications*, Kluwer Academic Publishers, Dordrecht, The Netherlands

Van Oijen, M., Cannell, M. G. R. and Levy, P. E. (2004) 'Modelling biogeochemical cycles in forests: state of the art and perspectives', in F. Andersson, Y. Birot and R. Päivinen (eds) *Towards the Sustainable Use of Europe's Forests: Forest Ecosystem and Landscape Research: Scientific Challenges and Opportunities*, European Forest Institute, Joensuu, Finland

Van Oijen, M., Rougier, J. C. and Smith, R. (2005) 'Bayesian calibration of process-based forest models: bridging the gap between models and data', *Tree Physiology*, vol 25, pp915–927

Van Waveren, R. H., Groot, S., Scholten, H., Van Geer, F. C., Wösten, J. H. M., Koeze, R. D. and Noort J. J. (1999) *Good Modelling Practice Handbook* (STOWA Report 99-05), Utrecht, The Netherlands

Vatti, B. R. (1992) 'A generic solution to polygon clipping', *Communications of the Association for Computer Machinery*, vol 35, pp56–63

Ventana Systems (2009) *VENSIM Software*, Harvard, MA

Vetterlein, D. and Hüttl, R. F. (1999) 'Can applied organic matter fulfil similar functions as soil organic matter? Risk–benefit analysis for organic matter application as a potential strategy for rehabilitation of disturbed ecosystems', *Plant and Soil*, vol 213, pp1–10

Villa, F. and Costanza, R. (2000) 'Design of multi-paradigm integrating modelling tools for ecological research', *Environmental Modelling and Software*, vol 15, pp169–177

Villa, F., Donatelli, M., Rizzoli, A. E., Krause, P., Kraslisch, S. and van Evert, F. K. (2006) 'Declarative modelling for architecture independence and data/model integration: a case study', in A. Voinov, A. J. Jakeman and A. E. Rizoli (eds) *Proceedings of the iEMSs Third Biennial Meeting, 'Summit on Environmental Modelling and Software'*, International Environmental Modelling and Software Society, www.iemss.org/iemss2006/sessions/all.html, accessed 16 September 2009

Vuorisalo, T. O. and Mutikainen, P. K. (1999) *Life History Evolution in Plants*, Kluwer Academic Publishers, Dordrecht, The Netherlands

Wackernagel, M. and Rees, W. (1996) *Our Ecological Footprint: Reducing Human Impact on the Earth*, New Society Publishers, Gabriola Island, BC

Wagener, T., Wheater, H. S. and Gupta, H. V. (2003) 'Identification and evaluation of watershed models', in Q. Duan, S. Sorooshian, H. V. Gupta, A. Rousseau and R. Turcotte (eds) *Calibration of Watershed Models*, American Geophysical Union, Washington, DC

Waldrop, T. A., Buckner, E. R., Shugan, H. H. and McGee, C. E. (1986) 'FORCAT: A single tree model of stand development on the Cumberland Plateau', *Forest Science*, vol 32, pp297–317

Walker, W. E. (1995) *The Use of Scenarios and Gaming in Crisis Management Planning and Training*, Rand, Santa Monica, CA

Walters, C. J. (1986) *Adaptive Management of Renewable Resources*, Macmillan, New York, NY

Walters, C. J. and Bunnell, F. (1971) 'A computer management game of land use in British Columbia', *Journal of Wildlife Management*, vol 35, pp644–652

Wang, Y. P. and Jarvis, P. G. (1990) 'Description and validation of an array model – MAESTRO', *Agricultural and Forest Meteorology*, vol 51, 257–280

Wang, J. R., Hawkins, C. D. B. and Letchford, T. (1998) 'Photosynthesis, water and nitrogen use efficiencies of four paper birch (*Betula papyrifera*) populations grown under different soil moisture and nutrient regimes', *Forest Ecology and Management*, vol 112, pp233–244

Wardlaw, I. F. (1990) 'Tansley review no. 27: the control of carbon partitioning in plants', *New Phytologist*, vol 116, pp341–381

Wei, X., Kimmins, J. P. and Zhou, G. (2003) 'Disturbances and the sustainability of long-term site productivity in lodgepole pine forests in the central interior of British Columbia: an ecosystem modelling approach', *Ecological Modelling*, vol 164, pp239–256

Weinstein, D. A. and Yanai, R. D. (1994) 'Integrating the effects of simultaneous multiple stresses on plants using the simulation model TREGRO', *Journal of Environmental Quality*, vol 23, pp418–428

Weinstein, D. A., Shugart, H. H. and West, D. C. (1982) *The Long-Term Nutrient Retention Properties of Forest Ecosystems: A Simulation Investigation* (ORNL/TM-84 72), Oak Ridge National Laboratory, Oak Ridge, TN

Weinstein, D. A., Yanai, R. D., Beloin, R. and Zollweg, C. G. (1992) *The Response of Plants to Interacting Stresses: TREGRO Version 1.74 – Description and Parameter Requirements*, Electric Power Research Institute, Palo Alto, CA

Weiskittel, A. R., Maguire, D. A. and Monserud, A. (2009) 'Development of a hybrid model for intensively managed Douglas-fir in the Pacific Northwest', in D. P. Dykstra and R. A. Monserud (eds) *Forest Growth and Timber Quality: Crown Models and Simulation Methods for Sustainable Forest Management*, US Department of Agriculture Forest Service, Pacific Northwest Research Station, Portland, OR

Welham, C. (2005) *Evaluating a Prescriptive Approach to Creating Target Ecosites Using D-Ecosites as a Test Case* (Project 2004 – 0014), Cumulative Environmental Management Association, Fort McMurray, AB

Welham, C. (2006) *Evaluating Existing Prescriptions for Creating Target Ecosites Using the Ecosystem Simulation Model, FORECAST: Implications for Ecosystem Productivity and Community*

Composition (Project 2005 – 0025), Cumulative Environmental Management Association, Fort McMurray, AB

Welham, C. (2009) *Evaluating Existing Prescriptions for Creating Target Ecosites Using the Ecosystem Simulation Model, FORECAST: Implications for Ecosystem Productivity and Community Composition in Reclaimed Overburden* (Project 2006 – 0030), Cumulative Environmental Management Association, Fort McMurray, AB

Welham, C., Seely, B. and Kimmins, H. (2002) 'The utility of the two-pass harvesting system: an analysis using the ecosystem simulation model FORECAST', *Canadian Journal of Forest Research*, vol 32, pp1071–1079

Welleck, S. (2003) 'Testing statistical hypothesis of equivalence', Chapman & Hall, London

Wells, R. and Nelson, J. (2006) 'Sustainable forest management basecase analysis: The Lemon Landscape Unit pilot project', Arrow IFPA Series: Note 4 of 8, *British Columbia Journal of Ecosystems and Management*, vol 7, pp67–75

Wells, R. W., Lertzman, K. P. and Saunders, S. C. (1998) 'Old growth definitions for the forests of British Columbia, Canada', *Natural Areas Journal*, vol 18, pp279–292

West, P. W. (1993) 'Model of above-ground assimilate partitioning and growth of individual trees in even aged forest monoculture', *Journal of Theoretical Biology*, vol 161, pp369–394

West, G. B., Brown, J. H. and Enquist, B. J. (1997) 'A general model for the origin of allometric scaling laws in biology', *Science*, vol 276, pp122–126

West, G. B., Brown, J. H. and Enquist, B. J. (1999) 'A general model for the structure and allometry of plant vascular systems', *Nature*, vol 400, pp664–667

Westoby, J. (1987) *The Purpose of Forests*, Blackwell, Oxford

Wilkie, D. S. and Finn, J. T. (1988) 'A spatial model of land use and forest regeneration in the Ituri forest of northeastern Zaire', *Ecological Modelling*, vol 41, pp307–323

Williams, K. S. (2000) 'Teaching and learning with EcoBeaker 2.0', *Ecology*, vol 81, pp1173–1174

Wilmking, M., Juday, G. P., Barbier, V. A. and Zald, H. S. J. (2004) 'Recent climate warming forces contrasting growth responses of white spruce at treeline in Alaska through temperature thresholds', *Global Change Biology*, vol 10, pp1–13

White, G. C. (2001) 'Statistical models: key to understanding the natural world', in T. M. Shenk and A. B. Franklin (eds) *Modeling in Natural Resource Management: Development, Interpretation, and Application*, Island Press, Washington, DC

Wolfram, S. (1984) 'Universality and complexity in cellular automata', *Physica*, vol 10D, pp1–35

Woodley, S., Kay, J. and Francis, G. (1993) *Ecological Integrity and the Management of Ecosystems*, St Lucie Press, Delray Beach, FL

Woodall, C. W., Oswalt, C. M., Westfall, J. A., Perry, C. H., Nelson, M. D. and Finlay, A. O. (2009) 'An indicator of tree migration in forests of the eastern United States', *Forest Ecology and Management*, vol 257, pp1434–1444

Woodward, F. I. (1987) *Climate and Plant Distribution*, Cambridge University Press, Cambridge

Woodward, F. I. and Osborne, C. P. (2000) 'The representation of root processes in models addressing the responses of vegetation to global change', *New Phytologist*, vol 147, pp223–232

World Commission on Environment and Development (1987) *Our Common Future*, Oxford University Press, Oxford

Wu, H., Sharpe, P. J. H., Walker, J. and Penridge, L. K. (1985) 'Ecological field theory: a spatial analysis of resource interference among plants', *Ecological Modelling*, vol 29, pp215–243

Wu, L., McGechan, M. B., McRoberts, N., Baddeley, J. A. and Watson, C. A. (2007) 'SPACSYS: integration of a 3D root architecture component to carbon, nitrogen and water cycling-model description', *Ecological Modelling*, vol 200, 343–359

Wykoff, W. R., Crookston, N. L. and Stage, A. R. (1982) *User's Guide to the Stand Prognosis Model* (General Technical Report INT-133), US Department of Agriculture, Forest Service, Intermountain Forest and Range Experiment Station, Ogden, UT

Yamasaki, S. H., Kneeshaw, D. D., Munson, A. D. and Dorion, F. (2002) 'Bridging boundaries among disciplines and institutions for effective implementation of criteria and indicators', *The Forestry Chronicle*, vol 78, pp487–491

Yan, H.-P., Kang, M. Z., de Reffye, P. and Dingkuhn, M. (2004) 'A dynamic, architectural plant model simulating resource-dependent growth', *Annals of Botany*, vol 93, pp591–602

Yan, X. and Zhao, S. (1996) 'Simulating Changbai Mt. forests with climate change', *Journal of Environmental Science*, vol 8, pp358–368

Yang., Y., Monserud, R. A. and Huang, S. (2004) 'An evaluation of diagnostic tests and their roles in validating forest biometric models', *Canadian Journal of Forest Research*, vol 34, pp619–629

Young, A. and Muraya, P. (1990) *SCUAF: Soil Changes Under Agroforestry*, ICRAF, Nairobi

Xi, W., Coulson, R. N., Birt, A. G., Shang, Z.-B., Waldron, J. D., Lafon, C. W., Cairns, D. M., Tchakerian, M. and Klepzig, K. D. (2009) 'Review of forest landscape models: types, methods, development and applications', *Acta Ecologica Sinica*, vol 29, pp69–78

Xiandong, Y. and Shugart, H. H. (2005) 'FAREAST: a forest gap model to simulate dynamics and patterns of eastern Eurasian forests', *Journal of Biogeography*, vol 32, pp1641–1658

Index

For Product Safety Concerns and Information please contact
our EU representative GPSR@taylorandfrancis.com Taylor & Francis
Verlag GmbH, Kaufingerstraße 24, 80331 München, Germany

T - #0070 - 230425 - C8 - 234/156/16 - PB - 9781138866942 - Gloss Lamination